Spontaneous Ordering
in Semiconductor Alloys

Spontaneous Ordering in Semiconductor Alloys

Edited by

Angelo Mascarenhas

National Renewable Energy Laboratory (NREL)
Golden, Colorado

Kluwer Academic / Plenum Publishers
New York, Boston, Dordrecht, London, Moscow

ISBN 0-306-46778-X

©2002 Kluwer Academic/Plenum Publishers, New York
233 Spring Street, New York, New York 10013

http://www.wkap.nl/

10 9 8 7 6 5 4 3 2 1

A C.I.P. record for this book is available from the Library of Congress

Contributors

S. Phillip Ahrenkiel, National Renewable Energy Laboratory, 1617 Cole Boulevard, Golden, Colorado, USA

Francesc Alsina, University of Barcelona, Bella Terra, Spain, and National Renewable Energy Laboratory, Golden, Colorado, USA

Hyeonsik M. Cheong, Department of Physics, Sogang University, Seoul, Korea, and National Renewable Energy Laboratory, Golden, Colorado, USA

Gottfried H. Döhler, Institut für Technische Physik I, Universität Erlangen-Nürnberg, Erwin-Rommel-Str.1 91058 Erlangen, Germany

Rebecca L. Forrest, University of California-Los Angeles, Department of Materials Science

Mark C. Hanna, National Renewable Energy Laboratory, Golden, Colorado, USA

Peter Kiesel, Institut für Technische Physik I, Universität Erlangen-Nürnberg, Erwin-Rommel-Str.1, 91058 Erlangen, Germany

Thomas Kippenberg, Institut für Technische Physik I, Universität Erlangen-Nürnberg, Erwin-Rommel-Str.1, 91058 Erlangen, Germany

Michael Kozhevnikov, Gordon McKay Laboratory of Applied Science, Harvard University, 9 Oxford Street, Cambridge, Massachusetts, USA

Peter Krispin, Paul-Drude-Institut für Festkörperelektronik, Hausvogteiplatz 5-7, 10117 Berlin, Germany

Jianhua Li, Physics Department, University of Houston, Texas, USA

Gerald Martinez, Grenoble High Magnetic Field Laboratory MPI/CNRS, 25, Av. des Martyrs, 38042 Grenoble Cedex 9, France (on leave from the Institute of Physics, Academy of Sciences, Prague, Czech Republic)

Angelo Mascarenhas, National Renewable Energy Laboratory, Golden, Colorado, USA

Dirk C. Meyer, Institut für Kristallographie und Festkörperphysik, Fachrichtung Physik, Technische Universität Dresden, D-01062 Dresden, Germany

Venkatesh Narayanamurti, Gordon McKay Laboratory of Applied Science, Harvard University, 9 Oxford Street, Cambridge, Massachusetts, USA

Andrew G. Norman, National Renewable Energy Laboratory, 1617 Cole Boulevard, Golden Colorado, USA

Peter Paufler, Institut für Kristallographie und Festkörperphysik, Fachrichtung Physik, Technische Universität Dresden, D-01062 Dresden, Germany

Ines Pietzonka, Dept. of Solid State Physics, Lund University, Box 118, S-221 00 Lund, Sweden

Kurt Richter, Institut für Kristallographie und Festkörperphysik, Fachrichtung Physik, Technische Universität Dresden, D-01062 Dresden, Germany

Torsten Sass, Dept. of Solid State Physics, Lund University, Box 118, S-221 00 Lund, Sweden

Maeng-Je Seong, National Renewable Energy Laboratory, Golden, Colorado, USA

Jeremy D. Steinshnider, Texas A&M University, Department of Physics, College Station, Texas, USA

Gerald B. Stringfellow, College of Engineering, University of Utah, Salt Lake City, Utah, USA

Tohru Suzuki, System Devices and Fundamental Research, NEC Corporation

Kazuo Uchida, Department of Electrical Engineering, The University of Electro-Communications, 1-5-1, Choufugaoka, Choufu, Tokyo 182, Japan

Gerald Wagner, Institut für Oberflächenmodifizierung Leipzig, Germany

Su-Huai Wei, National Renewable Energy Laboratory, Golden, Colorado, USA

Michael B. Weimer, Texas A&M University, Department of Physics, College Station, Texas, USA

Peter Y. Yu, Department of Physics, University of California, Berkeley and Materials Science Division, Lawrence Berkeley National Laboratory, Berkeley, California, USA

Jan Zeman, Grenoble High Magnetic Field Laboratory MPI/CNRS, 25, Av. des Martyrs, 38042 Grenoble Cedex 9, France (on leave from the Institute of Physics, Academy of Sciences, Prague, Czech Republic)

Yong Zhang, National Renewable Energy Laboratory, Golden, Colorado, USA

Preface

The phenomenon of spontaneous ordering in semiconductor alloys, which can be categorized as a self-organized process, is observed to occur spontaneously during epitaxial growth of certain ternary alloy semiconductors and results in a modification of their structural, electronic, and optical properties. There has been a great deal of interest in learning how to control this phenomenon so that it may be used for tailoring desirable electronic and optical properties. There has been even greater interest in exploiting the phenomenon for its unique ability in providing an experimental environment of controlled alloy statistical fluctuations. As such, it impacts areas of semiconductor science and technology related to the materials science of epitaxial growth, statistical mechanics, and electronic structure of alloys and electronic and photonic devices. During the past two decades, significant progress has been made toward understanding the mechanisms that drive this phenomenon and the changes in physical properties that result from it. A variety of experimental techniques have been used to probe the phenomenon and several attempts made at providing theoretical models both for the ordering mechanisms as well as electronic structure changes.

The various chapters of this book provide a detailed account of these efforts during the past decade. The first chapter provides an elaborate account of the phenomenon, with an excellent perspective of the structural and electronic modifications it induces. The second chapter focuses on ordering mechanisms, the third on the use of surfactants to control the phenomenon, the fourth and sixth on X-ray studies of the order parameter, the fifth and seventh on electron microscopy/electron diffraction, and the eighth on diffraction anomalous fine structure studies. The remainder of the chapters concern electronic properties, with the ninth focusing on ballistic electron emission microscopy studies, the tenth and eleventh on the physics of alloy statistical fluctuations, the twelfth on upconversion phenomena, the thirteenth on polarization-sensitive devices, the fourteenth on phonons, and the fifteenth on electronic structure calculations. The sixteenth chapter deals with the issue of spontaneous electric fields and the band alignment between ordered GaInP and GaAs, which has become especially important because of the widespread use of this system for heterojunction bipolar transistors used for cellular communications applications.

It is the dedicated efforts of the authors of the various chapters that have made this book possible, and I am deeply grateful for their contributions. The extensive research in this field conducted at NREL would not have been possible without the support of the Division of Materials Science/Basic Energy Sciences/Office of Science of the U.S. Department of Energy. The help with technical editing from Irene Medina and Don Gwinner is gratefully acknowledged. Finally, I wish to thank Dr. S. K. Deb for his unwavering encouragement of fundamental research at NREL, and especially of this project.

Angelo Mascarenhas
Golden, Colorado

Contents

2. The Nature and Origin of Atomic Ordering in Group III-V Antimonide Semiconductor Alloys
A.G. Norman

3. Effects of the Surface on CuPt Ordering During OMVPE Growth
G.B. Stringfellow

4. X-Ray Diffraction Analysis of Ordering in Epitaxial III-V Alloys
R.L. Forrest

5. Surface Morphology and Formation of Antiphase Boundaries in Ordered (GaIn)P—A TEM Study
T. Sass and I. Pietzonka

6. X-Ray Characterization of CuPt Ordered III-V Ternary Alloys
J. Li

7. Diffraction and Imaging of Ordered Semiconductors
S.P. Ahrenkiel

8. X-Ray Analysis of the Short-Range Order in the Ordered-Alloy Domains of Epitaxial (Ga,In)P Layers by DAFS of Superlattice Reflections
D.C. Meyer, K. Richter, G. Wagner, and P. Paufler

9. Ballistic Electron Emission Microscopy and Spectroscopy Study of Ordering-Induced Band Structure Effects in $Ga_{0.52}In_{0.48}P$
M. Kozhevnikov and V. Narayanamurti

10. Cross-Sectional Scanning Tunneling Microscopy as a Probe of Local Order in Semiconductor Alloys
J.D. Steinshnider, M.B. Weimer, and M.C. Hanna

13. Polarization Effects in the (Electro)absorption of Ordered GaInP and Their Device Applications
P. Kiesel, T. Kippenberg, and G.H. Döhler

14. Phonons in Ordered Semiconductor Alloys
A. Mascarenhas, H.M. Cheong, M.J. Seong, and F. Alsina

15. Effects of Ordering on Physical Properties of Semiconductor Alloys
S.-H. Wei

16. Polarization Charges at Spontaneously Ordered (In,Ga)P/ GaAs Interfaces
Peter Krispin

Spontaneous Ordering
in Semiconductor Alloys

Chapter 1

Basic Aspects of Atomic Ordering in III-V Semiconductor Alloys

Tohru Suzuki
System Devices and Fundamental Research, NEC Corporation

Key words: Atomic ordering, III-V semiconductor alloy, vapor-phase epitaxy, MOVPE, MBE, CuPt-B, CuPt-A, TP-A, surface reconstruction, surface step, GaInP, bandgap anomaly.

Abstract: Basic aspects of atomic ordering in III-V semiconductor alloys are described, with an emphasis on ordering mechanisms. We illustrate how a bandgap anomaly found in GaInP plays a crucial role in revealing CuPt-B type ordering, which is the most commonly observed type of ordering in III-V alloys. Various factors that influence the occurrence of CuPt-B ordering are depicted. In addition, properties of Triple-period-A (TP-A) and CuPt-A orderings, whose ordering directions and periods are different from those of ordering, are described. We describe the present understanding of how these representative types of ordering are generated during growth and demonstrate that atomic ordering in III-V alloys is a surface phenomenon that occurs under the influence of surface reconstructions and atomic steps at growing surfaces during vapor-phase epitaxy; this mechanism is completely different from the conventional atomic ordering in metal alloys. The implications of these studies of atomic ordering are also discussed.

1.   INTRODUCTION

The present information-oriented society has been using semiconductor devices extensively for computers, high-density optical data storage systems (CD-ROM and DVD[1]-ROM), and communication systems such as cellular phones, optical fiber, and satellite-broadcasting systems. Although most semiconductor devices used in these systems are ultra-large-scale integrated circuits (ULSIs) made of silicon, III-V semiconductor alloys (in short, III-V alloys) made of group-III and group-V elements such as AlGaAs, AlGaInP, and GaInPAs are used in optoelectronic devices (e.g., lasers, photo-

1

detectors) and high-speed, low-noise, and high-power electronic devices (e.g., FETs, HBTs). Although the production-scale of the III-V devices is rather small (in terms of world sales, around two orders of magnitude smaller than that for silicon devices), they are operating in key places in these systems, because of the unique properties of III-V alloys, e.g., their light-emitting capability, and the ability to tailor important material properties (such as bandgap and electron mobility) simply by varying alloy compositions in III-V ternary and quaternary alloys such as $III_xIII_{1-x}V$, IIIV_xV_{1-x}, or $III_xIII_{1-x}V_yV_{1-y}$. The applications of these III-V alloys have become extensive since the 1980s.

Until the early 1980s, it had been generally believed that, in these III-V alloys, group-III or group-V atoms distribute randomly on respective face-centered-cubic (f.c.c.) sublattices (Fig. 1(a)), even though spinodal decomposition of alloys into component binary compounds should be taken into consideration in certain cases.[2] When atoms of the alloys, e.g., Ga and In atoms in $Ga_xIn_{1-x}P$, distribute randomly on the group-III sublattice, material properties such as bandgap had been assumed to be a single-valued function of alloy composition (x). During the efforts in the early 1980s to realize red-light-emitting laser-diodes made of AlGaInP alloy, however, it was noticed that the bandgap of $Ga_{0.5}In_{0.5}P$, which is an end-ternary alloy used as the active layer of the laser diode and is lattice-matched to the (001) GaAs substrate, had an anomalous lower value of 1.85 eV (at room temperature)[3] than the then-most-reliable value (1.91-1.92 eV).[4,5] The growth method used was metalorganic vapor-phase epitaxy (MOVPE). This anomaly could significantly affect the laser wavelength of the red-light-emitting AlGaInP laser-diodes.[6] And it was later found to depend on growth conditions such as group-V source-gas pressure[7,8] and growth temperature.[7,8,9]

The searches for what kind of degree of freedom of the alloy system gives rise to this bandgap deviation revealed an atomic ordering called CuPt-B ordering.[10] The structure is a monolayer superlattice of Ga-rich planes and In-rich planes ordered in the [-111] or [1-11] directions (Fig. 1(b)). The degree of ordering corresponds qualitatively very well to the degree of bandgap anomaly.[10,11] The correlation between ordering and the bandgap anomaly was later theoretically confirmed by the calculations of band structure.[12,13] The theory clarified that the bandgap anomaly is due to the introduction of CuPt-B ordering, which introduces a periodicity double that of the zincblende structure for the random alloy. The introduction of this crystal field results in a Brillouin zone folding effect[14,15] on the conduction-band structure and a crystal-field-splitting effect on the valence band. The calculated properties of the bandgap anomaly, due to the perfect ordering calculated by the theory, was significantly larger than the values obtained experimentally. The reason for this was further studied experimentally by using photoluminescence polarization spectroscopy,[16,17] which has enabled quantitative estimations of the degree of ordering by measuring both the bandgap lowering and the valence-based splitting. Raman scattering spectroscopy has also detected this symmetry change in changes of the lattice vibration spectrum.[11,18]

Atomic ordering was found more or less independently at several institutions around the world in the mid 1980s.[19] Around 1985, the famatinite structure (Fig. 1(e)) was observed in GaInAs grown on (001) InP substrate by liquid-phase epitaxy (LPE).[20] CuAu-I ordering (Fig. 1(f))[21] was found in AlGaAs grown by MOVPE or molecular-beam epitaxy (MBE). CuAu-I in GaAsSb grown by MOVPE was also reported.[22] In the same study, chalcopyrite-type ordering (Fig. 1(g)) mixed with CuAu-I was also reported in GaAsSb. However, these types of ordering only very rarely occur or mostly occur on non-conventional substrate orientations, that is, other than (001); therefore, there are very limited studies on them. On the other hand, since the first discovery of the CuPt-type ordering in MBE-grown SiGe[23] in 1985, and in MBE-grown GaAsSb[24] in 1986, this type of ordering has also been reported for GaInP,[10,25] AlInP,[11] GaInAs,[26] AlInAs,[27] and *almost all* ternary and quaternary alloys with very few exceptions; even the exceptions have a logical explanation, as mentioned later in this chapter.

The epitaxial layers that show CuPt ordering are all grown by MOVPE, MBE, or chloride-transport vapor-phase epitaxy (Cl-VPE). A common feature in these growth methods is that *all* these epitaxial layers are *grown from vapor-phase*. However, CuPt-B ordering in crystals grown by LPE (which had been a major crystal-growth method before the 1980s) was never reported.[28] This fact is intimately related to the mechanisms of atomic ordering. During the study of ordering mechanisms, new types of ordering, triple-period (TP)-A[29] (Fig. 1(c)) and CuPt-A[30] (Fig. 1(d)), were found in AlInAs and AlInP, respectively, grown on (001) substrates by MBE under extremely group-V-rich conditions. The ordering directions are [11-1] (A_+) and [-1-1-1] (A_-) (Fig. 1), in which CuPt-B ordering is never observed. By correlations found between the peculiar structural properties of these types of ordering and characteristic surface reconstructions during growth, the atomic ordering in III-V alloys is now established to be a phenomenon that occurs under the influence of surface reconstruction during vapor-phase growth.

Among the several types of atomic ordering in III-V alloys, the CuPt-B type is the most important and most extensively studied. This is because it is the most prevalent type of atomic ordering, and it occurs under very ordinary crystal-growth conditions (including substrate orientation).

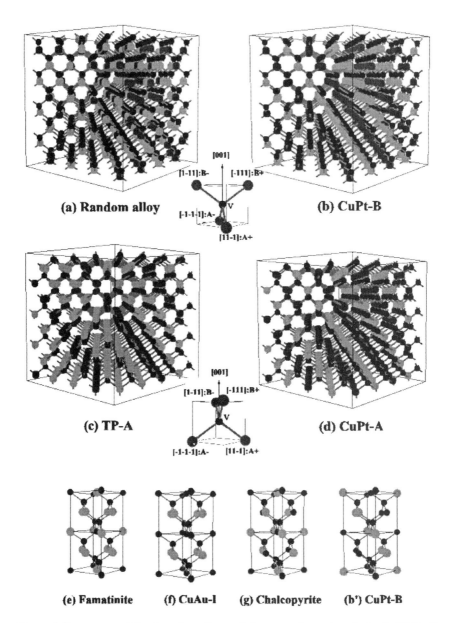

Figure 1. Structure models of random alloy and six types of atomic ordering in III-V alloys. The same CuPt-B structure with different scales is shown in (b) and (b') for reference. Red and green balls indicate group-III$_1$ and III$_2$ atoms for III$_1$-III$_2$-V type alloys. Black balls indicate group-V atoms. Crystal orientations for (a) and (b) are indicated between them, and those for (c) and (d) are also indicated between them. The four equivalent <-111> directions in the zincblende structure are called B$_+$, B$_-$, A$_+$, A$_-$ for convenience. CuPt-B (b) illustrates the ordering in the [-111] (B$_+$) direction. TP-A (c) and CuPt-A (d) illustrate the ordering in the [-1-1-1] (A$_-$) direction.

Regarding materials, GaInP has been the most thoroughly studied, because it shows a very clear bandgap anomaly that significantly influences the device characteristics of practical laser diodes.pensable.[31] Moreover, the bandgap anomaly itself is an indispensable too for studying the ordering. On the other hand, even though the A-type ordering, TP-A and CuPt-A, occurs under more restricted (unusual) growth conditions where electronic-device-quality crystals are difficult to grow, the fact that these new types of ordering have different superlattice orientations and periodicities from those for CuPt-B has provided a crucial role in determining the ordering mechanisms.

In section 2, we describe the basic properties of CuPt-B type atomic ordering. In section 3, TP-A and CuPt-A type ordering are described. Section 4 discusses the mechanism of atomic ordering, although we begin the discussions in the earlier sections, because it is a main thread of this chapter. We will add short remarks in section 5.

2.   CuPt-B ATOMIC ORDERING

Basic properties of CuPt-B atomic ordering and the factors that affect its formation are described in this section. We first introduce the bandgap anomaly, which was discovered in $Ga_{0.5}In_{0.5}P$ and resulted in the discovery of ordering in the alloy.

2.1   Bandgap Anomaly and CuPt-B Ordering

$Ga_{0.5}In_{0.5}P$[32] grown by MOVPE was found to show an anomalously low room-temperature photoluminescence (PL) peak energy, and the anomaly was later found to depend on MOVPE growth conditions. First, we look at a phosphine pressure dependence of PL spectra.[7,11] The epitaxial layers are lattice-matched to (001) GaAs substrates. The reactor pressure was 70 torr, source gases were TEGa (triethylgallium), TMIn (trimethylindium), and PH_3 (phosphine), the substrates were (001) GaAs, and the growth rate was 1μm/h. Because the flow rates of group-III source gases were kept constant in this series of growth runs, a change in V/III ratio (gas-flow-rate ratio used for group-V and group-III sources) means a change in group-V source-gas flow-rate. Figure 2 shows typical room-temperature PL spectra corresponding to the high (V/III = 412) and low (V/III = 62) phosphine pressures, respectively.[11] The PL peak energy corresponding to the lower phosphine pressure is 1.902 eV, which is rather close to the then-most-reliable values of 1.91-1.92 eV[4,5] reported for LPE crystals, whereas the PL peak energy corresponding to the higher phosphine pressure is an anomalous low value of 1.848 eV. Both PL spectra show a characteristic asymmetric feature of the band-to-band PL at room temperature, e.g., the high-energy sides of the peaks decay as $\exp(-h\nu/kT)$, where $h\nu$, k, and T are photon energy, Boltzmann's constant, and 300 K, respectively. Both crystals show comparable strong PL intensity and comparable threshold current densities when used as active layer,[33] although the full-width at half-maximum of the PL spectrum of the lower-bandgap crystal shows a slightly broader (~10 m eV) value than that for the crystal with the almost normal bandgap.

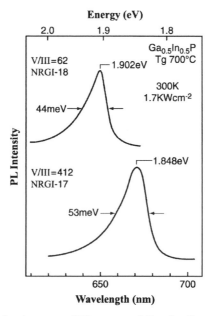

Figure 2. 300 K photoluminescence (PL) spectra of $Ga_{0.5}In_{0.5}P$ grown by MOVPE at 700°C under two different phosphine pressures. [Ref. 11]

Generally, the PL peak energy does not necessarily directly represent the bandgap of the semiconductor. However, according to studies[7,8,11] using X-ray diffraction, low-temperature PL, and photo-absorption spectroscopy, as well as the considerations mentioned above, the change in the PL peak energy in Fig. 2 was identified as being due to the change in *bandgap* and is not due to a change in alloy composition from sample to sample or fluctuation in alloy composition due to phase separation, or not due to recombination via deep levels that could be generated in the bandgap. This means that *the bandgap value of $Ga_xIn_{1-x}P$ with a fixed composition of x = 0.5 was not unique.*

A transmission-electron microscope (TEM) diffraction study revealed that samples with a bandgap anomaly show superstructures.[10] Figures 3(a) and (b) show [110]-zone and [-110]-zone diffraction (TED) patterns for $Ga_{0.5}In_{0.5}P$, with an anomalously low bandgap of 1.835 eV at room temperature. In the [110] zone, in addition to the basic spots coming from the zincblende structure, we see strong super-spots (-1/2,1/2,1/2; ...) halfway between the rows of basic matrix spots (-111; ...). Similar but relatively weak spots (1/2,-1/2,1/2; ....) are present along [1-11]. The [-110] zone TED, however, shows *no* super-spots at the 1/2,1/2,-1/2, and -1/2,-1/2,-1/2 positions. The spot at 0,0,1 in the [110] zone is due to a double diffraction from -1/2,1/2,1/2 and 1/2,-1/2,1/2. Figure 3(c) shows a high-resolution TEM lattice image of the [110] zone. A 6.5-Å periodicity in the [-111] direction, which is double that of the random alloy of zincblende structure, is clearly seen. Figure 1(b) is a structure model of CuPt-B ordering in the [-111] direction. Actually, the Ga plane and the In plane are not perfect, but are in fact the Ga-rich plane and the In-rich plane, respectively. In other areas of the crystal, the atomic ordering in the [1-11] direction is also seen. Directions of [-111] and [1-11] are two of the *four equivalent <-111>B directions* ([-111], [1-11], [11-1], and [-1-1-1]) in the zincblende structure (see the inset between (a) and (b) in Fig. 1). We call these [-111] and [1-11] directions, the B_+ and B_- directions, respectively, or B directions for both. We call the ordering in the [-111] and [1-11] directions the B_+ and B_- *variant*s, respectively, or B variants for both. Similarly, the [11-1], and [-1-1-1] are called A_+ and A_-, or A for both. The non-existence of A_+ and A_- variants is a characteristic feature of CuPt-B ordering.

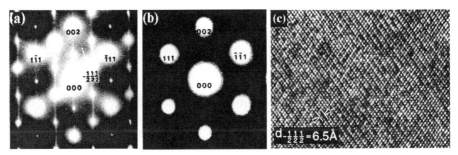

Figure 3. (a) [110]-zone and (b) [-110]-zone TED patterns of $Ga_{0.5}In_{0.5}P$ with an anomalous low bandgap value of 1.835 eV, and (c) a high-resolution [110]-zone TEM image of the same crystal. Right-side arrow in (a) indicates −1/2,1/2,1/2 super-spot and left-side arrow indicates 1/2,-1/2,1/2 super-spot. The double periodicity (6.5Å) in the [-111] direction of CuPt-B ordering is clearly seen in (c). [Ref. 10]

2.2   Growth Condition Dependence of CuPt-B Ordering

2.2.1   Group-V Source Pressure and Growth Temperature

The correlation between the bandgap and the degree of CuPt-B ordering is also seen in crystals grown under a wider range of conditions. The PL peak energy of $Ga_{0.5}In_{0.5}P$ is shown in Fig. 4 (left) as a function of growth temperature T_g, by taking V/III ratio as a parameter. This figure shows that bandgap takes minimum values around 650°–700°C, depending on the V/III ratio, and that at higher temperatures (e.g., 700°C), the bandgap decreases as V/III ratio increases, but at lower temperatures (e.g., 600°C) the bandgap increases as the V/III ratio increases. The GaAs substrates were (001) misoriented toward [011] by 2°, which is close to (001).

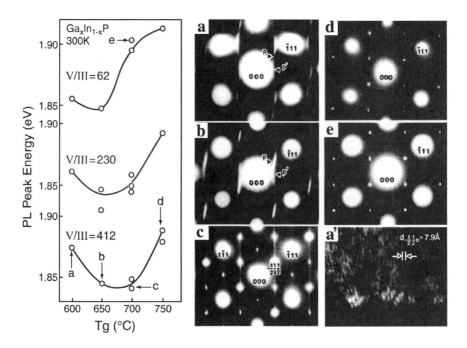

Figure 4. (Left) 300K PL peak energy vs. growth temperature of $Ga_{0.5}In_{0.5}P$ for three different phosphine-flow-rates and [110]-zone TEDs ((a) – (e)), corresponding to the data-points indicated in the left figure by 'a' – 'e', respectively. A typical [110]-zone dark-field- image corresponding to 'a' is shown in (a'). [Ref. 11]

[110]-zone TEM diffraction patterns from the crystals corresponding to the representative points ('a'-'e') in the bandgap vs. V/III ratio and T_g graph shown in Fig. 4 (left) are shown in Figs. 4(a) - (e),[11] respectively. Figure 4(a') shows TEM dark-field images corresponding to 'a' in Fig. 4 (left). In the TED patterns, the strongest super-spot intensity is seen in (c), which corresponds to the lowest E_g value (1.835 eV) among 'a'-'d' in Fig. 4 (left); this sample is identical to that for Fig. 3. In a higher T_g regime, as T_g increases (c: 700°C → d: 750°C) or phosphine pressure becomes lower (c: V/III 412 → e: V/III 62 at 700°C), the bandgap becomes higher and the super-spot intensity, by comparison with the intensities of the basic matrix spots, becomes weaker. In this regime, the degree of ordering is mainly limited by the degree of composition modulation (η, which is also called "order parameter"; see Fig. 5(a)),[34] even though there are still reductions in the degree of ordering due to the other imperfections, such as B_+/B_- domain boundaries and anti-phase boundaries (APBs), shown in Figs. 5(b) and (c).

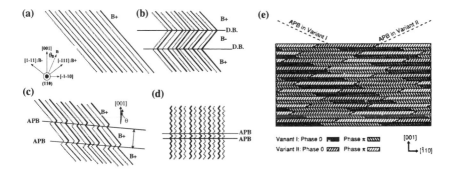

Figure 5. Imperfections in CuPt-B ordering: (a) CuPt-B ordering with incomplete composition modulation: $0 < \eta < 1$ (η: composition modulation or order parameter); Ga-rich $(0.5(1+\eta))$ plane (thick line) and In-rich $(0.5(1+\eta))$ plane (dotted line). $\eta = 1$ corresponds to complete modulation of Ga-plane and In-plane (Fig. 1(b)), and $\eta = 0$ corresponds to complete disordering (Fig. 1(a)). (b) B_+/B_- domain boundaries (D.B.). (c) Anti-phase boundaries (APBs). (d) Dense APBs or D.B.s. (e) A model [Ref. 35] to illustrate the streaks in TED patterns such as those seen in Figs. 4(a) – (e). 'Variant I' and 'variant II' indicated in (e) are identical to B_+ and B_- variants, respectively.

A characteristic feature in Figs. 4(a)-(e) is the streaks associated with the superstructures; they tend to become stronger as T_g decreases, although the streaks still exist at higher temperatures. Even in the streaky-TED regime, we can correlate the intensities at -1/2,1/2,1/2 or 1/2,-1/2,1/2 positions with the bandgap value. In this regime, reduction in η is also a cause of reduction in the degree of CuPt-B ordering. The B_+/B_- domain boundaries (Fig. 5(b))[35] and APBs within B_+ domains (shown in Fig. 5(c))[35,36] and within B_-

domains, respectively, are considered to be the causes of the streaks. A TEM dark-field image (Fig. 4(a')) has locally shown a quasi-periodicity of 7.9Å, which is double that of the zincblende structure and is modeled as in Fig. 5(d).[11] The double-period superlattice structure in the [-110] direction was proposed (Fig. 5(d)) in Ref. 8. This is interpreted as a kind of CuPt-B ordering with dense APBs (Fig. 5(c)) and/or dense B_+/B_- domain boundaries (Fig. 5(b)).

The slight tilt of streaks from the [001] direction, observed, e.g. at position 'p' (-1/2,1/2,1/2) in Fig. 4(b), is explained[25,35,36] by the same magnitude of slight tilt from the (001) plane of the anti-phase boundaries for B_+ domains (θ in Fig. 5(c) or the tilt angle from (001) of 'APB in variant-I' in Fig. 5(e)) and for B_- domains (tilt angles of APBs for B_- domains lie in the plane opposite of those for B_+ domains with respect to [001], an APB for B_- variant being indicated as 'APB in variant-II' in Fig. 5(e)). The elongation of the tilted streaks is correlated with the thickness of the domains between APBs (as indicated in Fig. 5(c)), and the streaks parallel to [001] are explained by the finite thicknesses of the B_+ and B_- domains whose normals are almost parallel to [001]. The whole feature of the streaks has been interpreted using a model shown in Fig. 5(e), in which plate-like B_+ and B_- domains are stacked consecutively in the [001] growth direction and anti-phase domain boundaries are interwoven.[35,37]

As above, the degree of CuPt-B ordering that influences the bandgap mainly depends on three factors: (1) the degree of composition modulation (η), (2) the density of the boundaries between B_+ and B_- domains, and (3) the density of anti-phase boundaries. From Fig. 4 (left) and Figs. 4(a)-(e), it is concluded that, qualitatively, the degree of ordering that affects bandgap is represented by the diffraction intensity at the -1/2,1/2,1/2 and 1/2,-1/2,1/2 positions; that is, the lower the bandgap value, the higher the degree of the CuPt-B type ordering. The correlation is reconfirmed abundantly by the other data shown below throughout section 2. We discuss related issues in this section (2.2) and in sections 2.4 and 4.1. Incidentally, the slightly broader full-width at half-maximum of PL spectra of ordered $Ga_{0.5}In_{0.5}P$ mentioned in section 2.1 is a general trend and may be due to the bandgap inhomogeneity arising from inhomogeneities of the above three factors.

$(Al_xGa_{1-x})_{0.5}In_{0.5}P$,[11,38] $Ga_{0.5}In_{0.5}As$,[39] and $InAsSb$[40] are the alloys that are also reported to show the bandgap anomaly. Their growth-condition dependencies appear similar to those in $Ga_{0.5}In_{0.5}P$.

2.2.2   Doping

Impurity doping also influences the degree of ordering. Figure 6 illustrates the bandgap of MOVPE-grown $Ga_{0.5}In_{0.5}P$ on (001) GaAs vs. p-type doping (Mg).[41] Under a growth condition of T_g = 730°C and V/III = 140, the bandgap (PL peak energy) shows almost no dependence on hole concentration up to about 2×10^{18} cm^{-3}. On the other hand, when T_g is reduced to 680°C, while the bandgap remains an anomalous value of ~1.85 eV up to 1×10^{18} cm^{-3}, the bandgap increases rapidly toward the normal value when hole concentration exceeds 1×10^{18} cm^{-3}. TEM studies revealed that the crystal grown at 680°C shows CuPt-B ordering when the doping level is less than 1×10^{18} cm^{-3}. However, when the doping level exceeds 1×10^{18} cm^{-3} and the bandgap increases, the crystal becomes disordered.

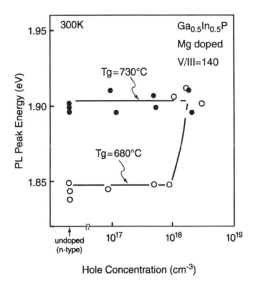

Figure 6. Mg-doping-level dependence of PL peak energy of $Ga_{0.5}In_{0.5}P$. [Ref. 41]

Similar disordering induced by impurity doping is observed in $Ga_{0.5}In_{0.5}P$ doped with Zn (p-type dopant),[41] Si, or Se (n-type dopants)[42]; also in these cases, TEM reveals the disordering of CuPt-B type ordering as the bandgap increases. There are two possible reasons for the disordering.[41] First, the dopant diffusion in the crystal during growth disorders the once-formed CuPt-B ordering, because the dopants enhance atomic diffusion,[43] and the threshold of occurrence of this impurity-enhanced diffusion is reported[43] to be ~ low 10^{18} cm^{-3}. Second, (2x1) surface reconstruction, which is responsible for the CuPt-B ordering (see section 3), is disordered by the doping during growth. Although which disordering process actually occurs during the doping of impurities such as Mg, Zn, Si, and Se is a question not yet studied in detail, there is at least an example where the first mechanism is active.[8,44] Regarding the second mechanism, an isoelectronic impurity of Sb has been reported to be capable of changing the surface reconstruction[45] (see section 3.1.2).

2.2.3 Growth Methods and Source Materials

Although crystals that show CuPt-B ordering described in sections 2.1 and 2.2 were grown by MOVPE, III-V alloys that show the CuPt-B type ordering are grown by several methods. We examine growth methods that were used for respective alloys:

III-III-V or III-III-III-V alloys: *AlInP* (MOVPE,[11] MBE[30]), *GaInP* (MOVPE,[10] MBE,[46] Cl-VPE,[47] HT-VPE[48]), $(Al_xGa_{1-x})_{0.5}In_{0.5}P$ (MOVPE[11]), *AlInAs* (MOVPE,[27] MBE[49,50]), *GaInAs* (MOVPE,[39,51,52] MBE, Cl-VPE[26]),
III-V-V alloys: *GaPAs* (MOVPE[53,54,55]), *GaAsSb* (MBE[24,56]), *InPAs* (MOVPE[57]), *InPSb* (MOVPE[58]), *InAsSb* (MOVPE,[59] MBE[60,61]).
III-III-V-V alloy: *GaInPAs* (Cl-VPE[26, 55]).

A common feature among the above-listed growth methods is that they are all VPE (vapor-phase epitaxy; MBE can be regarded as a kind of VPE, i.e., an extremely low-pressure VPE). And there is *no report on the CuPt-B type ordering in III-V alloys grown by LPE*.[28] In VPE, all sources are supplied is the form of isolated molecules or atoms. As explained in section 3, CuPt-B ordering is generated under the influence of group-V stabilized (2x1) surface reconstruction, which consists of dimer/dangling-bond pairs. Such a surface structure as (2x1) should be difficult to form at an LPE growth surface, which contacts a very dense (as dense as the solid surface) group-III-atom-rich liquid. On the other hand, crystal surfaces in VPE, including MBE, are quite different because they are in contact with gases of (at least three) orders-of-magnitude lower density compared to those of metal liquids. It is thus quite natural that there are no reports on CuPt ordering in any LPE-grown crystals.

Source materials, or precursors for the MOVPE growth of GaInP described in sections 2.1, 2.2, and 2.3 are the set: {TEGa, TMIn, and PH_3}.[7] Even if we focus only on GaInP, other combinations of source materials for which CuPt-B type ordering is formed are many, as follows: {TEGa, TEIn, PH_3},[6] {TMGa, TMIn, PH_3},[9] {TMGa, TMIn, TBP (tertiarybutylphosphine)}.[62] Cl-VPE[47] and HT-VPE,[48] in effect, use GaCl, InCl, and P_4. MBE uses {elemental Ga, elemental In, P_2}.[30]

As mentioned above, growth by the several different VPE methods using a variety of sets of source materials produces the same CuPt-B ordering in various III-V alloys. This fact indicates that CuPt-B ordering is a phenomenon common to these growth processes that is primarily related to Ga, In, and P themselves (in GaInP case) and is *irrelevant to the details of the reaction processes* of the source materials, i.e., the ordering should be the process that occurs after reactions of the source materials at the growth surface.

2.3   Bond-Length Difference

The list of alloys shown in section 2.2.3 is not complete in a sense that there are other alloys, such as GaInSb and GaPSb, that probably have not been examined yet but would show CuPt-B ordering. However, a most important missing alloy in the list is $Al_xGa_{1-x}As$.[28] This alloy system grown on (001) GaAs substrates is one of the most-studied materials in III-V alloys, which was grown by LPE since the 1960s and by VPE since the 1970s. However, no one has reported the existence of CuPt-B ordering in this alloy grown by MOVPE, MBE, or LPE; the only reported ordering is CuAu-I type.[21]

Bond lengths in III-V alloys tend to remain the same as those in their respective constituent binaries,[63] e.g., the lengths of the Ga-As bond and of the In-As bond in $Ga_xIn_{1-x}As$ are very close to those for the Ga-As bond and In-As bond in binary GaAs and binary InAs, respectively. Although the bond-length differences (ε)[64] in III_1-III_2-V, or III-V_1-V_2 alloys in the list in section 2.2.3, range from 3% (InPAs) to 10% (InPSb), $Al_xGa_{1-x}As$ has an extremely small difference between the bond lengths of Al-As and Ga-As (0.13%). For $Ga_{0.5}In_{0.5}P$, the bond-length difference between Ga-P and In-P is 7.4%. These data suggest that bond-length difference or covalent-atom-radius difference in alloys is a key factor that causes III-V alloys to form CuPt-B ordering.[28] If so, it is quite reasonable that there are no reports on CuPt-B ordering in AlGaP and AlGaSb, which have negligible or small length differences (0.004%, 0.6%, respectively). Although $(Al_xGa_{1-x})_{0.5}In_{0.5}P$ contains Al and Ga and the bond-length difference between Al-P and Ga-P is quite small, CuPt-B ordering can occur. Because there are large bond-length differences between Al-P and In-P (7.4%) and between Ga-P and In-P (7.4%), the alternate alignment of the (AlGa)-alloy-rich (-111) plane and the In-rich (-111) plane occurs in the [-111] direction (B$_+$ variant). The B$_-$ variant also occurs. However, CuPt-B ordering should not occur with respect to Al and Ga atoms.

2.4.1   Substrate Orientation Dependence

2.4.1   Two Kinds of Asymmetries in Variant Appearance

Two kinds of asymmetry in the super-spot intensities, i.e., in variant appearances, are noticed in Figs. 3(a) and (b). The first kind of asymmetry is the existence of B variants and the non-existence of A variants (*'BA asymmetry'*).[10,56] The second kind of asymmetry observed is that the intensity of the B_+ variant is stronger than that of the B_- variant,[10] as seen in Fig. 3(a). We call this asymmetry *'BB asymmetry.'*

The first kind of asymmetry (BA asymmetry) was observed consistently in $Ga_{0.5}In_{0.5}P$,[8,10,65] and in other III-III-V alloys such as $AlInP$[28] and $AlInAs$.[27] III-V-V alloys such as $GaAsSb$[66] and $GaPAs$,[53,54] also show the first kind of asymmetry; i.e., the appearance of only B variants.

The second kind of asymmetry (BB asymmetry) depends on the substrate misorientation from the (001) surface. The substrates on which the crystals shown in Figs. 3 and 4 were grown, have, to be precise, a misorientation of 2° from [001] toward [010]. This surface has a misorientation toward both [-111] and [111], by 1.4° each. The misorientation component toward [-111] introduces asymmetry in the surface structure with respect to both the B_+ and B_- directions, whereas the misorientation component toward [111] introduces no asymmetry with respect to the B_+ and B_- directions. The misorientation component of 1.4° toward [-111] is considered to introduce the BB asymmetry.[10] A similar second kind of asymmetry is also observed in $AlInP$[28] and $(AlGa)InP$.[28]

Figure 7 illustrates the effect of substrate misorientation on variant appearance.[67,68] Five kinds of substrate orientations are examined: $(001)2_A$, $(001)6_A$, $(001)2_B$, and $(001)6_B$, and the exact (001) surface ($\theta = 0°$). Here, $(001)\theta_{A\ or\ B}$ stands for (001) with a misorientation (θ in degrees) toward the A or B directions. $(001)2_B$ and $(001)6_B$ are the substrates that introduce asymmetry into the surface with respect to the B_+ and B_- directions. TED patterns (2b) and (6b) show that the larger the misorientation toward [-111], the stronger the -1/2,1/2,1/2 spot intensity. On the $(001)6_B$ substrate, almost solely the B_+ variant is observed, as is also seen in a TEM dark-field image (see Fig. 18(a) in section 4.1). The streaks parallel to the [001] direction observed in the TED pattern for $(001)_E$ almost disappeared, indicating the streaks are related to the coexistence of B_+ and B_- domains, as mentioned in section 2.2.1. With the 1/2,-1/2,1/2 diffraction having disappeared, it is noted that a diffraction at 0,0,1, which, in section 2.1, was ascribed to a double diffraction of -1/2,1/2,1/2 and 1/2,-1/2,1/2, has also disappeared.

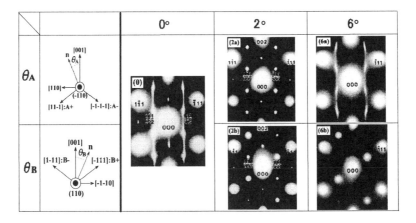

Figure 7. Misorientation dependence of CuPt-B ordering. [Adapted from Refs. 67 and 68.]

A surface with a few degrees of macroscopic misorientation from (001) toward [-111] (B+) may consist of one-molecular-layer-high (2.8Å) 'B+ step' arrays facing toward [-110], as illustrated in Fig. 8(a). A cross-section high-resolution TEM lattice image (Fig. 8(b)) of a typical (001) surface with a misorientation toward [-111] (B+) demonstrates this picture; that is, the B+ step arrays generate the B+ variant of CuPt-B ordering. Symmetrically, 'B- step' arrays facing toward [1-10] generate the B- variant (Fig. 8(c)).

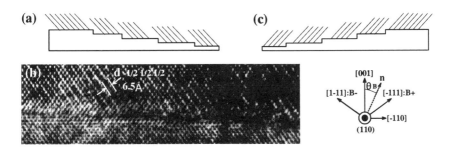

Figure 8. Relation between substrate misorientation and CuPt-B variant appearance: (a) B+ step array generates B+ variant, (b) a typical TEM image for a stepped surface with B+ step array shows B+ variant formation, and (c) B- step-array generates the B- variant.

On the other hand, the $(001)2_A$, $(001)6_A$, and $(001)_E$ substrates, which do not introduce asymmetry into the surface with respect to the B+ and B- directions, show equal intensity at the -1/2,1/2,1/2 and 1/2,-1/2,1/2 positions

(Figs. 7(2a) and (6a)), indicating that domains of B$_+$ and B$_-$ variants are equally generated.

Bandgap (PL peak energy) changes associated with an increase in θ_A and θ_B up to several degrees were observed,[69] which is described in detail in the next section.

2.4.2   Substrates with θ_B and θ_A Misorientations from (001)

The bandgap of Ga$_{0.5}$In$_{0.5}$P was found to depend on the misorientation from the [001] direction.[67,69,70,71,72] Figure 9 shows[72] the room-temperature bandgap (in terms of PL peak energy) as a function of misorientation (θ_B and θ_A) toward [-111] (B$_+$) and [-1-1-1] (A$_-$), respectively. As θ_B increases from 0° (the (001) surface), the bandgap first decreases slightly, has a minimum at around 6°, and then increases rapidly as the misorientation increases toward 15.8° (the (-115) surface). For $\theta_B > \theta_{(-115)B}$, the bandgap gradually approaches 1.918 eV at the (-111) surface ($\theta_B = 54.7°$, i.e., the so-called '(111)B' surface). TEM studies revealed that the Ga$_{0.5}$In$_{0.5}$P grown on (-111) GaAs substrates shows *no* ordering at all.[28,73] This means that the *highest bandgap* of 1.918 eV represents the room temperature bandgap

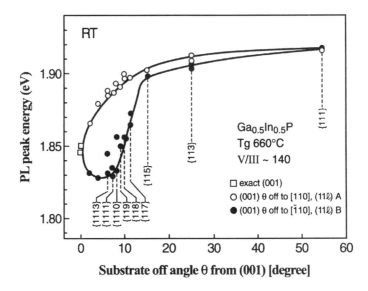

Figure 9. PL peak energy vs. substrate misorientation. Filled circle is for misorientation ($\theta = \theta_B$) from (001) towards (-111); open circle is for misorientation ($\theta = \theta_A$) from (001) towards (111). [Ref. 72]

corresponding to *complete disordering* on the group-III sublattice.[73] We refer to 1.918 (±0.002) eV as the *'standard bandgap'* of $Ga_{0.5}In_{0.5}P$ at room temperature. It is noted that 1.918 eV is close to the most reliable values reported for LPE-grown $Ga_{0.5}In_{0.5}P$ (1.921±0.004 eV,[4] 1.913±0.001 eV[5]). A more recent study on completely disordered $Ga_{0.5}In_{0.5}P$ reported a value of 1.910±0.008 eV[74] at 295 K.

The (-115) surface is a special surface with a *sharp kink* in the relation of bandgap vs. θ_B, as seen in Fig. 9. Figure 10(a) schematically illustrates a surface corresponding to $\theta_B \sim 6°$; several dimer/dangling-bond pairs can be accommodated on each (001) terrace, if the dimer/dangling-bonds exist as they do in the MBE growth surface. Figure 10(b), on the other hand, illustrates the (-115) surface ($\theta_B = 15.8°$) when each (001) surface is assumed to consist of one pair of dimer/dangling bonds, Figure 10(a) indicates that a surface with $\theta_B > 15.8°$ cannot even have one dimer/dangling-bond pair. From these observations, it was suggested that a (2x1) surface reconstruction consisting of an array of dimer/dangling-bond pairs (Fig. 10(a)) on each (001) terrace is responsible for the CuPt-B type ordering; in section 4, this is shown to be the case.

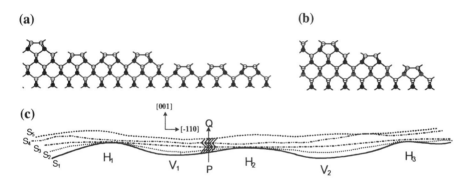

Figure 10. (a) Structure model of (001) surface with 6° misorientation toward (-111). (b) Structure model illustrating a special nature of the (-115) surface (see text). (c) Relation between local misorientations due to surface undulations and formation of CuPt-B variant domains. Here, undulation develops as s_1, s_2,..., growth proceeds. The B_+/B_- domain boundaries (as shown in Fig. 5(b)) are formed consecutively in the direction from P to Q, reflecting the local misorientation fluctuation due to the development of undulation. In (a) and (b), open circles with a horizontal bar are group-V atoms, and filled circles are group-III atoms.

$\theta_B = 0°$: The (001) surface usually develops slight surface undulation during growth.[10] Figure 10(c) schematically illustrates an unsteady development of the (110)-cross-section shapes of the undulation. The undulation introduces B_+ and B_- step arrays (Figs. 8(a) and (c), respectively), and thus, B_+ and B_- domains are generated[10] consecutively as growth proceeds (P $\rightarrow$ Q), according to the changes in the local misorientation, as shown in Fig. 10(c). A sample grown on an exactly (001)-oriented substrate has shown that kind of stacking of plate-like domains of B_+ and B_- with thicknesses of 2 to several nm.[35,36,37] The B_+/B_- domain boundaries shown in Fig. 5(b) may be formed in this way.

$0° < \theta_B < 6°$: When θ_B increases from $0°$ toward the [-111](B_+) direction, a shallow undulation may be insufficient to make a B_- step array appear; that is, this misorientation results in dominance of the B_+ step array and makes it difficult for the B_- variant to be generated. Thus, the number of B_+/B_- domain boundaries reduces and the bandgap decreases. The TED patterns of Figs. 7(2b) and (6b) show that the decrease in bandgap in this θ_B range ($0° < \theta_B < 6°$) is associated with an increase in the volume of B_+ domain and decrease in the volume of B_- domain. Although the surface undulation still makes B_- step arrays appear to some extent at $2°$ (Fig. 7(2b)), the B_- variant is almost completely suppressed at $6°$ (Fig. 7(6b)). The tilt angle (θ in Fig. 5(c)) of APBs of B_+ domains increases, and the APB density reduces with the increase in θ_B[75,76] for $0° < \theta_B < 6°$. The reduction in APB density, in addition to the reduction of the B_- variant domains, may also be contributing to the reduction of the bandgap in this regime. The mechanisms of the tilting of APBs have been proposed in Refs. 77 and 78. The relation between surface undulation and variant appearance has been further studied in Refs. 79 and 80.

$6° < \theta_B < 15.8°$ and $15.8° < \theta_B < 54.7°$: The decrease in the degree of ordering as θ_B increases from about $6°$ toward $15.8°$ is explained as follows. The average terrace width in units of dimer/dangling-bond pair reduces from 3 or 4 (Fig. 10(a)) to one (Fig. 10(b)). Due to a step-step interaction for such small step-step distances, the (2x1) reconstruction on each terrace is metastable and thus may change to some other structure that does not generate CuPt-B ordering. The decrease in the degree of ordering in this θ_B range was actually observed by TEM.[72] A bandgap of 1.90 eV at $\theta_B = 15.8°$ is slightly smaller than the standard bandgap of 1.918 eV for complete disordering at the (-111) surface. This difference can be ascribed to the slightly wavy surface morphology of $Ga_{0.5}In_{0.5}P$ during the growth, i.e., due to step bunching. When step bunching occurs, regions with $\theta_B < 15.8°$ and $\theta_B > 15.8°$ could appear, although the macroscopic average misorientation angle is $15.8°$. CuPt-B will be generated in the regions with $\theta_B < 15.8°$, and so, the regions will have a narrower bandgap. Actually, it is noticed that microscopically, nearly complete (-115) surfaces show no ordering,[81,82] which is in accord with the above discussion.

$0° < \theta_A < 54.7°$: Regarding the θ_A dependence of the bandgap, the bandgap increases monotonically from 0° to 54.7° (so-called '(111)A' surface). Similar to the θ_B-misorientation case, the (001) terraces almost disappear at 15.8° (the (115) surface), and the bandgap almost reaches that for complete disorder. A possible explanation for the increase of the bandgap, i.e., the decrease of the degree of ordering in this regime ($0° < \theta_A < 15.8°$), is as follows. The increase in θ_A in this region will enhance step flow growth to the A steps.[83] The increase in bandgap in this region indicates that A steps apparently do not contribute to CuPt-B ordering.

Because there is macroscopically no misorientation toward [-111] or [1-11] in this 'A-misorientation' ($\theta_A \neq 0$ and $\theta_B = 0$), there is no apparent mechanism that selects one CuPt-B variant between the two B-variants. Again, actual local-variant selection may be made by local misorientations with non-zero θ_B-component due to undulations, as shown in Fig.10(c). Generally, for an undulation with sufficient amplitudes (in the [001] direction) to develop, Ga and In atoms on the surface should migrate for sufficiently long distances in the ±[-110] directions before they adsorb at some (A or B) step edges. However, when θ_A increases, step flow into A-steps increases, on average. This shortens the migration distances of the group-III atoms and tends to prevent the surface undulations from developing sufficiently. On the other hand, the B_+ and B_- domain thicknesses will be on the order of the amplitude of the surface undulation. If so, the thicknesses of B_+ and B_- variants tend to reduce, because the undulation amplitude decreases. Then, thicknesses of the domains reduce and the density of B_+/B_- domain boundaries increases. The bandgap increase seems to be due to this increase in the number of the domain boundaries (Fig. 5(b)) nearly parallel to the (001) surface, because these boundaries work as defects with regard to the electronic state for the CuPt-B type ordering (Fig. 5(b)). The change in thickness of the domain boundaries occurs on a scale of the order of several nm. Thus, the size is able to affect the electronic structure. When θ_A approaches 16°, the (001) surface with dimer/dangling-bond pairs on each terrace may become metastable. These two effects will explain the monotonic increase of bandgap for $0° < \theta_A < 15.8°$.

At $\theta_A = 54.7°$ (the (111) surface), epilayers grown on this substrate show no trace of ordering. The layers show essentially the same 'standard bandgap' as that of $Ga_{0.5}In_{0.5}P$ on the (-111) GaAs substrates, as shown in Fig. 9. The small reduction in the bandgap between $16° < \theta_A < 54.7°$ may be explained by step-bunching of A-step arrays and the resulting appearance of a (001) terrace.

2.4.3   (110) Substrates

Ga$_{0.5}$In$_{0.5}$P epitaxial layers grown on *exactly* oriented (110) GaAs substrates did not show any trace of atomic ordering,[73] even if they were grown under the conditions under which CuPt-B ordering occurs when (001) GaAs is used as a substrate. The standard bandgap of 1.91-1.92 eV (the same as that of Ga$_{0.5}$In$_{0.5}$P on (-111) or (111) GaAs substrates) is observed.

A conclusion of section 2.4 is that the formation of CuPt-B ordering depends strongly on substrate orientation and *CuPt-B type ordering occurs only on (001) substrates with misorientation of θ < 15.8°*. This conclusion indicates that the (001) surface, in addition to the existence of B steps, plays an essential role in the CuPt-B ordering. Moreover, the assumption that there is (2x1) surface reconstruction comprising group-V dimer/dangling bonds on each (001) terrace during MOVPE growth of Ga$_{0.5}$In$_{0.5}$P has become quite justifiable.

Although Ga$_{0.5}$In$_{0.5}$As shows CuPt-B ordering when grown on (001) InP, the alloy shows CuAu-I ordering[84,85] when it is grown on (110) InP with slight misorientation toward the [001] or [00-1] direction. We briefly refer to the mechanism of CuAu-I ordering in section 4.2.

2.5   Degree of Ordering

In the previous sections, we showed that the lower the bandgap of Ga$_{0.5}$In$_{0.5}$P, the higher the degree of ordering. This relationship was later confirmed theoretically,[12,13] and thereby, quantitative estimation of the degree of ordering, in terms of composition modulation (η) (Fig. 5(a)) has become possible.[16,17] Figure 11[12,86,87,88] illustrates the theoretical interpretation for the effect of CuPt-B type ordering on the bandgap. The left part shows the band diagram for Ga$_{0.5}$In$_{0.5}$P random alloy. When CuPt-B type ordering occurs, the double-period crystal-field is introduced and the lattice periodicity in the [-111] direction doubles (from 3.26Å for a zincblende structure with cubic symmetry to 6.52Å for CuPt-B structure with trigonal symmetry). This causes the L-point of the Brillouin zone to fold back to the Γ point (indicated as 'C$_{folded}$' in the left figure). Due to the level repulsion between the two minimum conduction bands C$_{folded}$ and C$_{bottom}$, C$_{bottom}$ is pushed down, as indicated in the right part. The same double-period crystal-field introduced by the CuPt-B ordering lifts the three-fold degeneracy of the valence-band top level (V_{top}) into two levels, with a level spacing indicated by Δ_c. These two effects cause a narrowing of the bandgap, as shown in Fig. 11.

Polarization-dependent PL spectra[16] for $Ga_{0.5}In_{0.5}P$ due to a transition of A (E_g) (Fig. 11) and to a transition of B ($E_g +\Delta_c$) have been observed, being consistent with the theoretical analyses. The polarization dependence of PL spectra was used to obtain the crystal-field splitting (Δ_c); this splitting has been related to the degree of ordering due to composition modulation (η).[17] The obtained η values range from 0 to 0.4, depending on the bandgap anomaly and the magnitude of Δ_c. The maximum η value corresponds to a bandgap anomaly of 105 meV and Δ_c of 43.3 meV. A polarized piezo-reflectance study[87] has also been carried out for sigle variant samples grown on (001)16B samples to obtain $E_g +\Delta_{so}$, $E_g +\Delta_c$, and E_g. Elaborate theoretical treatments have been made for the relation between η and E_g, and for the relation between η and Δ_c.[89,90]

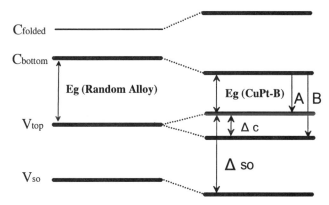

Figure 11.. Band diagram illustrating the bandgap reduction due to CuPt-B ordering [Adapted from Refs. 86 and 87.]

The nuclear magnetic resonance (NMR) of ^{31}P was used[91] to measure the degree of ordering of $Ga_{0.5}In_{0.5}P$ with a bandgap anomaly of ~100 meV. An order parameter (η) ≤ 0.6 was obtained. Another NMR study[92] (spin echo measurements of ^{71}Ga) gave the following: $\eta = 0(+0.1/-0)$ for $Ga_{0.5}In_{0.5}P$ grown by LPE and $\eta = 0.5(+0.15/-0.25)$ for an MOVPE-grown $Ga_{0.5}In_{0.5}P$ with a bandgap anomaly of 100 meV. These results are consistent with the optical measurement results mentioned above. An order parameter (η) was also measured by using X-ray diffraction,[93] and the results were consistent with the theoretical results. For a bandgap anomaly of 120 meV, the obtained η value was around 0.5. A calculation of the η value as a function of growth temperature has also been carried out[94] and are consistent with the experimental values.

From the results of the above studies, the maximum composition modulation (η) available so far for $Ga_{0.5}In_{0.5}P$ is < 0.6. The degree of ordering is reduced by the lattice defects of anti-phase boundaries and B_+/B_- domain boundaries, as well as by the composition modulation (η); this is mentioned in section 2.2.1.

3. TP-A AND CuPt-A ATOMIC ORDERING

Although CuPt-B ordering is observed almost ubiquitously in III-V alloys, other types of ordering are also found in III-V alloys grown on (001) substrates. Among these, two types of ordering, TP-A and CuPt-A, are formed by VPE under lower-than-normal growth temperature or under higher group-V pressure. This section describes the basic properties of TP-A and CuPt-A type atomic ordering.

3.1   TP-A Ordering

3.1.1   Structure and Variants

TP-A ordering was found in $Al_{0.5}In_{0.5}As$ grown by gas-source MBE on (001) InP substrates.[29] Figures 12(a) and (b) show TEM diffraction patterns of the [-110] and [110] zones, respectively. The crystal was grown at 460°C by gas-source MBE with AsH_3 cracked to As_2. In contrast to ordinary CuPt-B type ordering, super-spots are observed in the [-110] zone, but are not observed in the [110] zone. Moreover, the periodicity of the superstructure is triple that of the zincblende structure. Figure 12(c) shows a high-resolution TEM image of the [-110] zone. We see a superstructure with a triple periodicity in the [1-1-1] (A_-) direction, even though the ordering is rather short range (several nm in the domain size),[95] compared with CuPt-B ordering. The inset in Fig. 12(c) shows a Fourier transform of the optical image of Fig. 12(c), clearly indicating the triple periodicity. From these figures, it becomes evident that the periodicity in the [111] (i.e., -[-1-1-1]) or [-1-11] (i.e., -[11-1]) direction is triple that for the zincblende lattice. Figure 1(c) in section 1 illustrates the structure. A sequence of (111) planes with a triple periodicity consisting of Al-rich plane/ In-rich plane/ In-rich plane/ Al-rich plane/ In-rich plane/ In-rich plane/ etc., are formed in the [11-1] or [--1-1-1] directions (A_+ and A_- directions; see the inset figure between (c) and (d) of Fig. 1). CuPt-B ordering is not observed in the [110]-zone (Fig. 12(b)).

TP-A ordering is observed in $Ga_{0.5}In_{0.5}As$,[46,96] as well as in $Al_{0.5}In_{0.5}As$, grown by gas-source MBE. Solid-source MBE using As_4 also generates TP-A ordering in $Al_{0.5}In_{0.5}As$[50] and $Ga_{0.5}In_{0.5}As$.[50,97] MOVPE also generates TP-A ordering; $Al_{0.5}In_{0.5}As$ grown at a temperature of 500°C and a V/III ratio of 199 showed a very weak TP-A ordering,[98] while CuPt-B ordering was also observed in the same crystal and is still the main type of ordering. MOVPE-

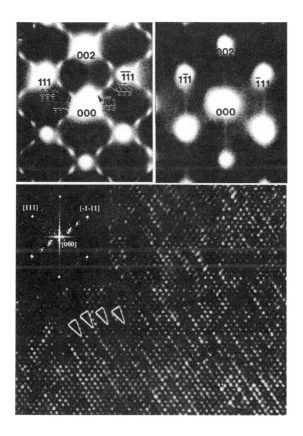

Figure 12. (a) [-110]-zone- and (b) [110]-zone TED patterns of MBE-grown $Al_{0.5}In_{0.5}As$. A high-resolution [-110]-zone TEM image (c), which shows triple-period (TP)-A ordering in the [-1-11] (= -[11-1]) direction. Inset shows a Fourier transform of image (c). [Adapted from Ref. 29]

grown $Ga_{0.5}In_{0.5}P$ doped with a small amount of Sb shows TP-A ordering (see section 3.1.2). The III-V-V-type alloy GaAsSb grown by MOVPE can generate triple periodicity in the [110] direction, though it is very weak.[99]

From the above observations, it became clear that there exists *'AB asymmetry'* in the TP-A type ordering; that is, A variants, but not B variants, appear. This asymmetry corresponds to the 'BA asymmetry' (section 2.4.1) observed in CuPt-B ordering.

'AA asymmetry' in TP-A ordering, corresponding to the 'BB asymmetry' (Fig. 3) observed in CuPt-B ordering, is also observed. Figure 13(a)[100] is a TEM dark-field image that illustrates how a TP-A ordering variant appears when the $Al_{0.5}In_{0.5}As$ is grown on a $Ga_{0.5}In_{0.5}As$ layer with slight misorientation from the (001) surface toward the A_+ and A_- directions. The image clearly indicates that a misorientation toward the [-1-11] (= $-A_+$) direction generates the A_+ variant in the left side of the image (area-I) and a misorientation toward [111] (= $-A_-$) direction generates the A_- variant (area-II).

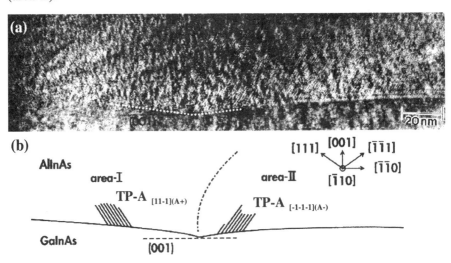

Figure 13. (a) A [-110]-zone TEM dark-field image of $Al_{0.5}In_{0.5}As$ grown on a $Ga_{0.5}In_{0.5}As$ undulating surface. (b) Schematic drawing of (a). Variants of TP-A ordering are formed on surfaces with A_+ and A_- misorientations. [Adapted from Ref. 100]

3.1.2   Surface Reconstructions

The surface phase-diagram of $Al_{0.5}In_{0.5}As$ is determined by observing the surface structure using RHEED (reflection high-energy-electron diffraction) *during* MBE growth[29,101](Fig. 14). At lower temperature or higher As pressure, a surface reconstruction with (2x3) symmetry is formed. Correspondingly, TP-A ordering is observed in the crystals when examined by TEM after growth. On the other hand, at higher temperature or lower As pressure, the surface reconstruction is (2x1). Corresponding to the (2x1) structure, CuPt-B ordering is formed in the crystals. The (2x1) structure is considered to be the surface covered with one monolayer of As and consisting of dimer/dangling-bond pairs of As (Fig. 15). The suggestion in section 2.4.2, based on the θ_B dependence of bandgap, that the (2x1) structure generates CuPt-B ordering is, by this correspondence, demonstrated to be true.

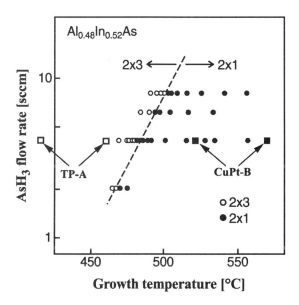

Figure 14. Surface phase diagram of MBE-grown $Al_{0.5}In_{0.5}As$. Four points are indicated, here TEM observations were made. [Adapted from Refs. 29 and 101]

The projection of the (2x3) structure onto the (-110) surface[29] is assumed to be that as illustrated in Fig. 15. The surface is probably covered with more than one monolayer of As, because the growth temperatures are low and As pressures are high in comparison with the growth conditions under which the (2x1) structure is observed. Note that the x3 structure of the surface in the [110] direction corresponds to the triple periodicity in the same direction of the TP-A type ordering. Therefore, the *periodicity and direction that TP-A type ordering has in the (001) plane* exactly correspond to the *periodicity and direction in the (2x3) structure.* For the top view of (2x3), a structure[94,102] slightly modified from Ref. 29 is assumed (Fig. 15). Together, the above two correspondences, (2x1) $\leftrightarrow$ CuPt-B and (2x3) $\leftrightarrow$

TP-A, clearly demonstrate that surface reconstructions, especially the directions and periods of the top-most dimers, are responsible for atomic ordering in III-V alloys.

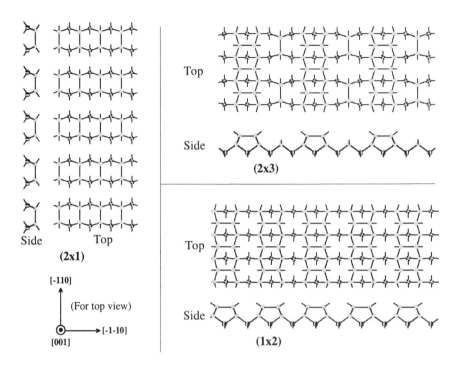

Figure 15. Top-view and cross-section view of structure models of (2x1), (1x2) [Refs. 30 and 50], and (2x3) [Refs. 29, 94, and 102]. Correspondingly, CuPt-B, CuPt-A and TP-A orderings are formed. Grey circles stand for group-V atoms, and gray-black circles stand for group-III atoms. Orientations (inset) are for top-views. (The top view of (2x3) shown here has been modified from Ref. 29 by adding one dimer per unit cell, according to Refs. 94 and 102.)

The crystals in which TP-A was observed (section 3.1.1) were all grown at lower temperature conditions than those commonly adopted for the alloys used for opto-electronic and electronic devices. However, even under normal 'high-temperature' conditions, TP-A ordering will be observed if the (2x3) structure is somehow formed during growth. As a matter of fact, $Ga_{0.5}In_{0.5}P$ grown at 650°C by MOVPE with a very small amount of Sb doping (1% of the PH_3 flow rate) in the gas phase was found to show TP-A ordering on the group III sublattice.[103,104,105] Without the Sb doping, the growth condition generates CuPt-B ordering. While only a very small amount of Sb (1.6%) is incorporated into the GaInP, a large number of Sb atoms seem to 'float' on the growing surface because of the low volatility of Sb; these atoms are thought to form a surface reconstruction of (2x3) or x3 structure made of double group-V layers (Fig. 15) in conjunction with P atoms. When the flow-rate ratio (Sb/P) decreases to 1.5×10^{-4}, a mixture of CuPt-B and TP-A appears. With an Sb/P flow-rate ratio less than 5×10^{-5}, only CuPt-B is observed; that is, in this very low Sb-doping regime, the surface should be (2x1),[104] which is a normal surface reconstruction under such growth conditions without Sb.

The occurrence of TP-A ordering in the slightly Sb-doped GaInP and in AlInAs (section 3.1.1) indicates that TP-A ordering is not a phenomenon peculiar to MBE-grown alloys, but it also occurs in MOVPE-grown alloys. However, the TP-A formation in MOVPE-grown alloys is, generally, rather difficult, because in MOVPE, the pyrolysis of source precursors is difficult at lower temperatures; these surfaces are required to have the (2x3) structure, unless group-V species with low vapor pressures such as Sb are used.

3.2   CuPt-A Ordering

3.2.1   Structure and Variants

$Al_{0.5}In_{0.5}P$ grown on (001) GaAs substrates at a lower temperature of 520°C by gas-source MBE shows CuPt-A ordering.[30] Figures 16(a) and (b) show TEM diffraction patterns of the [110] and [-110] zones. In both zones, superstructures coming from structures with 2x periodicity are observed. The high-resolution TEM image of the [-110] zone (Fig. 16(c)) shows the double periodicity in the A_ direction, in which atomic ordering with double periodicity was never observed before the report of Ref. 30. The atomic ordering with double periodicity and orientations in the A_+ and A_ directions is called CuPt-A ordering (Fig. 1(d)). The [110]-zone TEM image showed the conventional CuPt-B ordering (Fig. 16(d)), indicating that CuPt-A and CuPt-B ordering co-exist in the crystal in different domains.

Al-rich $Al_{0.9}In_{0.1}As$ grown by solid-source MBE at a low temperature of 430°C showed CuPt-A ordering.[50] CuPt-B was, however, *not* observed. Therefore, '*AB asymmetry*' is also present in CuPt-A ordering, just as it is in TP-A ordering.

The TEM image shown in Fig. 16(c) shows only the A_ variant of the CuPt-A type ordering. Although the macroscopic average orientation is exactly (001), there is a local misorientation from (001) toward the [111] (= -A_) direction in this place, due to step bunching. Therefore, '*AA asymmetry*' is present in CuPt-A ordering, too, as it is in TP-A ordering. In other places where local misorientation is toward the [-1-11] (= -A_+)

direction, the A_+ variant should be observed, as the diffraction pattern (Fig. 16(a)) indicates. CuPt-A ordering was also observed in $Ga_{0.5}In_{0.5}P$ grown by gas-source MBE.[46]

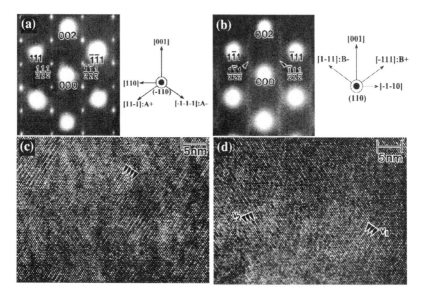

Figure 16. (a) [-110]-zone and (b) [110]-zone TED patterns of MBE-grown $Al_{0.5}In_{0.5}P$. TEM lattice images of respective zones are also shown in (c) and (d). Double-period-CuPt-A ordering is seen in (c). [Adapted from Ref. 30]

3.2.2   Surface Reconstructions

The surface reconstruction of $Al_{0.5}In_{0.5}P$, whose TEM diffraction patterns are shown in Figs. 16(a) and (b), was determined to be (2x2) by RHEED during growth (T_g = 520°C).[30] The (2x2) pattern is thought to consist of diffractions coming from (2x1) domains and (1x2) domains; that is, the (2x1) structure generates CuPt-B ordering (section 3-1-2) and the (1x2) structure generates CuPt-A ordering. This interpretation is confirmed by the following two observations. First, at a higher growth temperature of 560°C, only CuPt-B ordering was observed; correspondingly, the surface reconstruction was (2x1). Thus, the correspondence CuPt-B $\leftrightarrow$ (2x1) is again demonstrated in the different alloy system $Al_{0.5}In_{0.5}P$, as well as in $Al_{0.5}In_{0.5}As$ (section 3.1.2). The surface at the lower temperature of 520°C with the same phosphorous flux should therefore be more P-rich. Second, low-temperature growth of Al-rich $Al_{0.9}In_{0.1}As$ shows *only* CuPt-A ordering (section 3.2.1), and in this case, a (1x2) pattern is observed during growth.[50] Therefore, the correspondence CuPt-A $\leftrightarrow$ (1x2) is established. The (1x2) surface reconstruction is assumed to be a surface covered with two layers of phosphorus, as shown in Fig. 15. The periodicity and direction that CuPt-A ordering has in the (001) plane exactly correspond to the periodicity and direction of the (1x2) structure.

4. ORDERING MECHANISMS

4.1 CuPt-B, TP-A, and CuPt-A Ordering

The crucial inference that the CuPt-B type ordering occurs as a *surface phenomenon* and cannot be explained solely by bulk crystal thermodynamics was made based on the observation of two kinds of asymmetry in the variant appearances (the *'BA and BB asymmetries'* mentioned in section 2.4.1).[10,28] The fact that the ordering appears only when (001) or vicinal (001) substrates are used also supports the inference. A result of the thermodynamic first-principles calculation[106] that CuPt-B ordering is unstable with regard to disordering (i.e., the disordered state is more stable than the CuPt-B ordering) is consistent with the above inference.

If the ordering occurs during layer-by-layer crystal growth, two types of ordering processes should be explained (Fig. 17).[28] The first one is *intra-plane ordering*, shown in Fig. 17, in which [110]-oriented Ga columns and [110]-oriented In columns alternately align in the [-110] direction on the α_1 (001) plane. The second mechanism is *inter-plane ordering*, in which [110]-atom lines of particular atom species (e.g., Ga_1-line, Ga_2-line, Ga_3-line, ...) in the successive (001) planes (α_1, α_2, α_3, ...) have a fixed phase relation between adjacent layers (constant ϕ in Fig.17; *phase-locking mechanism*). Figure 18(a) shows a TEM dark-field image showing CuPt-B ordering in $Ga_{0.5}In_{0.5}P$ grown by MOVPE on $(001)6_B$ GaAs ((001) surface with a 6° misorientation toward [-111] (B_+)). The TED pattern is shown in Fig. 7(6b) (section 2.4.1). The average (001) terrace width is 27 Å (3–4 in units of dimer/dangling-bond pair), as shown in Fig. 18(b). Even in an area as large as 1000–2000 Å, few defects including anti-phase boundaries are observed. To explain this high coherency of the superlattice phase, a *'step-terrace reconstruction' (STR)* model was presented.[107,109] In this model, the phase relation between the group-V-atom dimer/dangling-bonds units of adjacent terraces remains fixed during the step-flow growth (Figs. 18(b) and (d)-(f)), and group-III atom incorporation occurs in a form of alternate periodic arrangement of [110]-Ga-columns and [110]-In-columns during step-flow growth, somehow reflecting the asymmetry on the surface structure (dimer/dangling bond) and the bond-length (or covalent radius) difference.

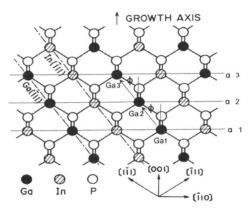

Figure 17. Structure model projected onto the (110) plane, illustrating CuPt-B ordering mechanism. [Ref. 28]

This model assumes that (1) there is surface reconstruction with 2x periodicity on each terrace due to the periodic arrangement of dimer/dangling-bond pairs[108] and (2) the step structure with a fixed phase relation is energetically favorable.

The place where Ga and In atoms (for the $Ga_{0.5}In_{0.5}P$ case) align in the (001) surface (intra-plane ordering) was first considered to be the top-most (001) surface at each step edge.[28,109] On the other hand, a theory using Monte Carlo simulation on Si_xGe_{1-x} predicted energetic stability of a near-surface structure with larger Ge atoms beneath dangling bonds and smaller Si atoms beneath dimers due to the compressive stress caused by the dimer formation.[110] A theoretical study on III-V alloy ordering provided several possible mechanisms on the intra-plane ordering[111]; (1) a mechanism similar to the SiGe case and (2) another different mechanism of Ga- and In-atom ordering at the top-most surface if the growth surface would show a reconstruction of group-III atoms. In any of these models, there are two problems to be solved: (1) what is the relevance of the surface reconstructions to the ordering phenomenon, and (2) where does the ordering occur?

The relevance of bond-direction anisotropy on the (001) surface with the appearance asymmetry of CuPt-B ordering[10] and the relevance of the (2x4) structure with CuPt-B ordering[56] had been suggested. Then, an experimental study on the relevance of (2x4) with CuPt-B ordering was carried out for

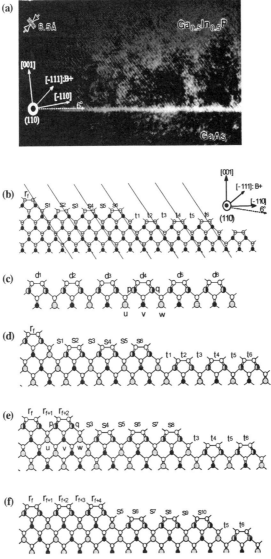

Figure 18. (a) TEM dark field image of $Ga_{0.5}In_{0.5}P$ grown on a $(001)6_B$ substrate, i.e., a (001) substrate with 6° misoriented towards [-111]; few defects are seen, suggesting (b) the step-terrace-reconstruction (STR) model [Refs. 107 and 109]. In (b), group-V atoms are represented by open circles with horizontal bars, and group-III atoms are represented by filled circles, respectively. (c) Relation between (2x1) reconstruction and CuPt-B ordering [Ref. 114]. (d)–(f): CuPt-B ordering processes [Refs. 121, 122, and 123] combining the STR model [Refs. 107 and 109] and the dimer-induced-stress model [Refs. 111 and 114]. In (c)–(f), open circles represent group-V atoms. Half-gray, half-filled circles represent In or Ga atoms. Gray circles and filled circles represent In and Ga atoms, respectively.

]BE-grown GaAsSb.[66] Relations of CuPt-B to (2x4) were further studied theoretically[111] and experimentally.[107,109] The relation was also reported for MBE-grown SiGe alloy.[112] The postulate that surface reconstruction on the (001) surface *during growth* plays a vital role has been perhaps most dramatically and definitively demonstrated by the correspondences between three characteristic surface reconstructions during growth and the three different types of ordering: (2x1) ↔ CuPt-B,[29,30] (1x2) ↔ CuPt-A,[29,50] and (2x3) ↔ TP-A.[29] These relations as a whole show that *the periodicities and the directions of arrangements of top-most group-V dimers* of the surface reconstruction correspond to the *periodicities and the directions of the ordered structures on the (001) plane of the respective types of ordering.*

These relations also provide the physical grounds for the asymmetric variant appearances (the 'BA asymmetry' in CuPt-B, and the 'AB asymmetry' in TP-A and CuPt-A) among the four equivalent variants, whereas the STR model provides a physical ground for the inter-plane ordering mechanism and for the 'BB-asymmetry' in CuPt-B ordering and the 'AA-asymmetry' in TP-A and CuPt-A ordering. This suggests that the ordering occurs at the step edges. As for BA (or AB) asymmetry, it is interesting to note that CuPt ordering in Si_xGe_{1-x} shows all the four <-111> variants. Thus, apparently, AB or BA asymmetry does not exist.[113] This is a natural consequence of (2x1) domains and (1x2) domains coexisting on a (001) SiGe surface.

The energetic stability of the respective structures (CuPt-B, CuPt-A, and TP-A) of the (001) intra-plane atomic ordering was clarified theoretically by the dimer-induced-stress model;[114] it was shown that, in the three types of ordering, smaller atoms (Ga for $Ga_{0.5}In_{0.5}P$) prefer to reside under the dimers and the larger atoms (In for $Ga_{0.5}In_{0.5}P$) prefer the sites not under the dimers (Fig. 18(c), Figs. 19(a) and (b)). The half-gray/half-filled circles in Fig. 18(c) represent the Ga or In atoms, for which the dimer-induced-stress model shows little site-preference. The experimental fact that the bond-length difference, or covalent radius difference, is a necessary condition (section 3.1) can be understood naturally.

It has also been reported that successive mono-layer by mono-layer growth of a Ga-(001) plane and an In-(001) plane produced CuPt-B ordering instead of the intended $(GaP)_1(InP)_1$ monolayer superlattice in the [001] direction.[115,116] This experimental result seems to favor the dimer-induced sub-surface ordering model[111,112] for the intra-plane ordering mechanism, rather than the top-most-surface selective adsorption model[107,109,117] of group-III atoms, or the group-III reconstruction model[111,118] at the top-most level on the (001) surface. The greatest merit of the dimer-induced-stress model is its simplicity. The model systematically explains major features of the intra-plane ordering, in spite of all the complications of the three different types of ordering. Simplicity, however, is not the sufficient condition; as for experimental verification of the dimer-induced-stress model, especially of the phase relation between the dimers and the ordered atomic arrangements, X-ray diffraction analyses have proved the phase relation between the atomic arrangements of Si and Ge and the positions of dimers and dangling bonds for CuPt ordering for SiGe.[119,120] An X-ray diffraction study on GaInAs[102] has also revealed the relation identical to the (2x3) structure shown in Fig. 15 and Fig. 19(a).

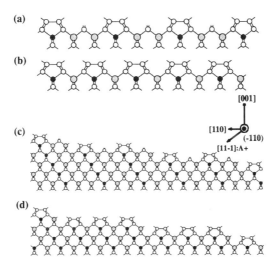

Figure 19. Dimer-induced-stress models: (a) for TP-A, and (b) CuPt-A [Ref. 114 for (a) and (b)]. STR models with dimer-induced-stress model: (c) for TP-A and (d) for CuPt-A. Symbols are the same as those for (c) - (f) in Fig. 18.

Based on the STR model[107,109] for inter-plane ordering and the dimer-induced-stress model[111] for the intra-plane ordering, a CuPt-B type ordering process has been proposed (Fig. 18(c)-(e)).[121,122,123] Starting from the surface structure of the STR model (Fig. 18(d)), adsorption of group-III atoms (e.g., p, and q) and group-V atoms (the dimer on p and q) may occur at the step edges, e.g., s_1, s_2 (Figs. 18(d)-(e)). Among various possible arrangements, the structure with the smaller atom Ga at v (beneath the dimer r_{f+2}) and the larger atom In at u between dimers (Fig. 18(e)) is energetically more stable than other structures, especially than that with In at v and Ga at u. When the crystal happens to take the latter structure or structures similar to it, such structures will find it difficult to survive and will tend to disintegrate to finally attain the structure with Ga at v and In at u before step-flow growth proceeds. Because step edges may be reactive places where crystal growth occurs, frequent restructurings such as mentioned above will occur more easily than on the flat (001) terrace surfaces. Similar processes may occur at each step edge and repeat the processes ((d) → (e) → (f), ...) to attain a coherent CuPt-B ordering such as that shown in Fig. 18(a). Although these figures illustrate the III-III-V-type alloy ordering, the III-V-V-type alloy ordering is also similarly explained.[123]

Similar processes should occur for TP-A and CuPt-A ordering, as shown in Figs. 19(c) and (d). The correctness of the detailed step-edge structures tentatively shown in these figures is not claimed; only the *fixed phase relation* for the dimers between adjacent terraces should be meaningful. The coherent phase-relation between upper-terrace dimers and lower-terrace dimers is expected to be a result of thermodynamic considerations at the surfaces, although theoretical treatments have not been reported so far.

In section 2.2.1, it was found that CuPt-B ordering best develops at around 650°–700°C, and the degree of ordering also depends on phosphorous pressure in MOVPE-grown crystals. These are well interpreted in terms of the ordering mechanism. For example, the fact that increasing T_g from 700° to 750°C while keeping phosphorous pressure constant reduces the degree of ordering (Fig. 4(c) → (d)) is explained by the change of surface reconstruction from (2x1) to a reconstruction with a lower phosphorous density,[107] which will give less stress due to a lower density of phosphorous dimers. In accord with this change of TED by the T_g increase, the TED intensity also reduces (Fig. 4(c) → (e)) when the phosphorous pressure is reduced under T_g being kept constant at 700°C. A reduction in P-dimer density with the increase in T_g is actually observed by surface photo-absorption spectroscopy.[124]

Increasing T_g may not only reduce the group-V coverage but may also increase the diffusivity of the atoms beneath the surface.[125] Although a moderate diffusivity of atoms near the surface is necessary for the restructuring at step edges for the ordering as mentioned above, excess diffusivity can reduce the degree of ordering. Further study is needed to determine how these two effects of diffusivity work, as T_g increases. However, another characteristic feature concerning the increase in bandgap at lower temperatures (e.g., 600°C) with the increase in V/III ratio in Fig. 4(left) is not yet well clarified. The general trend of the increase in bandgap in the lower T_g regime is due to an increase in the number of imperfections (Fig. 5) of CuPt-B ordering. An increase of these imperfections is likely to be explained by the lower mobility and diffusion coefficient of group-III atoms at and beneath the surfaces. The decrease in T_g reduces the diffusion near the surface and thus tends to make the restructuring at the step edges difficult, as mentioned above in the present section. In addition, lower atomic mobility at lower temperature tends to make atoms adsorb on each terrace rather than at step edges. Islands are then formed on (001) terraces. Because the islands have B_+ step arrays and B_- step arrays, B_+ and B_- domains are formed. This leads to the increase of the B_+/B_- domain boundaries and to a decrease in the degree of ordering. Pyrolysis of phosphine also becomes difficult as T_g decreases in this regime. Increase in adsorption of the insufficiently pyrolyzed phosphine on the growing surface may tend to disorder the (2x1) surface. This may be one of the reasons for the decrease in the degree of ordering in the lower-temperature regime.

4.2 Famatinite, Chalcopyrite, and CuAu-I ordering

Other types of ordering, famatinite, CuAu-I, and chalcopyrite (Fig. 1(e)-(g)), were reported in the early history of the atomic ordering study in III-V alloys. However, the mechanisms of these orderings seem to be quite different from those of CuPt-B, CuPt-A, and TP-A.

Famatinite-type ordering in LPE-grown GaInAs has been reported,[20] whereas there is no report on CuPt-A, CuPt-B, or TP-A ordering in LPE-grown alloys. CuAu-I ordering was first reported in AlGaAs.[21] Chalcopyrite ordering mixed with CuAu-I ordering was reported in MOVPE-grown GaAsSb.[22] TEM diffraction patterns of AlInAs and AlGaInAs grown by MOVPE similar to those in Ref. 20 and Ref. 22 have also been reported.[126] All the diffraction patterns that have appeared in Refs. 20, 22, and 126 are very similar. The diffractions are reported to come from small segregated domains. There are very few reports on famatinite and chalcopyrite ordering, and little is known about them.

CuAu-I ordering is, on the other hand, reproducibly observed in III-V alloys. Significant differences between CuPt-B (and CuPt-A and TP-A) and CuAu-I ordering are that (1) the CuAu-I ordering *occurs even in AlGaAs, in which the CuPt-B has never been reported,* and (2) CuAu-I is observed in alloys grown on (110) substrates with slight misorientation toward [001] or [00-1] by VPE methods. Exactly (110)-oriented substrates did not produce CuPt-B ordering[73] or produced little CuAu-I ordering.[127] It is noted that CuAu-I ordering is reported to occur in AlGaAs even on the (001) substrates.[21] These facts suggest that a different ordering mechanism is working in CuAu-I ordering.[73] RHEED observation during MBE growth of GaInAs indicated that a double-layer-height step array appears on the slightly misoriented (110) substrate, when growth conditions are such as that CuAu-I ordering occurs. A model of the ordering is proposed as follows.[128] If double-layer step-flow growth is assumed, Ga and In atoms can adsorb at each step edge. Because In has a stronger tendency to segregate to the surface, the double layers are formed with an In-rich layer on a Ga-rich layer. As the process repeats itself, CuAu-I ordering (Fig. 1(f)) appears. GaAsSb, a III-V-V type alloy, grown on slightly misoriented (110) InP by MBE, is also reported to show CuAu-I ordering.[129] The ordering mechanism may similarly be explained.

The CuAu-I ordering in AlGaAs grown on (001) GaAs substrates by MBE (classified as a VPE method) is interpreted as follows. When GaInAs is grown by VPE on substrates with near-(001) orientations under commonly used group-V-rich growth conditions, the (001) surface will show (2x1) (or (2x4)) reconstructions. Owing to the large difference (6.9%) between covalent radii (or bond lengths) of Ga and In, CuPt-B ordering will appear, as is actually observed.[26] However, in accordance with research experience, when AlGaAs is grown on (001) substrates, even if the surface takes the 2x surface reconstruction,[130] CuPt-B ordering does not occur because of the lack of the covalent-radius difference between Al and Ga atoms. Instead, if some growth conditions are fulfilled, e.g., formation of double-layer-high steps, CuAu-I should be formed even on the (001) surface, because Ga atoms may tend to segregate to the surface compared to Al atoms. This argument may explain the observed CuAu-I ordering in MBE-grown AlGaAs on (001) substrates[21] and explain why the observation is limited to AlGaAs.[131]

5. CONCLUDING REMARKS

Until the 1980s, LPE was used as a major practical growth technique for III-V alloys. In the 1980s, various demands appeared: high-precision thickness and composition control of ultra-thin heterostructure layers for advanced optoelectronic and electronic devices; new materials for new devices such as AlGaInP laser diodes that virtually cannot be grown by LPE; or the need for mass production. Such demand prompted the use of VPE growth techniques to a much greater degree than before. Techno-historically, these circumstances eventually provided researchers with the opportunity to discover the phenomenon of atomic ordering in III-V alloys in the mid-1980s.[132]

In the mid-1980s, the possibility of atomic ordering on the face-centered-cubic sublattices in III-V alloys was first discussed by theorists in terms of energetic stability of bulk crystals of III-V alloys by taking GaInP as a working material,[133] while CuPt-B ordering most prevalent today was not discussed. A possible relation between atomic ordering and bandgap deviation from that of disordered alloy was suggested, however.

Independently from the theorists, the experimentalists discovered famatinite, CuAu-I, and CuPt ordering and other types of ordering in III-V alloys.

CuPt-type ordering on the f.c.c. sublattice was originally reported in the $Cu_{0.5}Pt_{0.5}$ metal alloy many years ago,[134] but, interestingly, this has been the only metal alloy for which CuPt type ordering has been reported.[135] Its existence had been explained in terms of energetic stability of bulk $Cu_{0.5}Pt_{0.5}$ crystals, which was very recently proved to be the case.[135,136] On the contrary, in III-V semiconductor alloys grown by VPE, CuPt ordering is the most commonly observed type of ordering, as seen in section 2.2.3. And it has now become clear[29,30] that the CuPt(-B) ordering in III-V alloys occurs at a surface during VPE growth and is generated under the influence of the (2x1) surface reconstruction, which occurs under the most conventional VPE growth conditions that are used for obtaining high-quality epitaxial crystals. Although CuPt-B ordering in bulk crystals is calculated to be thermodynamically unstable,[106] the structure is thermodynamically stable at and near the growth surface.[111,114] Also, the embedded structure of CuPt-B ordering in the grown crystal after the successive growth of CuPt-B at the (001) growth surfaces by the intra-plane and inter-plane ordering mechanisms (section 4.1) is quite stable due to the high energetic barriers against atomic diffusion in bulk crystals at or below growth temperatures.

Atomic ordering is a conspicuous phenomenon that deserves extensive studies in its own right. But there is more to it. Epitaxial growth is a result of successive repetition of growth processes at surfaces, and thus, each growth process at the surface should largely determine the quality of the resultant crystal. Atomic ordering occurs just at and near the growing surface and is a very rare phenomenon in which surface reconstructions *directly* and *visibly* influence the atomic arrangement near the surface. Therefore the phenomenon reflects in detail the atomic processes at the growing surface. Thus, for studying the nature of atomic ordering, it should potentially provide an indispensable means for understanding the atomic processes at the growing surface. This would result in better control of crystal growth processes and improvements in crystalline quality and device characteristics.

MOVPE is an industrially important VPE growth method for III-V semiconductors, together with MBE. The surface processes of MBE have been relatively well studied due to the availability of *in-situ* monitoring methods during growth, e.g., RHEED; however, many of the surface processes of MOVPE have remained unknown due to the lack of easily available *in-situ* probing tools. In this sense, the insight obtained by the studies of atomic ordering in MOVPE-grown alloys is invaluable. Conclusions, such as that surfaces of almost all representative III-V alloys during MOVPE growth should have the (2x1) (or (2x1)-like) surface reconstruction under their typical growth conditions, were obtained for the first time from the studies of atomic ordering. This knowledge is of basic importance, because the first thing that we should do when we seriously study anything related to surfaces will be to know its atomic structure.

Another good example that shows the role of atomic ordering as a surface probing tool is that the observation of TP-A ordering in slightly Sb-doped $Ga_{0.5}In_{0.5}P$ grown by MOVPE at a high growth temperature of 650°C (section 3.1.2) quickly enabled us to suggest that the GaInP surface structure in this case should be the (2x3) reconstruction with double group-V layers in spite of the high growth temperature. Without the knowledge of atomic ordering, it would have been a difficult task in this case to infer the surface structure.

Although the ordering mechanisms and properties have been revealed to a large extent so far, there are still many issues to be elucidated. Studies on these problems may greatly contribute to deepen the understanding of the vapor-phase epitaxy processes.

ACKNOWLEDGEMENTS

The author would like to thank A. Gomyo, S. Iijima, K. Kobayashi, I. Hino, K. Makita, T. Ichihashi, and the members of the visible-laser group of NEC Corporation and the other present and past collaborators in works on atomic ordering. The author also would like to thank M. Ogawa and H. Watanabe for the support of this work

REFERENCES

1   Digital Versatile Disk.
2   H.Launois, M.Quillic, F.Glas, M.J.Treacy, *Inst.Phys.Conf. Ser.* No.**65**: Chap.6, 537(1983).
3   T.Suzuki, a progress report of the project: 'Visible Semiconductor Lasers' in the frame of the National Research and Development Project 'Optical Measurement and Control Systems', May 1983. The report was made to the Agency of Industrial Science and Technology, MITI of Japan.
4   H.Asai, and K.Oe, J.Appl.Phys. **53**, 6849 (1982).
5   M.Kume, J.Ohta, N.Ogasawara and R.Ito, Jpn.J.Appl.Phys. **21**, L424 (1982).
6   T.Suzuki, I.Hino, A.Gomyo, and K.Nishida, Jpn.J.Appl.Phys. **21**, L731(1982). In this paper, the operation wavelength of $Al_{0.5}In_{0.5}P/Ga_{0.5}In_{0.5}P/Al_{0.5}In_{0.5}P$ double-heterostructure by optical pumping was 6470Å at 90K. Since the active layer was precisely lattice-matched to GaAs, this wavelength is longer than the normal wavelength by about 250 Å.

7   A.Gomyo, K.Kobayashi, S.Kawata, I.Hino, T.Suzuki, and T.Yuasa, J.Cryst.Growth **77**, 367 (1986).

8   A.Gomyo, T.Suzuki, K.Kobayashi, S.Kawata, I.Hino and T.Yuasa, Appl.Phys.Lett. **50**, 673 (1987).

9   Y.Ohba, M.Ishikawa, H.Sugawara, M.Yamamoto, and T.Nakanishi, J.Cryst.Growth **77**, 374 (1986).

10   A.Gomyo, T.Suzuki and S.Iijima, Phys.Rev.Lett. **60**, 2645 (1988). 'B' in CuPt-B represents the ordering directions. See the caption of Fig.1. CuPt-B ordering and its correlation with the bandgap anomaly was first reported by T.Suzuki and A.Gomyo in the Invited Talk at the 7th Meeting Monbusho Special Project: 'Research on Alloy Semiconductor Physics and Electronics', Yamanashi, Japan, July 1987.

11   T.Suzuki, A.Gomyo, S.Iijima, K.Kobayashi, S.Kawata, I.Hino and T.Yuasa, Jpn.J.Appl.Phys. **27**, 2098 (1988).

12   S.-H.Wei, A.Zunger, Phys.Rev. **39**, 3279 (1989).

13   T.Kurimoto and N.Hamada, Phys. Rev. B **40**, 3889 (1989).

14   J.E.Bernard, S.-H.Wei, D.M.Wood, and A.Zunger, Appl.Phys.Lett. **52**, 311 (1987).

15   M.Kondow, H.Kakibayashi, S.Minagawa, Y.Inoue, T.Nishino, and Y.Hamakawa, J.Cryst.Growth **93**, 412 (1988). The zone-folding effect for CuPt-B was first suggested in this paper.

16   A.Mascarenhas, S.Kurtz, A.Kibbler, and J.M.Olson, Phys.Rev.Lett. **63**, 2108 (1989).

17   T.Kanata, M.Nishimoto, H.Nakayama, and T.Nishino, Phys.Rev. B **45**, 6637 (1992).

18   H.M.Cheong, A.Mascarenhas, P.Ernst, C.Geng, Phys.Rev. B **56**, 1882 (1997).

19   For reviews of atomic ordering from different points of view, see articles: T.Suzuki, Mater. Res. Bull. **22**, No.7, p.33, Materials Research Society (1997); A. Zunger, Mater. Res. Bull. **22**, No.7, p20 (1997); G. B. Stringfellow, Mater. Res. Bull. **22**, No.7, 27 (1997); For reviews until 1994, see also articles: A.G.Norman, T.-Y.Seong, B.A.Philips, G.R.Booker, and S. Mahajan, *Inst. Phys.Conf. Ser.* No**134**, Section 6, p.279, IOP Publishing Ltd. (1993); A.Zunger and S.Mahajan, *Handbook of Semiconductors*, Completely Revised Edition, Ed. by T.S.Moss, vol.3, ed. by S.Mahajan, Elsevier Science B.V., p.1399 (1994).

20   H. Nakayama, and H. Fujita, *Inst. Phys. Conf. Ser.* No.**78**, 289 (1986).

21   T.S.Kuan, T.F.Kuech, W.I.Wang, and E.W.Wilkie, Phys. Rev. Lett. **54**, 201(1985).

22   H.R.Jen, M.J.Cherng, and G.B.Stringfellow, Appl. Phys. Lett. **48**, 1603 (1986).

23   A.Ourmazd, and J.C.Bean, Phys.Rev.Lett. **55**, 765 (1985).

24   I.J.Murgatroyd, A.G.Norman, G.R.Booker, and T.M. Kerr, *Proc. Xth Int. Cong. On Electron Microscopy*, Kyoto, p.1497 (1986).

25   O.Ueda, M.Takikawa, J.Komeno, I.Umebu, Jpn.J.Appl.Phys. **26**, L1824 (1987).

26   M.A.Shahid, S.Mahajan, and D.E.Laughlin, Phys.Rev.Lett. **54**, 2567 (1987).

27   A.G.Norman, R.E.Mallard, I.J.Murgatroyd, G.R.Booker, A.H.Moore, and M.D.Scott, Inst. Phys. Conf. Ser., No.**87**, 77(1987).

28   T.Suzuki, A.Gomyo, and S.Iijima, "Strong ordering in GaInP alloy semiconductors: formation mechanism for the ordered phase," J.Cryst.Growth, **93**, 396(1988), Fig.1 used with permission of Elsevier Science.

29   A.Gomyo, K.Makita, I.Hino and T.Suzuki, Phys.Rev.Lett. **72**, 673 (1994).

30   A.Gomyo, M.Sumino, I.Hino and T.Suzuki, Jpn.J.Appl.Phys. **34**, L469 (1995).

31   The atomic ordering is conveniently and directly used in several ways in device studies and developments, for which we only refer to a review article by T. Suzuki in Ref. 19; some applications are discussed in the other chapters fof the present book.

32 $Ga_xIn_{1-x}P$ is lattice-matching to GaAs for x = 0.516. The value 0.516 is abbreviated to 0.5 in this chapter.

33 K.Kobayashi, I.Hino, A.Gomyo, S.Kawata, and T.Suzuki, IEEE J. Quantum Electron. QE **23**, 704 (1987).

34 A. Gomyo, F. Miyasaka, H. Hotta, K. Hukagai, K. Kobayashi, Appl. Surf. Sci. **130-132**, 469 (1998).

35 C.S.Baxter, W.M.Stobbs, J.H.Wilkie, "The morphology of ordered structure in III-V alloys: inferences from a TEM study," J.Cryst.Growth **112**, 373 (1991), Fig.14 used with permission of Elsevier Science.

36 E.Morita, M.Ikeda, O.Kumagai, and K.Kaneko, Appl.Phys.Lett. **53**, 2164 (1988).

37 D.M.Follstaedt, R.P.Scneider, Jr., and D.E.Jnes, J.Appl.Phys. **77**, 3077 (1995).

38 S.Yasuami, C.Nozaki, and Y.Ohba, Appl.Phys.Lett. **52**, 2031 (1988).

39 D.J.Arent, M.Bode, K.A.Bertness, Sarah R.Kurtz, and J.M.Olson, Appl. Phys. Lett. **62**, 1806 (1993).

40 S.R.Kurtz, L.R.Dawson, R.M.Biefeld, D.M.Follstaedt, and B.L.Doyle, Phys.Rev. B **46**, 1909 (1992).

41 T.Suzuki, A.Gomyo, I.Hino, K.Kobayashi, S.Kawata, and S.Iijima, Jpn.J.Appl.Phys. **27**, L1549 (1988).

42 A.Gomyo, H.Hotta, I.Hino, S.Kawata, K.Kobayashi, and T.Suzuki, Jpn.J.Appl.Phys. **28**, L1330 (1989).

43 W.D.Laidig, N.Holonyak, Jr., M.D.Camras, K.Hess, J.J.Coleman, P.D.Dapkus and J.Barden, Appl.Phys.Lett. **38**, 776(1981).

44 T.Suzuki, I.Hino, K.Kobayashi, A.Gomyo, and S.Kawata, OPTOELECTRONICS-Devices and Technologies, **4**, 317 (1989).

45 J.K.Schertleff, R.T.Lee, C.M.Fetzer, and G.B.Stringfellow, Appl.Phys.Lett. **75**, 1914 (1999).

46 A.Gomyo, T.Suzuki, K.Makita, M.Sumino and I.Hino, *Mat.Res.Symp.Proc.* **417**, p. 91, Materials Research Society (1996).

47 O.Ueda, M.Hoshino, M.Takeuchi, M.Ozeki, T.Kato, and T.Matsumoto, J.Appl.Phys. **68**, 4268 (1990). 'Cl-VPE' is abbreviation of Chloride-transport VPE.

48 Hydride-transport VPE uses hydrides (PH_3 in GaInP growth) for group-V sources. K.Nishi, and T.Suzuki, unpublished. PL peak energy had a lower value.

49 O.Ueda, T.Fujii, Y.Nakada, H.Yamada, and I.Umebu, J.Cryst.Growth **95**, 38 (1989).

50 T.Suzuki, T.Ichihashi and T. Nakayama, Appl. Phys. Lett. **73**, 2588 (1998).

51 S.N.Chu, R.A.Logan, and T.Tanbun-Ek, J.Appl.Phys. **72**, 4118 (1992).

52 T.-Y.Seong, A.G.Norman, G.R.Booker, A.G.Cullis, J.Appl.Phys. **75**, 7852 (1994).

53 T.Suzuki, and A.Gomyo, Abstracts for Sixth International Conf. On Molecular Beam *Epitaxy*, San Diego, USA, paper No. XII-3 (1990)

54 G.S.Chen, D.H.Jaw, and G.B.Stringfellow, Appl.Phys.Lett. **57**, 2475 (1990).

55 W.E.Plano, D.W.Nam, J.S.Major, Jr., K.C.Hsieh, and N.Holonyak, Jr., Appl.Phys.Lett. **53**, 2537(1988).

56 Y.-E.Ihm, N.Otsuka, J.Klem, H.Morkoc, Appl.Phys.Lett. **51**, 2013 (1987).

57 D.H.Jaw, G.S.Chen, and G.B.Stringfellow, Appl.Phys.Lett. **59**, 114 (1991).

58 Unpublished and referred to in H.R.Jen, D.S.Cao, and G.B.Stringfellow, Appl.Phys.Lett. **54**, 1890 (1989).

59 H.R.Jen, K.Y.Ma, G.B.Stringfellow, Appl.Phys.Lett. **54**, 1154 (1989).

60 T.-Y.Seong, G.R.Booker, A.G.Norman, and I.T.Ferguson, Appl.Phys.Lett. **64**, 3593 (1994).

61 S.R.Kurtz, L.R.Dawson, R.M.Biefeld, D.M.Follstaedt, and B.L.Doyle, Phys.Rev. B **46**, 1909 (1992-I).

62 S.Kurtz, J.M.Olson, A.Kibbler; J.Electron.Mater. **18**, 15 (1989).

63 J.C.Mikkelsen, Jr., and J.B.Boyce, Phys.Rev.Lett. **49**, 1412 (1982).

64 ε is defined as d(III$_1$-V) _d(III$_2$-V)]/(Average of d(III$_1$-V) and d(III$_2$-V)), where d(III$_1$-V) stands for a bond length for III$_1$-V. ε is similarly defined for III-V$_1$-V$_2$.

65 P.Bellon, J.P.Chevallier, G.P.Martin, E.Dupont-Nivet, C.Thiebaut, and J.P.André,Appl.Phys. Lett. **52**, 567 (1988).

66 I.J.Murgatroyd, A.G.Norman, and G.R.Booker, J.Appl.Phys. **67**, 2310(1990).

67 A.Gomyo, S.Kawata, T.Suzuki, S.Iijima, and I.Hino, Jpn.J.Appl.Phys. **28**, L1728 (1989).

68 T.Suzuki, A.Gomyo, S.Iijima, "Sublattice ordering in GaInP and AlGaInP: effects of substrate orientatons," J.Cryst.Growth **99**, 60 (1990), Fig.5 used with permission of Elsevier Science.

69 H.Hamada, M.Shono, S.Honda, R.Hiroyama, K.Yodoshi, and T.Yamaguchi, IEEE J.Quantum Electron. QE **27**, 1483 (1991).

70 S.Minagawa, and M.Kondow, Electron. Lett. **25**, 758 (1989).

71 M.Suzuki, Y.Nishikawa, M.Ishikawa, Y.Kokubun, J.Cryst. Growth **113**, 127 (1991).

72 A.Gomyo, T.Suzuki, K.Kobayashi, S.Kawata, H.Hotta, and I.Hino, NEC Res. Develop. **35**, 134 (1994). This work was first reported in the 1991 Spring Meeting of Jpn. Appl. Phys. Soc., Tokyo. (Extended Abstract, 30a-ZG-5).

73 A.Gomyo, T.Suzuki, S.Iijima, H.Hotta, H.Fujii, S.Kawata, K.Kobayashi, Y.Ueno, and I.Hino, Jpn.J.Appl.Phys. **27**, L2370 (1988).

74 M.C.Delong, D.J.Mowbray, R.A.Hogg, M.S.Skolnick, J.E.Williams, K.Meehan, S.R.Kurtz, J.M.Olson, R.P.Schneider, M.C.Wu, and M.Hopkinson, Appl.Phys.Lett. **66**, 3185 (1995).

75 A.Gomyo and T.Suzuki, unpublished.

76 L.C.Su, I.H.Ho, and G.B.Stringfellow, J.Appl.Phys. **75**, 5135 (1994).

77 M.Ishimaru, S.Matsumura, N.Kuwano, and K.Oki, Phys. Rev. B **51**, 9707 (1995).

78 S.Takeda, Y.Kuno, N.Hosoi, and K.Shimoyama, J.Cryst.Growth **205**, 11 (1999).

79 D.J.Friedman, J.G.Zhu, A.E.Kibbler, J.M.Olson, J.Moreland, Appl.Phys.Lett. **63**, 1774 (1993).

80 T-Y.Seong, A.G.Norman, G.R.Booker, and A.G.Cullis, J. Appl. Phys. **75**, 7852 (1994).

81 L.C.Su, S.T.Pu, and G.B.Stringfellow, J.Electron.Mater. **23**, 125 (1994) for GaInP.

82 G.S.Chen, and G.B.Stringfellow, Appl.Phys.Lett. **59**, 3258 (1991) for GaPAs.

83 'A step' indicates a monolayer-high step that will occur when (001) surface has a misorientation towards the [-1-1-1] or [11-1] directions.

84 T.S.Kuan, T.F.Kuech, and E.W.Wilkie, Appl.Phys.Lett. **51**, 51(1987).

85 O.Ueda, Y.Nakata, and S.Muto, J.Cryst.Growth **150**, 523 (1995).

86 A.Zunger, and S.Mahajan, *Handbook of Semiconductors*, Completely Revised Edition, Ed. by T.S.Moss, vol.3, ed. by S.Mahajan, Elsevier Science B.V., p.1399 (1994).

87 R.G.Alonso, A.Mascarenhas, G.S.Horner, K.A.Bertness, S.R.Kurtz, and J.M.Olson, Phys.Rev. B **48**, 11833(1993).

88 T.Nishino, Y.Inoue, Y.Hamakawa, M.Kondow, and S.Minagawa, Appl.Phys.Lett. **53**, 583 (1988). A possibility of the explanation is suggested on the analogy of the work for CuAu-I[14].

89 D.B.Laks, S.-H.Wei, and A.Zunger, Phys.Rev.Lett. **69**, 3766(1992).

90 S.-H.Wei and A.Zunger, Phys.Rev. B **57**, 8983(1998).

91   R.Tycko, G.Dabbagh, S.Kurtz, J.P.Goral, Phys.Rev. B **45**, 13452 (1992).

92   D.Mao, P.C.Taylor, S.Kurtz, M.C.Wu, W.A.Harrison, Phys.Rev.Lett. **76**, 4769 (1996).

93   R.L.Forrest, T.D.Golding, S.C.Moss, Z.Zhang, J.F.Geisz, J.M.Olson, A.Mascarenhas, P.Ernst, and C.Geng, Phys.Rev. B **58**, 15355 (1998).

94   S.B.Zhang, S.Froyen, and A.Zunger, Appl.Phys.Lett. **67**, 3141 (1995).

95   Considering the low coherency of the TP-A, the 80 meV bandgap-reduction reported in Ref.29 might be due to the large-scale composition-modulation, which has been reported by, e.g., F.Peiro et al., *Mat.Res.Soc.Symp.Proc.* **417**, p.265, Mater. Res. Soc (1996).

96   D.Shindo, A.Gomyo, J.-M.Zuo, J.C.H.Spence, J.Electron Micros. **45**, 99(1996).

97   B.A.Philips, I.Kamiya, K.Hingerl, L.T. Florez, D.E.Aspnes, S.Mahajan, and J.P.Harbison, Phys.Rev.Lett. **74**, 3640 (1995).

98   T.Suzuki, T.Ichihashi, and M.Tsuji, Proceedings of MRS Fall Meeting: Symposium I, paper I7-6, to be published.

99   T.Suzuki, T.Ichihashi, C.C.Hsu, and D.Shindo, unpublished.

100 A.Gomyo, K.Makita, I.Hino, T.Suzuki, "Effects of substrate misorientation on triple-period ordering in AlInAs," J.Cryst.Growth **150**, 533 (1995), Fig.4 used with permission of Elsevier Science.

101 K.Makita, A.Gomyo, "Gas source molecular beam epitaxy grown InGaAsP/InGaAlAs multi-quantum well structures with wide range continuum band-offset control," I.Hino, J.Cryst.Growth **150**, 579 (1995), Fig.4 used with permission of Elsevier Science.

102 M.Sauvage-Simkin, Y.Garreau, R.Pinchaux, M.B.Veron, J.P.Landesman, and J.Nagle, Phys. Rev. Lett. **75**, 3485 (1995).

103 T.Ichihashi, K.Kurihara, K.Nishi, and T.Suzuki, Jpn.J.Appl.Phys. **39**, L126 (2000).

104 T.Suzuki, T.Ichihashi, K.Kurihara, and K.Nishi, Abstract for the 10th Int. Conf. on MOVPE, Sapporo, Japan (2000). To be published in J.Cryst.Growth.

105 C.M.Fetzer, R.T.Lee, J.K.Schurtleff, G.B.Stringfellow, S.M.Lee, and T.Y.Seong, Appl.Phys.Lett. **76**, 1440 (2000).

106 J.E.Bernard, R.G.Dandrea, L.G.Ferreira, S.Froyen, S.-H.Wei, and A.Zunger, Appl. Phys. Lett. **56**, 731 (1990).

107 T.Suzuki, and A.Gomyo, in H.W.M.Salemink, and M.D.Pashley (eds.), *Semiconductor Interfaces at the Sub-Nanometer Scale*, Kluwer Academic Publishers, Netherlands, p.11 (1993).

108 The step-terrace structure illustrated in Fig. 18(b) was proposed for GaAs surface during MBE growth by Y.Horikoshi, H.Yamaguchi, F.Briones, and M.Kawashima, J.Cryst. Growth **105**, 326 (1990). However, there are two important differences. First, they considered that such a surface with group-V dimers does not exist during MOVPE growth, while the STR model assumes that the structure also exists for MOVPE growth. Second, their model for MBE growth of GaAs assumes that a whole reorganization of the group-V (2x1) structure on the terrace occurs every time atom adsorption occurs at each step; thus, the phase relation between upper-terrace dimers and lower-terrace dimers is not locked. On the contrary, the STR assumes that the terrace part of the (2x1) structure remain unchanged except at atep edges during step-flow growth to maintain the phase-relation, as illustrated in Figs.(d)-(f).

109 T.Suzuki, and A.Gomyo, J.Cryst.Growth **111**, 353 (1991).

110 P.C.Kelires and J.Tersoff, Phys.Rev.Lett. **63**, 1164 (1989).

111 J.E.Bernard, S.Froyen, and A.Zunger, Phys.Rev. B **44**, 11178 (1991).

112 F.K.LeGoues, V.P.Kesan, S.S.Iyer, J.Tersof, and R.Tromp, Phys.Rev.Lett. **64**, 2038 (1990).

113 F.K.LeGoues, V.P.Kesan, and S.S.Iyer, Phys.Rev.Lett. **64**, 40 (1990).

114 S.B.Zhang, S.Froyen, and A.Zunger, Appl.Phys.Lett. **67**, 3141 (1995).

115 B.T.McDermott, K.G.Reid, N.A.El-Masry, S.M.Bedair, W.D.Duncan, X.Yon, and F.H.Pollak, Appl.Phys.Lett. **56**, 1172 (1990).

116 Z.Spika, C.Zimprich, W.Stolz, E.O.Goebel, J.Jiang, A.Schaper, and P.Werner, J. Cryst. Growth **170**, 257 (1997).

117 G.S.Chen, D.H.Jaw, and G.B.Stringfellow, J.Appl.Phys. **69**, 4263 (1991).

118 S.Froyen, and A.Zunger, Phys.Rev.Lett. **66**, 2132 (1991).

119 N.Ikarashi, K.Akimoto, T.Tatsumi, and K.Ishida, Phys.Rev.Lett. **72**, 3198 (1994).

120 K.L.Whiteaker, I.K.Robinson, J.E.Nostrand, and D.G.Cahill, Phys.Rev. B **57**, 12410 (1998).

121 J.E.Bernard, *Materials Science Forum*, **155/156**, 131 (1994).

122 A.G.Norman, T.-Y.Seong, B.A.Philips, G.R.Booker, and S.Mahajan, *Inst. Phys.Conf. Ser.* No134, Section 6, p.279, IOP Publishing Ltd. (1993).

123 B.A.Philips, A.G.Norman, T.Y.Seong, S.Mahajan, G.R.Booker, M.Skowronski, J.P.Harbison, V.G.Keramidas, J.Cryst.Growth **140**, 249 (1994).

124 H.Murata, I.H.Ho, and G.B.Stringfellow, J.Cryst.Growth **170**, 219 (1997).

125 S.R.Kurtz, J.M.Olson, and A.Kibbler, Appl.Phys.Lett. **57**, 1922 (1990).

126 A.G.Norman, *NATO ASI Ser.*, Ser. B, Physics **203,** Plenum Press, New York, p.233 (1988).

127 O.Ueda, Y.Nakata, and S.Muto, J.Cryst.Growth **150**, 523 (1995).

128 O.Ueda, *Mat.Res.Symp.Proc.* **417**, p.31, Mater. Res. Soc. (1996).

129 O.Ueda, Y.Nakata, and S.Muto, Proc. 7th Int. Conf. InP and Related Materials, Sapporo, Japan, p.253 (1995).

130 L.Däweritz, and R.Hey, Surf. Sci. **236**, 15 (1990).

131 CuAu-I ordering may also occur in AlGaP and AlGaSb grown on (001) substrates by the same reason.

132 Among the published papers, the present authors notice a very carefully studied paper on $Ga_{0.5}In_{0.5}P$ by Olsen et al. in 1978 (G.H.Olsen, C.J.Nuese, and R.T.Smith, J. Appl. Phys. **49**,5523 (1978)). They reported a 25-meV-smaller bandgap of $Ga_{0.5}In_{0.5}P$, compared with the then-already-published value of crystals grown by LPE. However, they merely mentioned that their result was a 'refinement' of the already reported value. The crystals were grown by hydride-transport VPE (HT-VPE). Thus, the surface can have a reconstruction similar to that for MOVPE. The 'small' difference they noticed in their PL energy should be ascribed to the bandgap anomaly and the ordered structure should have been observed, if they would have made TEM observation. However, time had not matured in revealing the novel crystal structure that is being hidden behind the small difference in PL peak energy.

133 G.P.Srivastava, J.L.Martins, and A.Zunger, Phys.Rev. B **31**, 2561 (1985).

134 v.C.H.Johansson and J.O.Linde, Ann.Phys.(Leipzig) **82**, 29 (1927).

135 S.Takizawa, K.Terakura, and T.Mori, Phys.Rev.B **39**, 5792 (1989).

136 Z.W.Lu, S.-H.Wei, and A.Zunger, Phys.Rev.Lett. **66**, 1753 (1991).

Chapter 2

The Nature and Origin of Atomic Ordering in Group III-V Antimonide Semiconductor Alloys

A. G. Norman
National Renewable Energy Laboratory, 1617 Cole Boulevard, Golden CO 80401, USA.

Key words: atomic ordering, antimonides, III-V semiconductors, surface reconstruction, molecular-beam epitaxy, metal-organic vapour-phase epitaxy, transmission electron microscopy, reflection high-energy electron diffraction

Abstract: Group III-V antimonide semiconductor alloys exhibit several types of atomic ordering when grown by molecular-beam epitaxy and metal-organic vapour-phase epitaxy. This chapter describes in detail the ordered structures that are observed and discusses in depth the current understanding of the origin of the ordering. The atomic ordering is, in general, induced at the surface during growth. The type of ordering observed is shown to depend on the growth technique and the structure of the growth surface. For (001) surfaces, it is found that surface reconstruction, in particular, the formation of surface dimer bonds, plays a key role in the ordering process. Growth of layers with different surface reconstructions results in distinct types of atomic ordering. A segregation of different-sized atoms that is driven by dimer-induced subsurface stresses is believed to occur. This lowers the strain energy associated with the surface dimerisation and accommodation of the different-sized atoms at the reconstructed growth surface. Surface atomic steps play an important role in "phase-locking" consecutively ordered surface layers. It is concluded that this model is currently the one most able to explain the majority of the observed ordering behaviour in group III-V and group IV alloy semiconductor layers grown on near (001) orientation substrates.

1.   INTRODUCTION

Ternary and quaternary group III-V semiconductor alloys possess a wide range of optical and electrical properties and are thus important for a variety of devices.   Epitaxial layers of these alloys are grown by a number of techniques, including liquid-phase epitaxy (LPE), molecular-beam epitaxy (MBE), metal-organic vapour-phase epitaxy (MOVPE) and vapour-phase epitaxy (VPE).   These alloys, in general, have positive enthalpies of mixing and so were expected to exhibit miscibility gaps and be unstable toward clustering and phase separation, e.g., by spinodal decomposition, below a critical temperature.   However, it was calculated that for bulk alloys the coherency stresses associated with phase separation into regions having different lattice parameters would stabilise all the alloys against phase separation to extremely low temperatures.[1]   Despite this result, miscibility gaps have been experimentally observed and measured for the bulk form of some of these alloys, such as GaAsSb[2,3] and InAsSb[4], and evidence for phase separation was observed in epitaxial layers of III-V alloys such as GaInAsP[5-10] and InAsSb.[11-14]   Later work[15-17] indicated that when such alloys are grown in the form of epitaxial layers, relaxation of coherency stresses can occur at the free surface, an idea first suggested by Cahn,[18] and this leads to an increase in the critical temperatures toward values commonly used for epitaxial growth.

Simple bulk thermodynamic models, e.g., the regular solution model, suggested that they would not exhibit atomic ordering because of their positive enthalpies of mixing.   However, in a paper published in 1971, entitled "Thermodynamic Analysis of the III-V Alloy Semiconductor Phase Diagrams–1. InSb-GaSb, InAs-GaAs, and InP-GaP," Foster and Woods[19] deduced that the greater part of the excess free energy of mixing of the solids was in the excess entropy, rather than in the excess enthalpy.   This indicated that these materials might freeze with local ordering or structure, rather than as homogeneous random solids.   It was suggested that the excess free energy in the three size-mismatched semiconductor systems investigated resulted from strain-relief processes that occur during solidification to produce local ordering and a decrease in entropy.   They showed that these alloys are not regular solutions, and also, that they cannot be described by a quasi-chemical approach.   In their summary, it was suggested that "some of these alloys might not freeze as homogeneous random mixed crystals, but might exhibit some degree of local order in the nature of clustering, layering, or perhaps a vacancy-stabilised defect structure.   Such structures would be expected to have adverse affects on the performance of some semiconductor devices that are envisioned for these alloys."   Atomic ordering had already been observed in a variety of other bulk semiconductor alloys, e.g.,

$A^{II}B^{IV}C^{V}_{2}$ chalcopyrite-type alloys such as $ZnSnP_2$.[20,21] It was also well known that in metal alloys, phase separation, e.g., by spinodal decomposition, and atomic ordering can occur in the same alloy system, often in a cooperative or interdependent manner.[22] Therefore, the occurrence of atomic ordering in bulk III-V alloys could not be totally discounted. Furthermore, the majority of group III-V, II-VI, and group IV semiconductor alloys are grown epitaxially, which involves the incorporation of atoms at surfaces that are often reconstructed to have different bonding arrangements to the bulk. The nature of the surface and any atomic reconstruction would obviously be expected to play a critical role in how atoms of different types are incorporated into the growing alloy crystals. Therefore, surface thermodynamics, as well as bulk thermodynamics, needs to be considered for predicting the stability of epitaxially grown alloys toward phase separation and atomic ordering.

In 1974, Verner and Nichugovskii,[23] following the work of Khachaturyan[24] and Lifshitz,[25] established for III-V alloy substitutional solid solutions with the zinc-blende structure the possible ordered superstructures that may arise. This was achieved by analysing the conditions determining the thermodynamic stability of the ordered phases relative to the formation of antiphase domains. In 1975, Semikolenova and Khabarov,[26] after analysis of band gap and lattice parameter versus composition data for bulk $InAs_ySb_{1-y}$ alloys prepared by several methods, suggested that an ordered phase existed for the composition $InAs_{0.75}Sb_{0.25}$. Alloys of this composition exhibited the lowest band gap when prepared by the zone levelling method, and this was the only composition whose lattice parameter satisfied Vegard's law. X-ray diffraction measurements on synthesised solid solutions of this composition may have revealed the existence of superstructure reflections, indicating that ordering of the bulk alloy had occurred.

In 1984 and 1985, the first experimental evidence was obtained of atomic ordering in epitaxial layers of group III-V and group IV semiconductor alloys. Nakayama and Fujita[27,28] observed by transmission electron diffraction (TED) superlattice reflections in LPE $In_{1-x}Ga_xAs$ layers (size-mismatched alloy) grown on (001) InP substrates, which they interpreted as arising from the presence of famatinite ordering. Kuan et al.[29] found CuAu-I type ordering in MOVPE $Al_xGa_{1-x}As$ layers (size-matched alloy) grown on (110) GaAs substrates. Identical ordering was later reported for the size-mismatched $In_{1-x}Ga_xAs$ alloy grown by MOVPE on (110) InP substrates.[30] Ourmazd and Bean[31] and Murgatroyd et al.[32-35] discovered CuPt-type ordering on {111} planes in MBE SiGe and $GaAs_{1-y}Sb_y$ alloy layers, respectively, that were grown on (001) orientation substrates. Jen et al.[36]

reported a mixture of simple tetragonal CuAu-I type ordering and chalcopyrite ordering in (001) MOVPE $GaAs_{0.5}Sb_{0.5}$.

Simultaneously to these experimental discoveries, Srivastava et al.[37] performed first-principles, local-density, and total-energy minimisation calculations on both ordered and random models of bulk GaInP, and they concluded that certain ordered intermediate phases could be thermodynamically stable at low temperatures. The ordered phases considered were the $CuFeS_2$-type ($I\bar{4}2d$) chalcopyrite structure and simple tetragonal ($P\bar{4}m2$) with CuAu-I type cation sublattice for the alloy composition $Ga_{0.5}In_{0.5}P$ (Fig. 1). For the alloy compositions $Ga_{0.75}In_{0.25}P$ and $Ga_{0.25}In_{0.75}P$, the luzonite Cu_3AsS_4-type structure ($P\bar{4}3m$) with a Cu_3Au-type cation sublattice and the famatinite Cu_3SbS_4-type structure ($I\bar{4}2m$) with an Al_3Ti cation sublattice were considered (Fig. 1). The ordered phase with a CuPt-type cation sublattice ordered on {111} planes (R3m) was not considered in this work, although it has since been by far the most commonly observed ordered phase in epitaxially grown alloy layers on (001) orientation substrates. The ordered phases were concluded to be stable because they are strain reducing: they can simultaneously accommodate the different GaP and InP bond lengths in the alloy in a coherent fashion, thereby introducing less strain than would arise in a random alloy. Other work, however,[38-42] showed that the initial conclusions of this paper[37] were incorrect in that the ordered intermediate phases were not thermodynamically stable, but only metastable or unstable. For the majority of bulk III-V alloys the phase-separated state was in fact the lowest energy state. Epitaxial effects were found to change the stability of the ordered phases[40-42] because coherency strains associated with coherent epitaxial growth on a substrate acted to stabilise the alloys against phase separation. For example, chalcopyrite ordering was calculated to become stable in most epitaxial alloys, and the CuPt-type ordering was found to be less unstable.[42] These calculations, which ignored any effect of the free surface present during epitaxial growth, indicated consistently the chalcopyrite form of ordering to be the lowest-energy structure and the CuPt-type ordering the highest-energy structure for both the bulk and epitaxial forms of the majority of size-mismatched alloys, e.g., GaInP.

Since the initial experimental discovery of ordered structures in group III-V and group IV semiconductor alloys, several types of atomic ordering have been observed in a wide range of these alloy epitaxial layers, often present at the same time as phase separation. The most commonly observed ordered structures, i.e., CuPt-type on {111} planes, however, are not those calculated above to be the most stable in the bulk or epitaxial alloys, neglecting the free surface present during epitaxial growth. This is because the ordering observed is induced at the layer surface during epitaxial growth and subsequently frozen into the bulk of the layers. The extremely low rate of bulk diffusion in these alloys normally prevents the surface-induced ordering from rearranging to the lowest-energy structure calculated for the bulk. The occurrence of atomic ordering is often associated with significant changes in the electrical and optical properties of the layers such as band-gap reduction[43-45] and polarisation effects.[46] An understanding of the nature, origin, and effects of this atomic ordering is thus crucial to the optimisation of the performance of devices fabricated from these materials. In addition, if the atomic ordering can be controlled, the changed properties of the ordered material may be used in novel device structures.

In this chapter, we will first describe in detail the nature of atomic ordering and phase separation found in MBE and MOVPE antimonide alloys and then discuss the current understanding of the growth mechanisms operating to produce the observed microstructures.

2.   ATOMIC ORDERING IN GROUP III-V ANTIMONIDE SEMICONDUCTOR ALLOYS

2.1   Nature of Ordering

2.1.1   Possible ordered structures

Disordered ternary and quaternary III-V alloys have the zinc-blende crystal structure (Fig. 1). This consists of two interpenetrating, face-centred cubic (fcc) sublattices, one composed of group III atoms (cations) and the other of group V atoms (anions), displaced from each other by $a/4[111]$ (where a is the lattice parameter of the cubic zinc-blende alloy). For disordered alloys, the anions and cations are arranged randomly on the atomic sites of their respective sublattices. When atomic ordering occurs in a ternary alloy such as GaAsSb, the As and Sb atoms become arranged in an ordered fashion on the anion fcc sublattice. This leads to an increase in periodicity along certain crystal directions and the formation of a superlattice structure. In Fig. 1 are shown several possible superlattice structures for a perfectly ordered $AB_{1-y}C_y$ alloy such as $GaAs_{1-y}Sb_y$. In the superlattice structures, the lattice periodicity along certain crystallographic directions is increased over that in the zinc-blende random alloy. This means that normally forbidden superlattice reflections become allowed in X-ray and

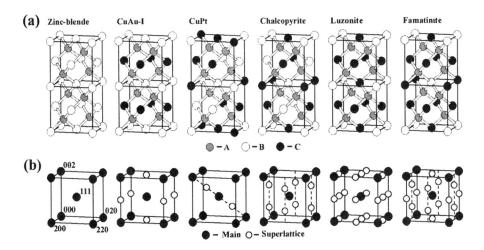

Figure 1. Atomic models and corresponding sections of reciprocal space for the random zinc-blende structure and various possible superlattice structures for $AB_{1-y}C_y$ alloys, e.g., GaAsSb.

electron diffraction patterns, and each superlattice structure has a characteristic array of superlattice reflections (Fig. 1), which enables the different superlattice structures to be distinguished. For the CuAu-I, CuPt, and chalcopyrite structures, perfect ordering can only occur at a composition of $AB_{0.5}C_{0.5}$, and the ordering occurs on {110} and {001} planes, {111} planes, and {210} planes, respectively. For the famatinite (Al_3Ti-type) and luzonite (Cu_3Au-type) structures, perfect ordering can only occur at compositions of $AB_{0.25}C_{0.75}$ or $AB_{0.75}C_{0.25}$. In the famatinite structure the ordering occurs on {100}, {110} and {210} planes, and in the luzonite structure the ordering occurs on {100} and {110} planes. The different superlattice structures are observed to occur over a wide range of compositions. Even at a composition for which perfect ordering is theoretically possible, only a partial ordering of the atoms to their correct sites on the ordered sublattice is observed normally. Antiphase boundaries (APBs) are also usually present in the ordered layers.

2.1.2   Experimental observations

2.1.2.1   CuPt-type ordering

During the early 1980s, A. G. Norman, whilst working for his D. Phil. in the research group of Dr. G. R. Booker in the Department of Materials, University of Oxford, observed evidence for the possible occurrence of phase separation, by spinodal decomposition, during the epitaxial growth of ternary and quaternary group III-V semiconductor alloys,[8,9,15] e.g., GaInAs and GaInAsP. In 1983, I. J. Murgatroyd, who had joined the group began to study if similar effects happened during the MBE growth of $GaAs_{1-y}Sb_y$ alloy layers.   In 1985, Prof. D. E. Laughlin of the Department of Metallurgical Engineering and Materials Science, Carnegie Mellon University, spent several months on sabbatical at Oxford, and was very interested in the results obtained by Murgatroyd and Norman; he is a specialist on phase transformations in metal alloys, in particular spinodal decomposition and ordering.   Despite the positive enthalpy of mixing observed for these alloys and their tendency toward phase separation, Prof. Laughlin suggested that they might also exhibit atomic ordering. The reason for this was that the simultaneous occurrence of phase separation and atomic ordering is quite often observed in metal alloys.[22] Following his remarks, Murgatroyd and Norman made a deliberate search for evidence of atomic ordering in these III-V alloy epitaxial layers and shortly afterwards, in 1985 and 1986, found the first evidence for atomic ordering in MBE GaAsSb layers[32-35,47] and MOVPE AlInAs and GaInAs layers[15,48] grown on (001) orientation substrates.   The type of ordering observed in these alloys was identical and corresponded to an ordering of atoms on {111} planes of the fcc mixed-atom sublattice of the zinc-blende structure of the alloys.   This type of ordering was called CuPt-type because ordering on {111} planes was previously observed in fcc CuPt alloys.   In both the mixed anion and mixed cation alloys, only two of the four possible variants were found, ordering on $(\bar{1}11)B$ and $(1\bar{1}1)B$ planes, referred to as the $CuPt_B$-type. This strongly suggested that the ordering was surface induced during growth and did not occur by diffusion in the bulk.   Since this initial discovery, CuPt-type ordering has been by far the most commonly observed type for III-V alloys grown on close to (001) orientation substrates.   It was shortly afterwards observed   independently   in   MBE   AlInAs,[49]   vapour-levitation   epitaxy GaInAs   and   GaInAsP,[50]   MBE   GaAsSb,[51]   and   MOVPE   GaInP layers.[44,45,52-57]

As described above, the first evidence of atomic ordering in antimonide III-V alloys was provided by Murgatroyd et al.[32-35,58] who found $CuPt_B$-type ordering on $(\bar{1}11)$ and $(1\bar{1}1)$ planes in MBE $GaAs_{1-y}Sb_y$ alloy layers, $0.25 \leq y < 0.71$, grown at 520°C at a rate of $\approx$ 1 µm/h. Fig. 2 shows [110] and $[\bar{1}10]$ transmission electron diffraction (TED) patterns taken from a MBE GaAsSb layer grown at 525°C. In the [110] pattern, Fig. 2(a), superlattice spots are present at the $1/2(\bar{1}11)$ and $1/2(1\bar{1}1)$ positions associated with ordering of As and Sb atoms on $(\bar{1}11)$ and $(1\bar{1}1)$ planes. The superlattice spots are connected together by weak rods of diffracted intensity running along [001]. This indicates the presence of monolayer disruptions in the ordering, e.g., antiphase boundaries or order twin boundaries,[59] lying on (001) planes. Fig. 3 shows a [110] projection of $GaAs_{0.5}Sb_{0.5}$ perfectly ordered on $(1\bar{1}1)$ planes. It can be seen that the perfectly ordered structure consists of a monolayer superlattice of GaSb and GaAs along the [1$\bar{1}$1] direction. A (001) monolayer of the ordered structure contains [110] rows of all As and all Sb atoms that alternate along the [$\bar{1}$10] direction giving a 2x periodicity in each ordered (001) layer along [$\bar{1}$10]. In the [$\bar{1}$10] TED pattern of Fig. 2(b), no superlattice spots are present at the 1/2(111) and 1/2($\bar{1}$$\bar{1}$1) positions, indicating no atomic ordering of As and Sb atoms occurring on these planes. This result strongly suggested that the atomic ordering originated at the surface during epitaxial growth because if the ordering occurred in the bulk, the four sets of {111} planes would be equivalent. The [$\bar{1}$10] TED pattern does, however, contain evidence for a modulation occurring in the crystal along [110] with a periodicity equal to $4d_{110}$. All the fundamental zinc-blende diffraction spots are extended by $\approx \pm$ $1/8g_{220}$ along [110], suggesting a modulation in the crystal along [110] of period $4d_{110}$. In addition, weak rods of diffracted intensity, which run along [001], are present at $\pm n/8g_{220}$ for all n except n = 0, 4, and 8. This may indicate the existence of weak ordering of periodicity $4d_{110}$ along [110] or quadruple period ordering on (111) and ($\bar{1}$$\bar{1}$1) planes, but with many monolayer disruptions in the ordering along the [001] growth direction. The 2x periodicity in the ordered crystal along the [$\bar{1}$10] direction and the 4x periodicity in the ordered crystal along the [110] direction correlate with the 2x and 4x periodicity of the (2x4) surface reconstruction observed during MBE growth of these layers by reflection high-energy electron diffraction (RHEED). Identical ordering behaviour was also reported shortly afterwards by Ihm et al.[51] in MBE GaAsSb alloy layers. CuPt$_B$-type ordering has been observed in MBE GaAsSb layers grown between $\approx$ 400° and 625°C.[13,32-35,51,60,61] However, it becomes very weak both at the low and high growth temperatures. At low growth temperatures, < 475°C, the alloy undergoes strong phase separation concomitant with the atomic ordering,[13,60,62] as also observed in MBE InAsSb alloy layers grown at low temperatures.[11-14,60,62,63] At the high growth temperature of 625°C, a different form of atomic ordering occurs,[61] as described later in this chapter. Surprisingly, CuPt$_B$-type ordering has not so far been reported for MOVPE GaAsSb layers.

CuPt$_B$-type ordering has also been reported for other antimonide alloys grown by both MOVPE and MBE. Jen et al.[64] first reported CuPt$_B$-type ordering in MOVPE InAs$_{1-y}$Sb$_y$ alloys grown at temperatures between 450° and 480 °C. The ordering was observed over a wide composition range, from y = 0.22 to 0.88, with a maximum degree of ordering for y $\approx$ 0.5. Seong et al. first observed CuPt-type ordering in MBE InAs$_{1-y}$Sb$_y$ layers.[11,63] The ordering was observed in layers grown at 370 °C and a rate of $\approx$ 1 μm/h, spanning the composition range from y = 0.2 to y = 0.8. A maximum in the

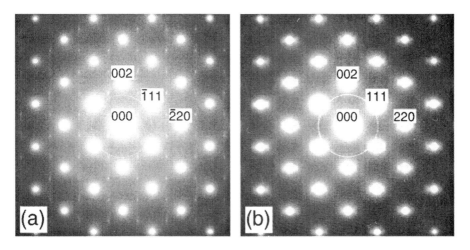

Figure 2. (a) [110] and (b) [$\bar{1}$10] TED patterns of MBE GaAsSb layer grown at 525°C showing evidence of atomic ordering on ($\bar{1}$11) and (1$\bar{1}$1) planes, and modulation along [110] direction of period $4d_{110}$, respectively.

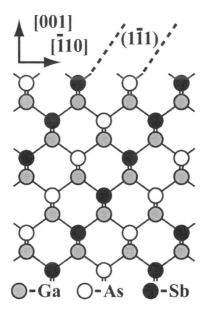

Figure 3. [110] projection of GaAs$_{0.5}$Sb$_{0.5}$ layer perfectly ordered on (1$\bar{1}$1) planes.

degree of ordering was observed for $y \approx 0.5$. Concomitant with the CuPt-type ordering, strong phase separation was observed in layers in the

composition range $0.2 \leq y \leq 0.6$.[11-14,60,62,63]   The phase separation
produced a microstructure consisting of tetragonally distorted platelets of
InAs-rich and InSb-rich material described by detail in Seong et al.[14]   In
later work, Seong et al.[63,65] studied CuPt-type ordering in MBE InAs$_{1-y}$Sb$_y$
layers $(0.4 < y < 0.6)$ grown over the temperature range 295° to 470°C and a
rate of ≈ 1 µm/h. Layers grown between 295° and 400°C exhibited phase
separation, whereas layers grown in the temperature range 430° to 470°C
were homogeneous in composition. The strength of the ordering was found
to be a maximum at growth temperatures of 370° and 400°C when the layers
had also undergone phase separation. The ordering became very weak at
low growth temperatures and appeared to be absent in layers grown at 450°C
and above. Micro-domains of ordered material were found in both the InAs-
rich and InSb-rich plates in the phase-separated samples, but were more
pronounced in the InAs-rich plates. [$\bar{1}$10] TED patterns contained lines of
diffuse intensity along the [001] growth direction that were uniformly spaced
with a separation of $1/6$ $g(220)$. These showed a maximum intensity in
layers grown at 430°C. These lines indicate a modulation in the layers along
the [110] direction of periodicity $3d_{110}$, where d is the lattice spacing along
[110]. RHEED patterns observed during growth of such layers revealed
intensity streaks associated with a periodicity of $2d_{110}$ along the [$\bar{1}$10]
direction and $3d_{110}$ along the [110] direction, corresponding to a (2x3)
surface reconstruction. The precise correlation of the periodicities observed
for the layer growth surface and the atomic ordering in the layers is further
evidence that the ordering is induced at the layer growth surface. Analogous
behaviour was described earlier for MBE GaAsSb layers, where a $4d_{110}$
periodicity in [$\bar{1}$10] TED patterns from the bulk layers was the same as the
$4d_{110}$ periodicity observed in the RHEED patterns of the (2x4) reconstructed
growth surface of the layers.

Kurtz et al.[66,67] subsequently reported CuPt$_B$-type ordering in MBE
InAs$_{1-y}$Sb$_y$ alloy layers $(y \approx 0.4)$ and superlattices, grown at 425°C with a
growth rate of 1 µm/h. A significant band-gap reduction, in comparison to
disordered alloy layers of the same composition, was found to accompany
the ordering.   This suggested that CuPt$_B$-type ordered InAs$_{1-y}$Sb$_y$ alloys
could effectively span the 8-12 µm window for long-wavelength infrared
devices. Earlier pseudopotential calculations indicated that significant bandgap reduction should occur in CuPt$_B$-type ordered InAs$_{1-y}$Sb$_y$ layers and that
a semimetal may result for fully ordered InAs$_{0.5}$Sb$_{0.5}$ layers.[68] A similar
band-gap reduction was also measured in MOVPE InAs$_{1-y}$Sb$_y$ alloy layers
and InAs$_{1-y}$Sb$_y$/InAs strained-layer superlattices grown at temperatures
between 475° and 525°C and growth rates between 0.75 and 3 µm per
hour.[69,70] These layers were also found to contain CuPt$_B$-type ordering,
even at Sb concentrations as low as $y = 0.07$.

As described above, $CuPt_B$-type ordering has been observed in both MOVPE and MBE $InAs_{1-y}Sb_y$ alloy layers. The degree of ordering was found to be greater and occurred at higher temperatures, with larger ordered domains, in the MOVPE layers compared to the MBE layers grown at similar rates. Similar behaviour has also been observed for $CuPt_B$-type ordering in AlInAs and GaInP alloy layers. The differences between the ordering behaviour in MOVPE- and MBE-grown alloys is poorly understood. Seong et al.[71] suggested several possible reasons for this difference in ordering behaviour observed in MBE and MOVPE $InAs_{1-y}Sb_y$ alloy layers. One reason may be that the surface reconstruction is different during MOVPE and MBE growth. If the surface reconstruction is the same, e.g., (2x3), it may occur over a higher temperature range for MOVPE growth as compared to MBE growth due to the effect of the different gaseous environments and molecular species present. In the MBE material, the ordered domains are observed to be smaller and poorly ordered in comparison to the ordered domains in MOVPE material. The main reason for this may be related to a difference in surface topography between the material grown by MBE and MOVPE, as revealed by atomic force microscopy. For the MOVPE material, there are surface ridges along the [110] direction, giving rise to surface steps along [110] that are favourable for producing large well-ordered domains of the two different variants at opposite sides of the ridges. For the MBE material that also exhibited phase separation, the ridges are along [$\bar{1}$10], with atomic steps along [$\bar{1}$10], a configuration unfavourable for $CuPt_B$-type ordering. As a result, poorly ordered small domains of the two variants form in a random manner on the sides of the ridges in the MBE material. There may also be a difference in the rate of disordering of the CuPt ordering in MOVPE and MBE layers during growth. The disordering occurs during subsequent growth, once the growth surface-induced ordering is buried in the bulk of the layer, because $CuPt_B$-type ordering is predicted to be unstable in the bulk. A difference in the rate of disordering could arise, for example, due to different concentrations of point defects and impurities in MOVPE and MBE-grown material.

$CuPt_B$-type ordering has also been observed in MOVPE $GaP_{1-y}Sb_y$ (very weak),[72] $InP_{1-y}Sb_y$ (grown at 450°C),[72] and $Ga_xIn_{1-x}Sb$ (grown at 525°C)[73] alloys by the research group of G. B. Stringfellow.

2.1.2.2 CuAu-I and chalcopyrite ordering

Simple tetragonal CuAu-I atomic ordering was first reported for antimonide alloys in 1986 by Jen and coworkers,[36] in MOVPE $GaAs_{1-y}Sb_y$ layers (y ≈ 0.5) grown at 580° and 600°C at a rate of ≈ 0.09 μm/min onto (001) InP substrates offcut 3° toward (110). Only two of the three possible variants of CuAu-I ordering were present. These were the two variants containing alternating planes of As and Sb atoms along the [100] and [010] directions that lie perpendicular to the [001] growth direction. At least two variants of six possible variants of chalcopyrite ordering were also simultaneously present in the layer. The simultaneous presence of both these types of ordering resulted in the (001) TED pattern shown schematically in Fig. 4. This pattern contains superlattice spots at {100} and {1 1/2 0} positions arising from the CuAu-I and chalcopyrite ordering, respectively. Further work[74,75] suggested that the two variants of chalcopyrite with the c-axis parallel to the growth direction were missing. TED patterns obtained from $GaAs_{0.25}Sb_{0.75}$ alloys contained {001} and {012} superlattice spots consistent with the occurrence of the $L1_3$ ordered structure, as has also been reported for the Cu_3Pt system. The orthorhombic structure of this ordering consists of {001} layers of the anion fcc sublattice composed alternately of all Sb atoms and of a 50/50 mixture of Sb and As atoms. A $Ga_{0.63}In_{0.37}As_{0.98}Sb_{0.02}$ sample contained a 2–3-nm-scale layered structure along [110], as well as a mixture of CuAu-I-type and chalcopyrite-type ordering.

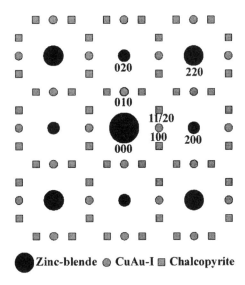

● Zinc-blende   ◉ CuAu-I   ▦ Chalcopyrite

Figure 4. Diagram of the (001) TED pattern observed for MOVPE $GaAs_{1-y}Sb_y$ layers by Jen et al.[36] containing a mixture of CuAu-I and chalcopyrite ordering.

A later paper[76] examined the effect of kinetics on the atomic ordering in (001) MOVPE $GaAs_{1-y}Sb_y$ ($y \approx 0.5$) layers by investigating layers grown at different growth temperatures and rates. Strong CuAu-I and chalcopyrite ordering were observed in $GaAs_{0.5}Sb_{0.5}$ layers grown in the range 580° to 660°C at both low and high growth rates. It was reported that the layer composition y must be very close to 0.5 in order to observe the ordered structures. At a growth temperature of 550°C, only weaker, more short-range, CuAu-I type ordering was observed for a growth rate of 0.02 μm/min. Reducing the growth rate by a factor of 2 led to the reappearance of weak chalcopyrite ordering, in addition to the more short-range CuAu-I type ordering. These results indicate that kinetics plays an important role in determining both the degree of ordering and selection of the type of ordering. In all the (001) $GaAs_{0.5}Sb_{0.5}$ samples examined in this work, a new superlattice structure was observed with a periodicity 3 times the normal zinc-blende periodicity along the [110] direction. The effect of substrate orientation was also investigated. For growth on (110) substrates, only the CuAu-I ordering was observed. It was observed in all layers grown in the temperature range 550° to 600°C, but was found to be weaker than that observed in the (001) layers. In addition, evidence was also found for a superlattice structure with 4 times the normal zinc-blende periodicity along the [001] directions. Layers grown on (221) and (311) orientation substrates showed virtually no atomic ordering. Ueda et al.[77,78] reported CuAu-I type ordering in MBE $GaAs_{1-y}Sb_y$ layers grown on (110) orientation InP substrates. In the temperature range of 470° to 530°C, the degree of ordering was found to increase with growth temperature. Off-cutting the substrate toward the [00$\bar{1}$] direction also led to an increase in the degree of ordering. Evidence was also observed for compositional modulations along the <001> and <110> directions, suggesting that phase separation and atomic ordering processes may compete with each other at the growth surface.

2.1.2.3   CuAu antiphase superlattice ordering

In 1990, when working jointly at the Interdisciplinary Research Centre for Semiconductor Materials, Imperial College of Science, Technology and Medicine, University of London and the Department of Materials, University of Oxford, on a fellowship, A.G. Norman investigated the connection between the surface reconstruction present during MBE growth and the nature of atomic ordering induced in ternary III-V alloys. The nature and origin of the phase separation that had been discovered in $InAs_{1-y}Sb_y$ and $GaAs_{1-y}Sb_y$ alloy layers grown at low temperatures[11] was also studied. During the course of this work, the effect of changing the surface reconstruction present during growth on the nature of the atomic ordering induced in $GaAs_{1-y}Sb_y$ was studied by growing layers of different compositions y over a wide range of temperatures. The growth rate (Ga flux) was kept constant and the alloy composition was varied by changing the Sb flux while keeping the As flux, supplied in excess, constant.

As mentioned earlier, CuPt$_B$-type ordering was observed in layers grown at temperatures between $\approx$ 400° and 625°C.[13,32-35,51,60,61] A weak modulation was also present in the layers grown at $\leq$ 600° C along the [110] direction of period four times the (110) plane spacing.[34,35,51,58,60,61] The CuPt$_B$-type ordering became very weak at both high and low growth temperatures and appeared to exhibit a maximum strength between 525° and 575°C. At growth temperatures below 475°C, the CuPt$_B$-type ordering occurred simultaneously with strong phase separation, as described in section 2.1.2.1. At a growth temperature of 625°C, detailed TED studies revealed the presence of a different form of atomic ordering in the GaAs$_{1-y}$Sb$_y$ layers,[61,79-81] described in detail below, simultaneously with very weak CuPt$_B$-type ordering. In-situ RHEED studies revealed that a (2x4) reconstruction was present during growth of the layers, with y $\leq$ 0.7, at all the growth temperatures investigated. At growth temperatures < 625°C, this is thought to be the β2(2x4) surface reconstruction[82,83] as observed for GaAs layers grown at similar temperatures and group V to group III flux ratios. At a growth temperature of 625°C, the intensity of the 2/4 and 3/4 [01] streaks, arising from the 4x periodicity of the reconstructed surface along [110], were very weak, as shown in Fig. 5. This suggests that the surface reconstruction may have changed to the α(2x4) reconstruction,[83-85] previously reported for (001) GaAs surfaces grown at high temperatures, or to the δ(2x4) surface reconstruction thought to occur for GaAs (001) surfaces in the presence of Sb at high temperatures.[86,87]

All the GaAs$_{1-y}$Sb$_y$ layers grown at 625°C contained virtually no CuPt$_B$-type ordering, as revealed by the virtual absence of 1/2($\bar{1}$11) and 1/2(1$\bar{1}$1) superlattice spots in [110] TED patterns, e.g., see Fig. 6(a). However, in [$\bar{1}$10] TED patterns (Fig. 6(b)), relatively strong superlattice spots were present (e.g., arrowed) at ±[3/4 3/4 0] from each of the fundamental zinc-blende diffraction spots, indicating that a new type of atomic ordering had occurred. In some samples of higher Sb content grown at this temperature, weaker diffraction spots were also observed at ±[1/4 1/4 0] from each of the fundamental diffraction spots. The superlattice spots were extended along the [001] direction, indicating disruptions in the ordered structure along the [001] growth direction. The strength of the superlattice spots was observed to be a maximum in layers with 0.1 < y < 0.2, and to decrease at higher Sb-contents simultaneously with the appearance of more continuous [001] rods

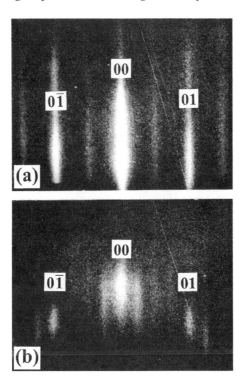

Figure 5. RHEED patterns obtained from surface of MBE $GaAs_{0.81}Sb_{0.19}$ layer growing at 625°C, showing streaks characteristic of (2x4) surface reconstruction. (a) [110] azimuth, (b) [$\bar{1}$10] azimuth.

of diffracted intensity connecting the spots. The (001) TED patterns taken from plan-view samples also contained relatively strong superlattice spots at ±[3/4 3/4 0] from each of the fundamental zinc-blende diffraction spots (Fig. 7). In some cases, weaker superlattice spots were also present at ±[1/4 1/4 0] from each of the fundamental diffraction spots.

The array of the strong superlattice spots observed in these samples can be viewed as consisting of pairs of spots (e.g., arrowed in Figs. 6(b) and 7), split by [1/2 1/2 0] along the [110] direction. These pairs of spots are arranged in the fingerprint pattern of {110} and {001} spots expected for simple tetragonal, CuAu-I type ordering on (001) planes of the group V (anion) atom fcc sublattice in the zinc-blende $GaAs_{1-y}Sb_y$ alloy (Fig. 1). Such characteristic splitting of superlattice spots is observed in some ordered metal alloys that contain a periodic array of antiphase boundaries (APBs) forming an antiphase superlattice.[88,89] Antiphase boundaries correspond to boundaries at which a displacement of the atom type occurs on the ordered planes in the superlattice structure. A classic example of such an antiphase

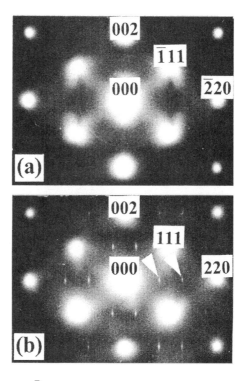

Figure 6. (a) [110] and (b) [$\bar{1}$10] cross-section TED patterns of GaAs$_{0.89}$Sb$_{0.11}$ layer grown at 625°C showing pairs of superlattice spots in (b) arising from antiphase superlattice along [110].

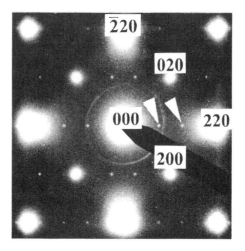

Figure 7. (001) TED pattern of GaAs$_{0.89}$Sb$_{0.11}$ layer grown at 625°C showing pairs of superlattice spots arising from antiphase superlattice along [110].

superlattice is CuAu-II. The crystallographic unit cell of CuAu-II is orthorhombic, and it is formed by stacking 10 CuAu-I tetragonal unit cells along the [010] direction and switching the content of the (001) planes from all Au to all Cu halfway along the long cell (i.e., after five CuAu-I unit cells). This results in an APB halfway along the cell and at subsequent similar intervals along the [010] axis. This structure can be referred to as a one-dimensional antiphase superlattice along the [010] direction. The periodicity of the APBs leads to a characteristic splitting of the superlattice reflections associated with the CuAu-I ordered crystal along a direction perpendicular to the APBs.[88,89] To create the splitting observed for the superlattice spots of the $GaAs_{1-y}Sb_y$ layers grown at 625°C, APBs need to be present along [110], with a periodicity of $2d_{110}$. This forms an antiphase superlattice structure of the CuAu-I type ordering with an orthorhombic unit cell of length $4d_{110}$ along [110] of the zinc-blende structure, as shown in Fig. 8. The structure is drawn for an alloy containing 11% Sb, i.e., identical to the alloy composition of the sample whose TED patterns are shown in Figs. 6 and 7, with all the Sb atoms segregating to the atomic columns

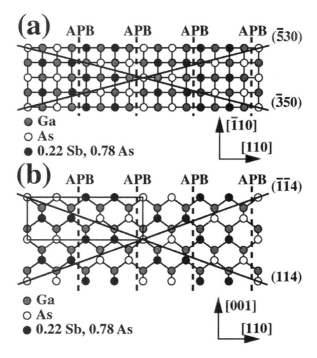

Figure 8. (a) [001] and (b) [$\bar{1}$10] projections (zinc-blende) of proposed antiphase superlattice structure in $GaAs_{0.89}Sb_{0.11}$ layer grown at 625°C. Unit cell outlined by box in (b).

shown producing the ordered structure. Confirmation of this ordered structure in MBE $GaAs_{1-y}Sb_y$ layers grown at 625°C was recently provided by quantitative synchrotron x-ray diffraction.[90] Kinematic diffraction calculations of the [$\bar{1}$10] and [001] TED patterns expected for the proposed [$\bar{1}$10] ordered structure, Fig. 9, reproduce well the experimentally obtained patterns of Figs. 6 and 7.

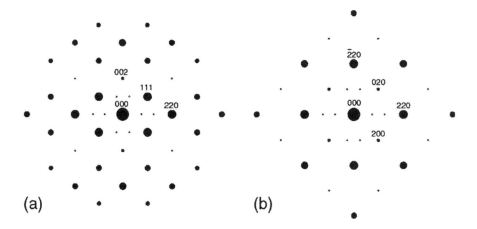

Figure 9. Kinematic diffraction calculations of (a) [$\bar{1}$10] and (b) [001] pole TED patterns expected from the ordered structure of Fig. 8.

Examination of the [$\bar{1}$10] and [001] projections of the ordered structure shown in Fig. 8 reveals that this ordering also corresponds to the simultaneous occurrence of long-period superlattices along [114] and [$\bar{1}\bar{1}$4] directions and [$\bar{3}$50] and [$\bar{5}$30] directions. This ordering on {114} and {350} planes can clearly be seen in the Fourier filtered high resolution lattice images of Fig. 10 together with the periodic array of antiphase boundaries along the [110] direction.

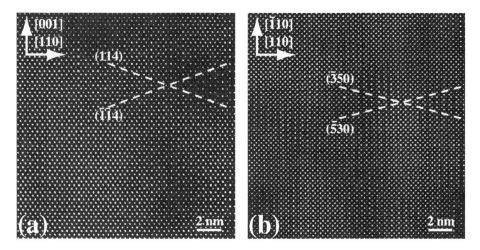

Figure 10. Fourier filtered lattice images of (a) [$\bar{1}$10] cross-section and (b) [001] plan-view samples of layer grown at 625°C showing ordering on {114} and {350} planes and periodic array of antiphase boundaries along [110].

3.   ORIGINS OF ATOMIC ORDERING IN SEMICONDUCTOR ALLOYS

3.1   CuPt-Type Atomic Ordering

3.1.1   Important characteristics of the observed CuPt-type ordering

CuPt-type ordering on {111} planes is observed in group IV semiconductor alloys such as GeSi and in III-V and II-VI semiconductor alloys grown on close to (001) orientation substrates. In III-V alloy layers grown on close to (001) substrates, $CuPt_B$-type ordering is by far the most commonly observed type of ordering. Since its first observation in MBE GaAsSb alloys,[32-35] it has been reported in virtually all III-V ternary alloys where a significant difference in atomic sizes exists for the atoms on the mixed anion or cation sublattice. It has not been observed in AlGaAs alloys, where the Al and Ga atoms have virtually identical sizes, strongly suggesting that atomic size differences play an important role in the ordering mechanism. Only the two $CuPt_B$-type variants of {111} ordering are normally observed for III-V alloys, indicating that the ordering is surface-induced because all four variants of CuPt ordering would be equally expected if the ordering occurred in the bulk of the layers. Evidence for monolayer abrupt changes in the ordering along the [001] growth direction such as APBs and order-twin boundaries[59] were observed in the diffraction patterns, also suggesting that the ordering was induced at the surface during growth. Annealing experiments indicated that the CuPt$_B$-type ordering in III-V alloys does not occur by bulk diffusion[15] and that in fact the ordered structure is thermodynamically unstable in the bulk of the layers.[91] CuPt-type ordering is found only in layers grown by MBE, MOVPE, and VPE techniques, where it is observed that surface reconstruction occurs during growth. It has not been observed in layers grown by LPE where reconstruction at the solid/liquid interface is not thought to occur.

CuPt-type ordering is only observed in layers grown on close to (001) orientation substrates under conditions that favour (2x1), (2x4), (1x2), and c(4x4) surface reconstruction in SiGe and III-V alloys, indicating that the presence of {110}-oriented surface atom dimers plays a key role in the ordering mechanism. Surface photoabsorption experiments by Murata et al.[92,93] suggested that the degree of CuPt$_B$-type ordering in MOVPE GaInP layers is linked to the concentration of [$\bar{1}$10]-oriented surface atom dimers. The critical role of surface reconstruction during growth in the origin of the atomic ordering is beautifully illustrated by experiments where the type of reconstruction during growth of III-V alloys is changed. This leads to the generation of different ordered structures to the normally observed CuPt$_B$-type ordering such as triple-period ordering on {111}A planes,[94] CuPt$_A$-type ordering,[95] and CuAu-I antiphase ordering.[61,79-81] Other work has shown that CuPt-type ordering can be almost completely eliminated during epitaxial growth of SiGe and III-V alloys by the use of surfactants, e.g., Sb and Bi,[96-100] that destroy or modify the surface reconstruction. Surface atomic steps are also found to play an important role in the evolution of CuPt-type atomic ordering. For example, growth of GaInP on suitably offcut (001) substrates leads to the preferential selection of a single variant of CuPt-type atomic ordering. Growth on (001) surfaces offcut toward [$\bar{1}$10] or [1$\bar{1}$0], with arrays of B-type monolayer steps whose edges run along [110], that descend toward [$\bar{1}$10] or [1$\bar{1}$0], promote extended domains of the ($\bar{1}$11) and (1$\bar{1}$1) CuPt$_B$-type variants, respectively.[101-103] Growth on (001) surfaces offcut toward [110] or [$\bar{1}$$\bar{1}$0], with arrays of A-type steps with edges running along [$\bar{1}$10], are found to suppress CuPt$_B$-type ordering. This indicates that the direction of step flow growth is important in determining the variant of CuPt-type ordering produced in the layers.

The above experimental observations on the nature of CuPt-type ordering in semiconductor alloys clearly demonstrate that atomic size differences, and the presence of dimer bonds at reconstructed growth surfaces, play key roles in the origin of the ordering. It was clear to experimentalists very early on that the $CuPt_B$-type ordering observed in epitaxial layers of III-V alloys was almost certainly induced at the surface during growth and that surface reconstruction probably played a vital part in the ordering mechanism. It is somewhat surprising, therefore, that the early theoretical calculations of the relative stability of ordered structures in III-V alloys totally ignored the key effect of the presence of a reconstructed free surface during MBE and VPE growth. These calculations only considered the bulk and epitaxial stability of the different ordered structures and neglected the presence of the free surface during epitaxial growth. Indeed, the most commonly observed type of ordering, CuPt-type on {111} planes, was not considered in the first theoretical papers on ordering in III-V alloys.[37,104] In later theoretical papers, which still ignored the effect of the free surface present during growth, it was shown that the CuPt-type of ordering was the most energetically unfavourable form for both bulk and epitaxial ordering in III-V alloys.[41,42] The widespread occurrence of CuPt-type ordering in epitaxial layers, when it was theoretically calculated to be the most unstable form of ordering in the bulk and epitaxial case, clearly indicated that the free surface present during growth needed to be included in models to explain the origin of this ordering.

If one examines closely the $CuPt_B$-type ordered structure in III-V alloys, e.g., for $GaAs_{1-y}Sb_y$ as shown in Fig. 3, it can be seen that both variants contain (001) ordered monolayers that consist of alternating [110] rows of the different types of atoms on the mixed atom sublattice. To form extended domains of either of the two variants, consecutively ordered (001) monolayers of this type need to be stacked in phase with each other to avoid either the formation of antiphase boundaries or order-twin boundaries between variants. Models to explain this ordering therefore need a mechanism to create the ordered (001) monolayers of the correct type, and also require a phase-locking mechanism to phase-lock consecutive (001) ordered monlayers to produce extended regions of the two different variants.

3.1.2   "Bond length" models

In 1987 and 1988, Norman[15] and Suzuki et al.[101] independently proposed surface mechanisms to explain the origin of the $CuPt_B$-type ordering observed in MOVPE mixed-cation alloys such as AlInAs, GaInAs, and GaInP. From extended x-ray absorption fine-structure (EXAFS) measurements, it was well known that in ternary III-V alloys such as GaInAs, the bond lengths and bond angles between Ga and As atoms and between In and As atoms remain very close to their values in the binary compounds.[105] Norman and Suzuki et al. therefore independently suggested that a two-dimensional (2D) ordering of the group III surface atoms occurred at the (001) growth surface to minimise the strain energy associated with incorporating the different-sized atoms and their different binary bond lengths into the growing crystal. Due to the anisotropy of bonding at the (001) surface of group III-V zinc-blende alloys, group III atoms bonding to a group V atom terminated (001) surface form chains of bonds along the [$\bar{1}$10] direction. A surface arrangement of group III atoms corresponding to a (001) monolayer of the two variants of CuPt$_B$-type ordering was concluded to be the lowest in energy, because in this arrangement the different-sized group III atoms alternate along the [$\bar{1}$10] direction. This enables the different bond lengths to be incorporated with a minimum in strain energy because of relaxation of the underlying group V atoms in both the [001] and [$\bar{1}$10] directions (Fig. 11). Some support for this hypothesis was provided by the valence force field (VFF) model calculations of the energies of different 2D ordered unreconstructed Ga and In cation (001) surfaces of GaInP published by Boguslawski.[106,107] The ordered cation (001) surfaces corresponding to a monolayer of the observed CuPt$_B$-type variants and to a monolayer of CuAu-I type ordering along [100] are composed of [$\bar{1}$10]-oriented chains of alternating Ga and In atoms. These ordered cation surface arrangements were found to have negative surface formation enthalpies, and hence to be stable against 2D surface segregation. This stabilisation was due to the effective lateral relaxation of the subsurface group V atoms described above. In contrast, the ordered cation surface corresponding to the two CuPt$_A$-type variants that are not normally observed, and for which this strain relaxation mechanism cannot occur, was shown to have a negligibly small surface formation energy, and hence, to be much less stable. Hamada and Kurimoto[108,109] also calculated the energies of different 2D ordered surface arrangements of Ga and In cations on GaAs and found that the (2x1) structure corresponding to a monolayer of CuPt$_B$-type ordering was the most stable. The c(2x2) arrangement corresponding to a monolayer of the CuAu-I or chalcopyrite ordered structure was also found to have a low energy. However, Froyen et al.[110-112] pointed out that the energies of the different ordered cation arrangements on unreconstructed (001) GaInP surfaces were too close to lead to a preference for any particular pattern at the temperatures typically used for epitaxial growth.

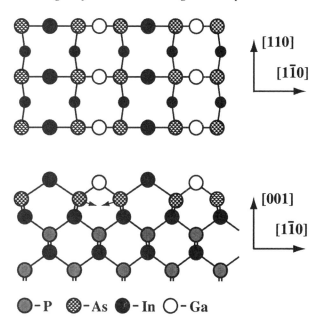

○ - P ⊕ - As ● - In ○ - Ga

Figure 11. Schematic diagram showing the lowest strain-energy arrangement of different-sized group III atoms on the unreconstructed (001) surface. It is in the form of [110] rows of the different-sized atoms that alternate along the [1̄10] direction. Arrows indicate displacements of subsurface anion atoms.

For GaInP, Suzuki et al.[101] also suggested that [110] B-type step arrays, descending in the [1̄10] or [11̄0] direction, could act as "phase lockers" to produce extended domains of the two CuPt$_B$-type variants. This was associated with a preferential incorporation of Ga atoms that was proposed to occur at {111}B atomic-scale micro-facets thought to be present at the [110] step arrays. Bellon et al.[102] also proposed a growth-induced ordering mechanism at step edges to explain CuPt$_B$-type ordering in the mixed-cation alloy GaInP. They considered the attachment of group III atoms to atomic steps on the growing surface. They proposed that the lowest-energy group-V centred tetrahedron at step edges on the surface would correspond to a tetrahedron with either a base of three Ga atoms and an apex of an In atom or a base of three In atoms and an apex of a Ga atom. This assumption enabled the system to accommodate bond-length differences at the surface. Hamada and Kurimoto[108,109] calculated the energies of different configurations of the cations adsorbed at B-type atomic steps on the unreconstructed (001) GaInP growth surface. The stablest arrangement was found to be with a row of Ga atoms first absorbed at the step edge followed by a row of In atoms. This supported the idea that CuPt$_B$-type ordering may be selected in the initial stages of growth by a preferential absorption of one of the cations at the step edges and the subsequent flow of the step edges along the [$\bar{1}$10] direction. The above "bond length" models, however, wrongly predicted that the CuPt$_A$-type variants should occur in mixed group V alloys such as GaAsSb. This is contrary to the observation of only the CuPt$_B$-type variants in these alloys by, e.g., Murgatroyd et al.,[35,58] Ihm et al.,[51] Jen et al.,[64] Seong et al.,[11,63] and Chen et al.[113]

3.1.3   Surface reconstruction models

It is now generally believed that the origin of CuPt-type ordering in semiconductor alloy epitaxial layers is related to the occurrence of surface reconstruction during epitaxial growth. On an unreconstructed group V terminated (001) surface of a III-V semiconductor, e.g., GaAs, each surface group V atom has two dangling bonds aligned along the [$\bar{1}$10] direction. The surface can lower its energy by reconstructing such that pairs of group V atoms form [$\bar{1}$10]-oriented dimer bonds, thus reducing the number of dangling bonds. Similar dimerisation of group III atoms occurs on group III atom-rich surfaces, but with the dimer bonds now normally being oriented along [110]. Depending on the semiconductor alloy, surface orientation, surface temperature, and surface group V/III atom ratio, a wide variety of surface reconstruction structures are observed that differ in the arrangement, direction, and number of the group V and group III atom dimers. For example, for (001) MBE GaAs surfaces, the surface reconstruction changes progressively in the order c(4x4), (2x4), (2x6), (4x2), and (4x6) as one goes from As-rich to Ga-rich surfaces.[114]   The As-terminated (2x4) reconstruction, under which MBE growth of (001) GaAs is normally performed, is also thought to occur as three different phases, α, β, and γ, depending on the group V/III surface atom ratio and temperature.[83,84,114] Such reconstructed surfaces have been widely studied in-situ under ultrahigh-vacuum MBE growth conditions by techniques such as low-energy electron diffraction and RHEED, and ex-situ on quenched samples by scanning tunnelling microscopy (STM). More recent studies using grazing incidence X-ray scattering,[115,116] in-situ optical techniques such as reflectance difference spectroscopy[117,118], and ex-situ STM[119-121] have indicated that similar surface reconstructions also occur under MOVPE growth conditions for selected systems, i.e., GaAs and InP. Figure 12 shows the currently accepted structure of the β2(2x4) surface reconstruction for (001) GaAs that is commonly present during MBE growth.[82,83,114] The surface structure consists of blocks of pairs of [$\bar{1}$10]-oriented As dimers followed by blocks of pairs of missing As dimers that alternate along the [110] direction. A single [$\bar{1}$10] row of As dimers is present at the next layer of As atoms beneath the missing blocks of As dimers. The 2x periodicity of the surface along [$\bar{1}$10] arises from the dimerisation of the surface group V atoms, whereas, the 4x periodicity along [110] arises from the regular spacing of the rows of missing dimers. The blocks of pairs of As dimers lie in phase across the rows of missing dimers in the (2x4) reconstructed surface. If they were out of phase, the reconstruction would correspond to c(2x8). The formation of dimers at reconstructed surfaces is found to induce sizeable subsurface strains (and associated stresses) in the atomic sites in the first few atomic layers beneath the reconstructed surface, as reported for (2x1) reconstructed (001) Si surfaces by Appelbaum and Hamann.[122] These subsurface strains (and associated stresses) are now believed to play a very important role in the ordering mechanisms, as described in detail below in section 3.1.3.2.

3.1.3.1　Surface models

In 1987, Murgatroyd[35] and Ihm et al.[51] independently suggested a link between the (2x4) surface reconstruction present during growth and the appearance of CuPt$_B$-type atomic ordering and [110] modulations in mixed anion alloy GaAs$_{1-y}$Sb$_y$ MBE layers grown on (001) substrates. Murgatroyd[35] applied the (2x1) and (2x4) asymmetric dimer surface reconstruction models proposed for (001) GaAs surfaces by Larsen et al.[123] to the growth of GaAs$_{1-y}$Sb$_y$ alloys on (2x4) reconstructed surfaces. From an analysis of RHEED and angle-resolved photoemission data, Larsen et al.[123]

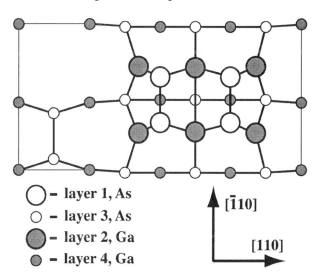

○ – layer 1, As
○ – layer 3, As
● – layer 2, Ga
● – layer 4, Ga

[$\bar{1}$10]

[110]

Figure 12. Diagram showing the structure of the β2(2x4) surface reconstruction for (001) GaAs.

proposed that the (2x1) and (2x4) reconstructed (001) surfaces of GaAs contain asymmetric dimer bonds, between pairs of group V atoms that are all aligned along the $[\bar{1}10]$ direction. (The (2x4) surface reconstruction model of Larsen et al.[123] for (001) GaAs was later shown to be incorrect by the scanning tunnelling microscopy results of Pashley et al.[124] on (2x4) reconstructed (001) GaAs.) Murgatroyd hypothesised that the As and Sb atoms might preferentially occupy different sites on the group V stabilised, reconstructed surface during layer growth. For example, the Sb atoms might preferentially occupy either the "up" or "down" atom sites of the asymmetric dimers of the Larsen model because of the $\approx 15\%$ difference in tetrahedral covalent radii between the Sb (0.1405 nm) and As (0.1225 nm) atoms and/or because of differences in their electronegativities (As = 1.57, Sb = 1.31).[125] Such a preferential occupation of the "up" and "down" atom sites of asymmetric dimers on a (2x1) reconstructed surface would result in an ordered surface monolayer of As and Sb atoms that is identical to that of an ordered (001) monolayer of the two $CuPt_B$-type variants observed in the layers. However, a (2x4) surface reconstruction was observed by RHEED during MBE growth of the layers, and this has a periodicity of $4d_{110}$ along the [110] direction, the same periodicity as the [110] modulation observed in the layers. Murgatroyd therefore suggested that during growth, alternate surface layers of the group V atoms reconstructed with the (2x1) and (2x4) structures of Larsen et al.[123] A lamellar domain structure of the two observed $CuPt_B$ variants having a periodicity of $4d_{110}$ along [110] would result from the preferential occupation of sites at the reconstructed surfaces. This model could not, however, explain the $CuPt_B$-type ordering observed in mixed-cation alloys, e.g., AlInAs and GaInP, and it did not include any phase-locking mechanism that would produce extended domains of the two $CuPt_B$ variants.

This model was later modified[58] to incorporate the missing dimer structure of (2x4) surface reconstruction revealed by the calculations of Chadi[82] and the STM experiments of Pashley et al.[124] It was again suggested that As-Sb dimers would preferentially form at the growing reconstructed surface due to chemical interactions. It was assumed that these dimers would all be oriented in the same sense, forming the alternating [110] rows of As and Sb atoms at the surface required for a (001) monolayer of $CuPt_B$-type ordering, although no mechanism was suggested for this. The origin of the $4d_{110}$ modulation observed in the layers was linked to the rows of missing dimers present on the (2x4) reconstructed surface. The model was able to predict the occurrence of only the two $CuPt_B$-type variants, the $4d_{110}$ modulation, and also, the monolayer abrupt disruptions along the [001] growth direction of both the $CuPt_B$-type ordering and the $4d_{110}$ modulation.

However, the model included no phase-locking mechanism to give extended domains of the two observed variants and was unable to explain the occurrence of CuPt$_B$-type ordering in mixed-cation alloys such as AlInAs and GaInP. Also, no calculations were performed to test the validity of the model.

Chen et al.[113] extended these ideas in attempting to explain the origin of CuPt$_B$-type ordering in MOVPE GaAsP alloy layers. They assumed that similar (2x1) or (2x4) surface reconstruction would occur at the group-V-rich MOVPE growth surface resulting in the formation of [$\bar{1}$10]-oriented, group V atom surface dimers. Experimental evidence for the presence of such group V atom surface dimers on MOVPE GaAs growth surfaces was later provided by the reflectance difference spectroscopy work of Kamiya et al.[117,118]. Chen et al. suggested that a thermodynamic selectivity of the incorporation of the different group V atoms at B-type [110]-oriented steps occurred as they moved in the [$\bar{1}$10] direction of the dimerised 2x reconstructed surface. The selectivity of incorporation of the group V atoms was proposed to occur because the bonding of a group V atom at a [110] step moving across the dimerised reconstructed surface alternates between two different configurations. In one configuration, the group V atom bonds to two group III atoms on the plane underlying the adatom only. In the other, it bonds to the two underlying group III atoms, in addition to forming a dimer bond to the group V atom adjacent to the adatom on the propagating step. It was thus suggested that the two different group V atoms in GaAsP would preferentially be incorporated at these different bonding sites. This process would then form rows of [$\bar{1}$10]-oriented As-P surface dimers all arranged in the same sense to give an ordered (001) monolayer corresponding to a (001) monolayer of CuPt$_B$-type ordering. Chen et al.[113] also introduced a mechanism for phase locking of consecutively ordered (001) monolayers to give extended regions of the CuPt$_B$-type variants. They suggested that dimers in the surface layer would preferentially form directly above the rows of smaller group V atoms in the buried lattice in order to minimise the dimer-induced subsurface strains described previously.[122] For mixed group III sublattice alloys, it was suggested that a selectivity in bonding of the different-sized group III atoms would be introduced by the motion of a kink down the [110] steps on the reconstructed surface, although no more details were given to support this model. Again, no calculations were performed to test these ideas.

In a series of papers, Suzuki et al.[126-128] re-examined their earlier formation mechanism of CuPt$_B$-type ordering in mixed-cation ternary III-V alloys. They proposed the following model from an analysis of the ordering behaviour observed during the MOVPE growth of GaInP on different offcut

(001) GaAs substrates and assuming that the growing surface was 2x reconstructed to form [$\bar{1}$10]-oriented group V atom dimers. It was first noted that the existence of an array of monolayer-high B-type steps, whose step edge extends along the [110] direction, was essential to obtain extended domains of the CuPt$_B$-type ordering variants in the layers. The concept of step-terrace-reconstruction (STR) was then introduced. For STR, it was suggested that on a non-planar 2x reconstructed group V atom terminated (001) surface containing [110] B-type step arrays descending toward either the [$\bar{1}$10] or [1$\bar{1}$0] directions, the terrace widths between step edges would contain an even number of group III atom sites. This enables all the group V atoms on the terraces to form dimers, as shown in Fig. 13, which is thought to be the most stable surface configuration. It was then shown that step-flow growth on such a reconstructed stepped surface would lead to the formation of large domains of the two observed CuPt$_B$-type variants if the following processes occurred: firstly, preferential incorporation of one of the group III atoms at the B-type step edges, e.g., Ga due to the stronger bonding energy of GaP in comparison to InP, or the larger In atoms due to a steric effect; secondly, if this was followed by the incorporation alternately of [110] rows of Ga and In atoms due to bond energy and bond length differences, and a steric effect. Providing that in local areas the direction of motion of the [110] B-type step edges is the same, phase locking of consecutively ordered (001) monolayers was shown to automatically occur for growth on a step-terrace-reconstructed surface, producing extended domains of the two CuPt$_B$-type variants without introducing antiphase boundaries. It was suggested that it would be most likely that the larger In atoms would be incorporated at the B-type step edges, due to a steric effect, rather than the incorporation of the smaller Ga atoms, that is favoured by the higher bonding energy of Ga-P in comparison to In-P. However, the selective incorporation of the larger In atoms at the B-type step edges would lead to the larger In atoms being situated directly under the P dimers and the smaller Ga atoms situated beneath the gaps between the dimers of the following monolayer overgrowth. This would be a very high strain-energy situation because the microscopic stresses introduced by the surface dimerisation would favour the opposite arrangement of these different-sized atoms.

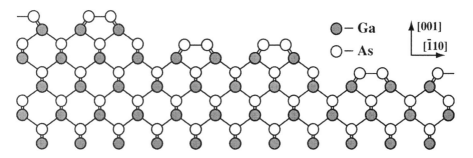

Figure 13. Step-terrace-reconstruction model of a 2x reconstructed vicinal (001) GaAs surface.

Suzuki et al.[127,128] also considered the origin of the $CuPt_B$-type ordering experimentally observed in mixed group V anion alloys. This cannot be simply explained by the mechanism described above, which predicts the occurrence of the two $CuPt_A$-type variants in these alloys. Similar to Murgatroyd et al.[35,58] and Chen et al.,[113] they suggested that there would be an energetic preference for the formation of mixed-anion atom dimers at the (001) growth surface. The mixed dimers were assumed to be composed of the two different kinds of group V atoms, all aligned in the same sense. It was proposed that the driving force for the formation of the mixed group V atom dimers was a reduction in the strain energy stored in the back bonds of the dimers. The alignment of the different group V atoms in the mixed atom dimers was associated with the bonding-energy difference between the two binary components of the III-V ternary alloy. It was suggested that as the B-type steps move across the growing surface, this would promote preferential incorporation of the group V atoms with the highest bonding energy at the atomic sites where bonding occurred to the two underlying group III atoms only. This was to be followed by incorporation of the other kind of group V atoms to form the mixed group V atom surface dimers, all aligned in the same sense, to minimise the strain energy of the dimerised surface. This is very similar to the model proposed by Chen et al.[113] No calculations were performed by Suzuki et al. to test these ideas. However, Philips et al.[129] performed valence force-field calculations to determine the minimum strain-energy atomic configurations during the growth of GaAsSb or GaAsP on (2x4) reconstructed (001) GaAs surfaces. The configuration of arsenic and antimony (phosphorus) atoms in the dimerised first-atom layer was found to result in no significant difference in the strain energy. Therefore, it was concluded that the strain-energy reduction in the dimerised surface layer, proposed by Suzuki et al.,[127,128] does not provide a driving force for the formation of 2D surface ordering, and hence, $CuPt_B$-type ordering in these mixed-anion alloys.

Froyen, Bernard, Osório, and Zunger[110-112,130,131] performed first-principles total-energy calculations to investigate the origin of spontaneous surface-induced CuPt$_B$-type ordering in (001) GaInP$_2$ layers. For unreconstructed cation-terminated surfaces, the calculations revealed that size differences between the cations could be readily accommodated by relaxation of the atoms perpendicular to the growth surface. However, the energy differences between the possible configurations of the surface cations were found to be insufficient to lead to any surface ordering at normal growth temperatures. On the other hand, an electronically driven (2x2) surface reconstruction (dimerisation with buckling and tilting) of a cation-terminated surface was shown to be energetically favourable. This stabilised the surface topology, corresponding to a monolayer of CuPt$_B$-type ordering over less favourable surface configurations by $\approx$ 90 meV per surface atom. However, as growth of III-V alloys is normally performed under group V (anion)-rich surface conditions, it is unlikely that such cation atom surface reconstruction could occur significantly during growth, thus severely limiting any ordering by this mechanism.

3.1.3.2   Dimer-induced subsurface stress ordering

In a key paper in 1989, Kelires and Tersoff[132] simulated the equilibrium alloy properties at the (2x1) reconstructed (001) SiGe alloy surface. The composition was found to vary with depth in a complex oscillatory way. In the following discussions on ordering mechanisms, the convention is adopted that the surface layer of atoms is layer one, the first subsurface layer is layer two, the next subsurface layer is layer three, etc., as presented in the original paper of Kelires and Tersoff. A strong segregation of Ge to the surface (layer one) was found to occur related to the lower surface energy of Ge with respect to Si. The Ge concentration in layer two was strongly reduced relative to the bulk. Even more unusual was the finding that layers three and four showed strong deviations from the bulk composition. In layers three and four, a lateral ordering of Si and Ge atoms was shown to occur to the two different sites of the (2x1) unit cell. These surprising results were explained by considering the local stress field beneath the dimerised surface. Subsurface strains beneath the (2x1) reconstructed dimerised (001) surface of Si were studied previously by Appelbaum and Hamann.[122] Some atomic sites could be viewed as being under compression, whereas, others could be viewed as being under tension. For the (2x1) reconstructed (001) SiGe alloy surface, Kelires and Tersoff[132] found that sites under compression would favour occupation by the smaller Si atoms, whereas, sites under tension would favour occupation by the larger Ge atoms to lower the strain energy. Layer two of the (2x1) reconstructed (001) SiGe surface was found to be under a large compression, and so, it favours occupation by the smaller Si atoms. In layers three and four, the atomic sites in the (2x1) cell are not equivalent. The atomic sites directly below the dimer are under compression, whereas, the other sites under the gap between dimers are under tension. The sites under compression thus favour occupation by the smaller Si atoms, and the sites under tension favour occupation by the larger Ge atoms at low temperatures. The effect was found to be slightly stronger in layer four than in layer three.

LeGoues et al.[96] used these findings to develop a model to explain the CuPt-type ordering on all four sets of {111} planes observed in MBE grown (001) SiGe alloy layers. They proposed that this ordering, previously thought to be a bulk effect, was in fact a result of a stress-induced local segregation at the dimerised (2x1) reconstructed surface, as observed in the calculations of Kelires and Tersoff.[132] The basis of their model (KTL model) is illustrated in Fig. 14. In atomic layers three and four, the smaller Si atoms tend to segregate to the atomic sites under compression directly beneath the dimers; the larger Ge atoms tend to segregate to the sites under tension beneath the gaps between dimers in order to lower the strain energy. In their model, they assumed that growth proceeded by the motion of double-height atomic steps across the growth surface, along either the [110] or [$\bar{1}$10] directions, resulting in the (Si-Si)–(Ge-Ge) {111} plane-type ordered structure, termed **RS2**, shown in Fig. 14. The other assumption of this model was that sufficient atomic diffusion could occur in the atomic layers three and four to give rise to the lateral ordering. Diffusion deeper in the crystal was assumed to be insufficient to allow disordering of the metastable ordered structure in the bulk. At the typical growth temperatures employed, bulk diffusion is negligible. However, it was suggested that the near-surface diffusion is enhanced by several orders of magnitude due to the considerable stresses induced by the surface reconstruction.

A different form of CuPt-type ordering will also result from the above stress-induced lateral segregation beneath a (2x1) dimerised surface by the motion of single-height atomic steps across the surface.[133] As each mono-layer is deposited, the dimer direction will rotate by 90°. Growth of successive monolayers forms a CuPt-type ordered structure that simultaneously contains ordering on two sets of {111} planes that intersect the (001) surface at 90° to each other. This structure, termed **MS2**, was also presented later by Jesson et al.,[134] who proposed an alternative ordering mechanism based on a kinetic segregation occurring as monolayer-height steps move across the (001) growth surface. Jesson et al.[135] later introduced another more general structure for {111} ordering in SiGe layers termed **RS3**. In this structure, the <111> stacking sequence was proposed to be (α-β)–(γ-δ), where α, β, γ, δ represent the probabilities of a Ge atom occupying

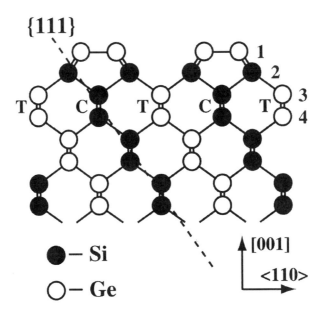

Figure 14. (110) projection of (2x1) reconstructed (001) SiGe surface illustrating the dimer-induced subsurface-stress ordering mechanism of Kelires and Tersoff[132] and LeGoues et al.[96] Numbers refer to atomic layers of the reconstructed surface. Sites under tension labelled T, sites under compression labelled C.

a site in each of these {111} planes. The structure was derived from a bilayer step-flow growth model, with the ordering resulting from a combination of surface segregation and atomic exchange processes at kink sites on the steps

TEM and TED experiments by LeGoues et al.[136] and x-ray diffraction measurements by Tischler et al.[137] indicated that the **MS2** structure was unlikely to be correct for the {111} ordering in SiGe layers. The x-ray diffraction measurements of Tischler et al.[137] revealed a best fit for the sample examined in detail, corresponded to values of 0.59, 0.54, 0.46, and 0.41 for α, β, γ, and δ, respectively, in the **RS3** structure. This structure was described as being similar to, but distinct from, a weakly ordered **RS2** structure. They thus concluded that the <111> (Si-Si)–(Ge-Ge) is a dominant stacking sequence in these ordered films. As pointed out by LeGoues et al.,[138] such a structure is still consistent with the dimer-induced subsurface strain ordering mechanism proposed by them[96] from the original calculations of Kelires and Tersoff.[132] These calculations also predict a stacking sequence of four inequivalent {111} planes of different alloy compositions. The surface x-ray diffraction experiments of Whiteaker et al.[139] on thin $Si_{0.5}Ge_{0.5}$ layers revealed the nature of the atomic ordering at the initial stages of growth beneath a (2x1) reconstructed surface. The ordering observed in the subsurface layers under the dimerised surface was consistent with the **RS2** structure and confirmed the predictions of Kelires and Tersoff that form the basis of the KTL ordering mechanism. The high-resolution TEM studies of ordered structures at Si on (001) Ge interfaces of Ikarashi et al.[140] also provided support for the KTL mechanism. Recent work by Uberuaga et al.[141] studied the diffusion of Ge into the subsurface layers of the dimerised Si (001) surface. They found, by Auger electron diffraction measurements, evidence for Ge present throughout the top four atomic layers after submonolayer deposition of Ge at temperatures as low as 500°C. A higher Ge occupation was observed in the atomic layers three and four in the sites between the dimer rows than beneath them, consistent with the KTL model. Density functional theory also predicted Ge distributions in layers three and four, consistent with the KTL model, and identified a surprisingly low-energy diffusion pathway, resulting from low interstitial formation energy in layers three and four. Strain-energy calculations by Araki et al.[142] showed that the **RS2** structure is stable under a (2x1) reconstructed surface, and they also proposed that atomic exchange processes were involved in the formation of the ordering. Thus, despite some controversy existing between the KTL ordering mechanism[96,132,136,138] and the alternative mechanisms proposed by Jesson et al.,[134,135] the majority of the experimental and theoretical data published seem still to be supportive of the KTL model.

The papers of LeGoues et al.[96] and Kelires and Tersoff,[132] led Norman et al.[13,60,63,143] and Mahajan et al.[144-146] to conclude that the model proposed to explain the CuPt-type ordering in SiGe alloys could be extended to the $CuPt_B$-type ordering observed in mixed-anion and -cation ternary and quaternary III-V alloys. The basis of the model[13,60,61,63,129,143-146] applied to III-V alloys is shown in Fig. 15. It was first assumed that the layers are grown under group-V-rich surface conditions and that the surface group V atoms reconstruct to form dimers, as shown in Fig. 15. The surface dimerisation of the group V atoms induces subsurface stresses in the III-V crystal, similar to those observed for (2x1) reconstructed (001) Si.[122] Atomic sites in layers three and four of the reconstructed surface are placed alternately under compression directly below the dimers and under tension or dilated below the gaps between dimers along the $[\bar{1}10]$ direction, as shown in Fig. 15. Now, if we consider the growth of a mixed-cation alloy such as GaInAs on such a reconstructed surface, it is proposed that the larger In atoms will tend to segregate to the [110] rows of atomic sites under tension in layer four. The smaller Ga atoms will tend to segregate to the [110] rows of sites under compression in layer four. This segregation minimises the strain energy associated with the dimerisation and the incorporation of the different-sized cation atoms at the reconstructed surface. This mechanism acting on a (2x1) or β2(2x4) reconstructed GaInAs alloy surface will produce the [110] rows of Ga and In atoms that alternate along the [$\bar{1}$10] direction required for an ordered subsurface (001) monolayer of the observed CuPt$_B$-type ordering. Disordering of the metastable CuPt-type ordering in the bulk during further growth is assumed kinetically limited, freezing in the surface-induced atomic ordering during further growth. For the growth of a mixed-anion alloy, e.g., GaAsSb, on such a reconstructed surface, it is proposed that a similar ordering process will occur. The larger Sb atoms will tend to segregate to the [110] rows of atomic sites under tension and the smaller As atoms to the [110] rows of sites under compression in layer three of the reconstructed surface to minimise the strain energy. This again leads to the required atomic arrangement for a (001) monolayer of the observed CuPt$_B$-type atomic ordering in mixed-anion alloys.

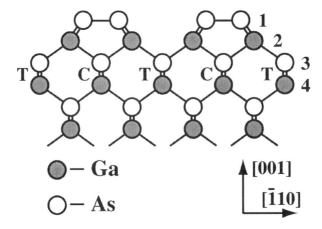

Figure 15. [110] projection of 2x reconstructed As-terminated (001) GaAs surface showing the nature of dimer-induced subsurface stresses. Sites under tension labelled T, sites under compression labelled C. Numbers refer to atomic layers of the reconstructed surface.

These proposals were later supported by the results of valence force-field calculations[61,129] that were performed to calculate the strain energy associated with the (2x4) surface reconstruction and the incorporation of the different-sized atoms in different configurations beneath the reconstructed surface. Simulation of the growth of GaInAs on (2x4) reconstructed (001) GaAs indicated that incorporation of In atoms in layer two of the reconstructed surface produced negligible strain-energy differences between different configurations. Incorporation of the In atoms in layer four of the reconstructed surface, however, produces significant strain-energy differences for different configurations. Preferential occupation of the [110] rows of atomic sites under tension beneath the gaps between dimers by the larger In atoms reduces the strain energy by ≈ 100 meV / dimer site over less favourable arrangements. Simulation of the growth of GaAsSb or GaAsP on (2x4) reconstructed (001) GaAs gave similar results. The configuration of arsenic and antimony (phosphorous) atoms in the dimerised first atom layer does not produce large differences in the strain energy. Therefore, strain-energy reduction in the dimerised surface layer proposed by Suzuki et al.[127,128] as a possible driving force for the origin of CuPt$_B$-type ordering in mixed-anion alloys such as GaAsSb does not seem to be responsible for the ordering. In atomic layer three of the reconstructed surface, a large difference between the strain energies of different configurations was found. The larger group V atoms tend to occupy the [110] rows of atomic sites under tension, whereas, the smaller group V atoms tend to occupy the [110] rows of sites under compression. A difference in strain energy between the favourable and unfavourable configurations of ≈ 120 meV / dimer site was calculated for the GaAsSb and GaAsP epitaxial layers. The valence force-field calculations therefore indicate that the CuPt$_B$-type atomic ordering in mixed-anion and -cation III-V alloy epitaxial layers most likely arises from a preferential occupation of atomic sites in atomic layers three and four of the group V stabilised reconstructed surface. This occurs to minimise the strain energy associated with surface dimerisation and the incorporation of different-sized atoms at the reconstructed surface.

Phase-locking of consecutively ordered layers to form extended domains of the two variants was attributed[13,61,63,129,144] to the effect of surface steps on the reconstructed surface. This was achieved by using the step-terrace-reconstruction model for a reconstructed stepped surface, proposed by Suzuki et al.[126-128] and illustrated in Fig. 13 for a vicinal (001) GaAs surface slightly offcut toward [$\bar{1}$10]. It is assumed first that the [110] steps on 2x reconstructed vicinal surfaces preferentially occur with the unbonded riser configuration shown in Fig. 13, so that the terraces between steps contain an even number of surface group V atoms. This configuration is thought to be the lowest in energy because all the group V atoms on the terraces between steps can form dimers. Now consider the growth of GaAsSb on such a surface. The dimer-induced subsurface stress ordering mechanism described above will lead to the segregation of the smaller As atoms to the compressed sites directly under the dimers and the larger Sb atoms to the dilated atomic sites under the gaps between dimers (Fig. 16(a)). It is assumed that step flow growth proceeds by the incorporation of atoms at the step edges, which continuously move toward the [$\bar{1}$10] direction. Each newly attached pair of surface anion atoms reconstruct to form dimers, and hence, cause further subsurface segregation of the different-sized group V atoms to occur. The situation after deposition of the next monolayer of crystal is shown in Fig. 16(b). The newly deposited anion dimers are shifted by $a/(2\sqrt{2})$ in the $[\bar{1}10]$ direction from the position of those that were in the underlying monolayer. Consequently, the [110] rows of dilated and compressed sites beneath the newly dimerised surface are also displaced similarly with respect to those in the lower terrace. This leads to the generation of CuPt$_B$-type ordering of the As and Sb atoms on $(\bar{1}11)$ planes, as observed experimentally for this type of vicinal surface. If a terrace width corresponding to an odd number of group V atoms was present on the reconstructed surface, the above mechanism would lead to the formation of an antiphase boundary in the ordered structure. The presence of fairly large domains containing few APBs in ordered samples suggests that the majority of terraces are an even number of group V atoms wide.

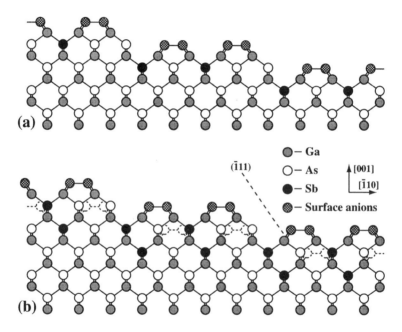

Figure 16. (a) Step-terrace-reconstruction (STR) model of 2x reconstructed vicinal (001) GaAsSb surface showing dimer-induced subsurface segregation of As and Sb atoms. (b) Crystal after step-flow growth of next monolayer showing phase-locking effect of STR.

The above model is able to easily explain why the same two variants of CuPt-type ordering, the B variants, are normally observed in both mixed group III and mixed group V alloys. Previous models that proposed to explain the origin of CuPt$_B$-type ordering in III-V alloys were not satisfactorily able to do this. The model also explains why CuPt$_B$-type ordering is not observed in AlGaAs layers. The atomic sizes of Al and Ga are almost identical, and hence, there is little driving force for the surface reconstruction stress-induced ordering. It would also explain why CuPt-type ordering is not found in alloy layers grown by LPE because surface reconstruction probably does not occur at the solid/liquid interface. It also can explain why the $CuPt_B$-type ordering is only found in samples grown on substrates oriented within a few degrees of (001). The presence of [110] rows of [$\bar{1}$10]-oriented group V atom surface dimers is essential for the ordering to occur and these are not present on surfaces such as (111), (110), and (311) that have very different surface structures. The model is also able to explain the ordering behaviour observed in layers grown on vicinal (001) substrates.

Concurrent with our work, Froyen, Bernard, Osório, and Zunger[110-112,130,131] performed first-principles total-energy calculations to investigate the origin of spontaneous surface-induced $CuPt_B$-type ordering in (001) $GaInP_2$ layers. Their results for unreconstructed and reconstructed cation-terminated surfaces were discussed previously in section 3.1.3.1. For a phosporous anion-terminated reconstructed surface, expected under normal growth conditions, little driving force was found for ordering in the cation layer, layer one, immediately beneath the surface.

Valence force-field calculations were performed[110-112,130,131] to investigate the energies of different arrangements of the cations when buried in deeper layers beneath both cation and anion reconstructed surfaces and for different topologies of cation atoms in atomic layers closer to the surface. For cation-terminated reconstructed surfaces, a weak preference for the topology corresponding to one of the $CuPt_A$-type variants was indicated for the first subsurface cation layer. In the case of the second subsurface cation layer, the topology required to produce true 3D $CuPt_B$-type ordering was found to be the lowest in strain energy for a cation-terminated reconstructed surface. For anion-terminated reconstructed surfaces, with buckling of the P dimers allowed, the cation arrangement corresponding to $CuPt_B$-type ordering was preferred in layer four, the second subsurface cation layer, by about 35 meV over the next most favourable arrangement (a $CuPt_A$ variant) regardless of the arrangement in layer two, the first subsurface cation layer. For a (2x1) surface reconstruction with unbuckled P dimers, the energy selectivity of the $CuPt_B$-type arrangement of cations in atomic layer four was increased to about 90 meV. The origin of this large preference for the $CuPt_B$-type arrangement of cations in layer four is the subsurface stresses induced by the surface P dimers. This favours occupation in the [110] rows of dilated sites under the gaps between dimer rows by the larger In atoms and occupation of the compressed rows of sites under the dimers by the smaller Ga atoms to lower the strain energy (Fig. 15). This is identical to the ordering mechanism first proposed by Kelires and Tersoff,[132] and LeGoues et al.[96] to explain the origin of CuPt-type ordering in SiGe alloy layers and independently extended by Mahajan et al.[129,144-146] and Norman et al.[13,60,61,63,129,143] to III-V alloys. Several scenarios were presented by Froyen and Bernard et al.[110-112,131] to account for the occurrence of the long-range CuPt$_B$-type ordering in GaInP$_2$ layers by both cation and anion surface reconstruction induced mechanisms, despite it being unstable in the bulk. One of these scenarios is the anion dimer-induced subsurface stress ordering mechanism that is identical to the work of Mahajan et al. and Norman et al. described in detail above. No calculations were performed for mixed-anion alloys such as GaAsSb, but it was suggested that a buckled dimer structure, which would tend to favour CuPt$_A$-type ordering, might be favoured in the surface anion layer. Subsurface strain induced by the dimerisation, however, was expected to promote the anion arrangement corresponding to CuPt$_B$-type ordering in layer three, the first anion subsurface layer, as suggested above (Fig. 15). This suggests that the dominant mechanism in III-V alloys giving rise to CuPt$_B$-type ordering is the dimer-induced subsurface stress mechanism, the extension of the KTL model for SiGe layers, described in detail above. It would be beneficial in determining the exact ordering mechanism if detailed first-principles total-energy calculations, combined with valence force-field calculations, were also performed for the near-surface layers of a mixed-anion alloy such as GaAsSb. In a later paper, Bernard[147] presented a dimer-induced subsurface stress model for the ordering in III-V alloys coupled with a step-terrace-reconstruction model to explain phase locking of consecutively ordered layers that is essentially identical to that presented previously[13,60,61,63,129,145,146] and described above.

If the CuPt$_B$-type ordering is a result of dimer-induced subsurface stresses, then if one changes the pattern and/or orientation of the surface dimers present during growth, one would expect also the type of ordering induced to change. This was experimentally confirmed by the work of Gomyo et al.,[94,95,148] Philips et al.,[149-152] and Norman,[60,61,79-81] who demonstrated that changing the surface reconstruction present during growth leads to the formation of different types of atomic ordering. Gomyo et al.[94] investigated the gas-source MBE growth of (001) Al$_x$In$_{1-x}$As layers as a function of growth temperature. At high temperatures (520°–570°C), a (2x1) surface reconstruction was present during growth and the expected CuPt$_B$-type ordering occurred, correlating with the double periodicity along the [$\bar{1}$10] direction of the (2x1) reconstructed surface. At lower growth temperatures (415°–460°C), a (2x3) surface reconstruction was observed by RHEED. This anion-rich surface reconstruction is thought to consist of a bilayer surface structure of As atoms, with the uppermost chemisorbed As atoms forming [$\bar{1}$10] rows of [110]-oriented dimer bonds separated by

[$\bar{1}$10]-oriented As dimers on the atomic layer below, giving rise to a triple periodicity along [110]. These samples contained a new form of atomic ordering with a triple period along the <111>A directions that correlates with the [110] direction of the triple period of the uppermost [110]-oriented As dimers in the (2x3) reconstructed surface. Identical triple-period <111>A ordering was also observed in GaInAs layers grown with (2x3) surface reconstruction.[148-152] Mahajan and coworkers suggested an explanation of its origin in terms of the subsurface stresses induced under the dimerised surface. The smaller Ga atoms were expected to preferentially occupy the [$\bar{1}$10] rows of atomic sites in the first subsurface cation layer directly under the rows of [110]-oriented uppermost As dimers. The larger In atoms were expected to occupy the neighbouring [$\bar{1}$10] rows of atomic sites under tension between the uppermost As dimers, thus giving rise to a 2D ordered arrangement of cations corresponding to a (001) monolayer of the triple-period <111>A ordering. Experimental evidence for this dimer-induced subsurface stress ordering mechanism was provided by the grazing incidence x-ray diffraction studies of (2x3) reconstructed (001) GaInAs alloy surfaces of Sauvage-Simkin et al.[153,154] Their results clearly indicated that the smaller Ga atoms preferentially occupied the [$\bar{1}$10] rows of compressive sites directly beneath the [110]-oriented uppermost As dimers, whereas, the larger In atoms occupied the adjacent [$\bar{1}$10] rows of sites that are under tension. However, similar work by the same authors on a (2x4) reconstructed (001) GaInAs alloy surface revealed a new form of (2x4) surface reconstruction and suggested an arrangement of Ga and In atoms in the near surface atomic layers that did not seem consistent with the dimer-induced subsurface stress ordering mechanism.[155] Recent STM and RHEED studies by Ohkouchi and Gomyo[156] linked the occurrence of triple-period <111>A ordering in MBE InGaAs layers to the presence of an anion-rich (4x3) surface reconstruction during growth. The proposed structure contained blocks of three [110]-oriented As dimer bonds in the uppermost chemisorbed As layer arranged with a three-fold periodicity along the [110] direction correlating with the triple period of the <111>A ordering.

Gomyo et al.[95,148] also discovered that CuPt$_A$-type ordering with double the periodicity in the crystal along the <111>A directions could be produced by gas-source MBE growth of AlInP and GaInP at low temperatures. Growth was performed on anion-rich (1x2) or (2x2) reconstructed (001) surfaces as deduced from RHEED. These surfaces are again thought to have a group V atom coverage of more than one monolayer. The uppermost chemisorbed P atoms form dimer bonds aligned along the [110] direction. The double periodicity on the reconstructed surface along [110] correlates with the double periodicity of the CuPt$_A$-type ordering. Recently, Suzuki et al.[157] also reported CuPt$_A$-type ordering in MBE AlInAs layers grown at low

temperatures with an As-rich (1x2) surface reconstruction as deduced from
RHEED. CuPt$_A$-type and triple-period <111>A ordering have also been
observed recently in MOVPE AlInAs layers grown at low temperatures.[158]
In contrast to the above work, Norman[60,61,79-81] investigated growth of
GaAsSb alloys at high temperatures with a less anion-rich surface
reconstruction present. At normal growth temperatures, the β2(2x4) surface
reconstruction, containing [$\bar{1}$10]-oriented group V surface dimers, was
present and led to CuPt$_B$-type ordering in the layers. At high growth
temperatures, a different form of (2x4) reconstruction, containing [110]-
oriented Ga dimers in addition to [$\bar{1}$10]-oriented group V dimers, was
present. This produced a one-dimensional antiphase superlattice along the
[110] direction, based on CuAu-I type ordering, as described previously in
section 2.1.2.3 and later in section 3.2.

Zhang et al.,[159,160] using valence force-field and ab initio
pseudopotential calculations, examined the influence of dimer-induced
subsurface stresses for different anion-rich surface reconstructions on the
nature of ordering in GaInP alloys. The results indicated that the orientation
of the uppermost surface phosphorous dimers correlates with the type of
subsurface ordering induced. The anion-terminated (2x1) reconstruction
with [$\bar{1}$10]-oriented P dimers stabilises subsurface 2D CuPt$_B$-type ordering,
the (1x2) bilayer P reconstruction with [110]-oriented uppermost P dimers
strongly stabilises subsurface 2D CuPt$_A$ ordering, and the (2x3) bilayer P
reconstruction with [110]-oriented uppermost P dimers stabilises subsurface
2D triple-period CuPt$_A$-type ordering. This is in very good agreement with
the experimental findings of Gomyo et al.,[94,95,148,156] Suzuki et al.,[157,158]
Philips et al.,[149,150,152] Mahajan et al.,[151] and Sauvage-Simkin et al.[153,154]
More extensive calculations by Froyen and Zunger[161] for the more realistic
surface reconstructions β2(2x4) containing [$\bar{1}$10]-oriented surface P dimers
and the c(4x4) reconstruction containing [110]-oriented P dimers in the
uppermost P layer revealed strong preferences for subsurface ordering with
the CuPt$_B$ and CuPt$_A$ structures, respectively, caused by the subsurface
stresses induced by the surface P dimers.

If dimer-induced subsurface stresses are responsible for the ordering,
then if one can completely remove the surface dimers, reduce their density,
or increase the length of the surface dimer bonds and so reduce the driving
force for the ordering, the ordering should be eliminated or the degree of
ordering reduced. LeGoues and coworkers[96] demonstrated the above effect
for SiGe layers by growing layers using Sb as a surfactant to destroy the
(2x1) surface reconstruction normally present during growth of CuPt-type
ordered layers. The presence of the Sb surfactant induced a (1x1) surface
reconstruction not thought to contain surface dimer bonds and resulted in no
CuPt-type ordering. A similar effect was also achieved by growing at low

temperatures using ultra-high-vacuum chemical vapour deposition,[162] where significant hydrogen absorption also leads to a (1x1) reconstruction and no CuPt-type ordering in the layers. Stringfellow and coworkers have extended these studies to III-V alloys. The degree of CuPt-type ordering in MOVPE GaInP alloys was reduced by decreasing the [$\overline{1}$10]-oriented P surface dimer concentration present during growth[92,93] by increasing the growth temperature and by decreasing the partial pressure of the group V precursor. Recently, they have investigated the use of isoelectronic surfactants such as Sb[97-99] and Bi[100] to modify the surface structure present during MOVPE growth of GaInP and so control the atomic ordering without increasing the bulk Ga/In interdiffusion. As small amounts of Sb are added to the system during growth, it was suggested that Sb dimers with longer dimer bonds replace P dimers on the surface. This reduces the driving force for the CuPt$_B$-type ordering resulting in a decrease in the degree of ordering in the layers.[97-99] Higher concentrations of Sb lead to a change in the surface reconstruction resulting in the formation of triple-period <111>A-type ordering in the layers.[163,164] This suggests that a change in the surface reconstruction as Sb is added to the surface might also contribute to the observed decrease in the degree of CuPt$_B$-type ordering. The addition of small amounts of Bi during growth similarly leads to large reductions in the degree of ordering.[100] It is not known how the presence of the surfactant atoms influences processes associated with surface atomic steps.

In summary, the dimer-induced subsurface stress mechanism for ordering, first proposed by Kelires and Tersoff[132] and LeGoues et al.[96] and extended by others to III-V alloys, currently seems to be the model best able to explain the widest range of ordering behaviour observed in SiGe, III-V, and II-VI alloys that are grown on near (001) orientation substrates.

3.2   Antiphase Superlattice in MBE GaAsSb Grown at High Temperatures

As described in detail in section 2.1.2.3, a novel form of ordering is observed in MBE GaAsSb alloy layers grown at high temperatures (625°C) on (001) substrates.[61,79-81] It is a one-dimensional antiphase superlattice based on simple tetragonal CuAu-I -type ordering, as shown in Fig. 8. In the previous section, it was shown that the best model currently to explain a wide variety of atomic ordering in size-mismatched alloys grown on (001) substrates is the dimer-induced subsurface stress ordering mechanism. It will now be explored whether the origin of the antiphase superlattice ordering in the GaAsSb layers can also be explained using this mechanism.

The growing surface of these layers exhibited a (2x4) surface reconstruction as revealed in-situ by RHEED, Fig. 5, although the intensity of some of the diffraction features indicated a different form of (2x4) reconstruction to the β2(2x4) structure expected at lower growth temperatures. At high temperatures, the surface reconstruction of (001) GaAs is thought to change to the less anion-rich α(2x4) structure,[83-85,114] Fig. 17, that contains two [$\bar{1}$10]-oriented As surface dimers and two [110]-oriented Ga surface dimers per unit cell. As described previously, the presence of the two [$\bar{1}$10]-oriented As surface dimers would be expected to cause a lateral ordering of As and Sb atoms along [$\bar{1}$10] in layer three of the reconstructed surface by the dimer-induced subsurface stress ordering mechanism. This would form a (001) monolayer of CuPt$_B$-type ordering. However, the α(2x4) surface reconstruction also has highly strained [110]-oriented surface Ga-dimers whose presence will induce sizeable subsurface compressive and tensile stresses, that alternate along the [110] direction, in layer five of the reconstructed surface. In section 3.1.3.2, it was described that the presence of subsurface stresses associated with chemisorbed [110]-oriented anion dimers in the uppermost surface layer of the anion-rich (2x3), (1x2), and c(4x4) reconstructions leads to a subsurface lateral ordering of different-sized atoms along this direction to lower the strain energy.[151,152,159-161] Growth of further phase-locked ordered layers, e.g., by the action of steps on the reconstructed surfaces, results in domains of triple-period <111>A and CuPt$_A$-type ordering, respectively. Therefore, it is highly likely for the α(2x4) surface reconstruction that the presence of the highly strained [110]-oriented surface Ga-dimers will also give rise to a tendency for a subsurface ordering of the different-sized As and Sb atoms in layer five along [110] to lower the strain energy. And so, during further growth, as the CuPt$_B$-type ordered (001) monolayers, induced in layer three by the [$\bar{1}$10]-oriented As surface dimers, pass through layer five it is believed that the Sb and As atoms will reorder along [110] to reduce the stresses induced by the [110]-oriented surface Ga-dimers. However, the [$\bar{1}$10] projection of this reconstruction (Fig. 17) reveals that the subsurface stresses induced by the dimers of this reconstruction in layer five of the reconstructed surface (anions) probably form a pattern of three [$\bar{1}$10] rows under compression adjacent to one [$\bar{1}$10] row under tension. The rows under compression would favour occupation by the smaller As atoms: the rows under tension favouring occupation by the larger Sb atoms. This would not produce an ordered (001) monolayer of the proposed antiphase structure of Fig. 8 that is the best fit to the electron diffraction data for these ordered samples.

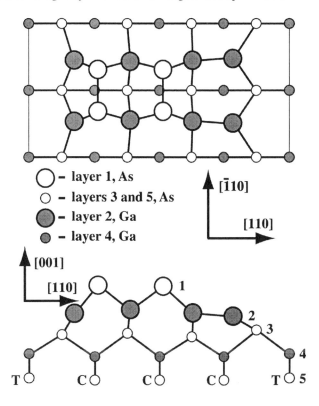

Figure 17. Diagram showing the proposed structure of the high-temperature, α(2x4) surface reconstruction of (001) GaAs. Subsurface sites thought to be under compression are labelled C, and subsurface sites under tension are labelled T.

To produce an ordered (001) monolayer of the proposed antiphase ordered structure, Fig. 8, requires a pattern of subsurface stresses, in layer five, of two [1̄10] rows of atomic sites under compression adjacent to two [1̄10] rows of atomic sites under tension. To give the required pattern of subsurface stresses required in the fifth atom layer for the antiphase ordering, it was suggested[79-81] that the surface reconstruction shown in Fig. 18 is present during growth. This reconstruction, termed δ(2x4) by Schmidt and Bechstedt[87], contains single [1̄10]-oriented group V atom surface dimers on the uppermost and third atomic layers and two [110]-oriented Ga dimers on the second surface layer per unit cell. As expected for the high growth temperature used, this reconstruction has a reduced group V coverage of 0.5 monolayer, in comparison to the β2(2x4) structure present at lower temperatures, which has a group V coverage of 0.75

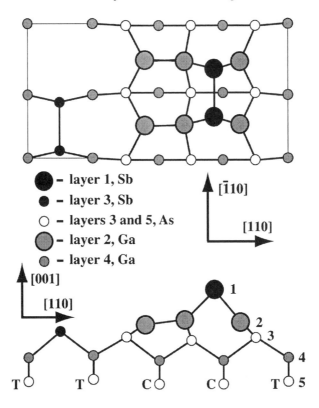

Figure 18. Diagram showing the Sb-terminated δ(2x4) reconstructed GaAs (001) surface. Subsurface sites thought to be under compression are labelled C, and subsurface sites under tension are labelled T.

monolayer. Some evidence for the existence of this surface reconstruction for Sb-stabilised (001) GaAs surfaces has been reported. Moriarty et al.[86] suggested a surface reconstruction with a unit cell containing only a single [1̄10]-oriented Sb dimer in the uppermost layer and two [110]-oriented Ga dimers in the second layer from scanning tunnelling microscopy studies of Sb-terminated (001) GaAs. Schmidt and Bechstedt[87] performed first-principles pseudopotential calculations of various Sb-stabilised (001) GaAs (2x4) reconstructions. Two structures, termed δ_1 and δ_2, with a single [1̄10]-oriented Sb dimer in the top surface layer and two [110]-oriented Ga dimers in the second layer, were found to be stable for the low Sb and high Ga chemical potentials expected for the high growth temperature of the ordered layers. This is also consistent with reflectance anisotropy spectroscopy measurements that revealed a coexistence of Sb and Ga dimers on a Sb-stabilised (2x4) reconstructed (001) GaAs surface.[165] Very recent theoretical work by Schmidt et al.[166] indicates that the equivalent single As dimer (2x4) reconstruction for (001) GaAs, termed α2, is more stable than the two-As-dimer α structure assumed previously,[83-85,114] due to its lower electrostatic energy. Pseudopotential and valence force-field calculations for the minimum strain-energy configurations of different-sized anion or cation atoms beneath the less anion rich α(2x4), α2(2x4), and δ(2x4) reconstructions would be of great interest.

In the antiphase ordering described above (Fig. 8), consecutively ordered (001) monolayers are laterally displaced by $3a/(2\sqrt{2})$, with alternate (001) monolayers being displaced in opposite <110> directions. This is different to the vertical stacking of ordered layers in CuPt$_B$, CuPt$_A$, and triple-period <111>A ordering where, consecutively ordered (001) monolayers are laterally displaced by $a/(2\sqrt{2})$ along either the [$\bar{1}$10] or [110] directions. This lateral displacement of $a/(2\sqrt{2})$ in these structures is associated with the identical lateral displacement that occurs for the surface dimers in consecutive monolayers and so it is easily understood. The origin of the different lateral displacement observed for adjacent ordered (001) monolayers of the antiphase ordering, however, is not understood at present.

In summary, the formation of a 2D ordered (001) monolayer of the antiphase ordering is explainable by the dimer-induced subsurface stress ordering model if growth occurs on a δ(2x4) or α2(2x4) reconstructed surface. The phase-locking mechanism for the vertical stacking of the ordered (001) monolayers is not known at present.

3.3   CuAu-I Type Ordering in Layers Grown on (110) Substrates

The experimental observations on this form of ordering suggest that it is also surface induced. However, the (110) surface is formed of an equal mix of group III and V atoms, and no reconstruction by dimerisation of surface atoms is thought to occur. Growth on different offcut substrates is seen to influence the degree of ordering, suggesting that surface steps play an important role in the ordering process.[78,167] The strongest ordering is observed for offcuts toward the [001] or [00$\bar{1}$] direction, with the degree of ordering increasing with the angle of offcut from (110) up to a value of $\approx 5°$. Atomic size differences do not appear to be important because it has been reported in both size-matched (e.g., AlGaAs)[29] and size-mismatched alloys (e.g., InGaAs).[30,168] Therefore, it may be that bond-energy difference effects play an important role in the ordering mechanism. It is also found in mixed-anion alloys such as GaAsSb.[77,78] Ueda et al.[78,167] proposed several possible mechanisms based on step flow growth to explain the origin of this ordering. Growth experiments suggested that the presence of two-monolayer-high atomic steps on the growth surface was required for strong ordering.[78,167] Recent calculations by Kangawa et al.[169] on the origin of

CuAu-I ordering in (110) InGaAs layers suggest that ordered InGaAs clusters are stabilised at kinks on two-monolayer-high steps and that the CuAu-I ordering results from the propagation of these kinked surface steps.

3.4 CuAu-I and Chalcopyrite Ordering in Layers Grown on (001) Substrates

The origin of this type of ordering (section 2.1.2.2, Fig. 4), observed by Jen and coworkers[36,74-76] in MOVPE antimonide alloys, remains uncertain. Chalcopyrite ordering is calculated to be thermodynamically stable in coherent epitaxial layers of Ga_2AsSb if one neglects any surface effects such as reconstruction.[41,42] CuAu-I ordering, however, was found to be unstable. The results of Jen et al. suggest that the CuAu-I and chalcopyrite-type ordering thought to be present in their layers occurred at the growth surface. This is because only the two variants of CuAu-I ordering, with alternating {100} planes along the two <100> directions lying in the (001) plane of the growth surface, were present. The two variants of chalcopyrite ordering, with the c-axis parallel to the growth direction, were also missing. Identical TED patterns to those of Jen et al.[36,74-76] (Fig. 4) have been reported from LPE InGaAs layers by Nakayama et al.,[27,28] but interpreted as arising from the presence of famatinite-type ordering. Because these InGaAs layers were grown by LPE, it is unlikely that the ordering was associated with any reconstruction of the growing surface. Norman also obtained similar TED patterns from MOVPE AlInAs and GaInAlAs layers grown at high temperatures (650°C).[15,59] High-resolution electron microscopy suggested that this ordering may also have been famatinite type. Further work needs to be done to determine the exact nature and origin of the ordering observed in these GaAsSb, InGaAs, and GaInAlAs samples.

4.   CONCLUSIONS

Several types of atomic ordering have been observed in group III-V antimonide alloys grown by molecular-beam epitaxy and metal-organic vapour-phase epitaxy. The ordering occurs at the growing surface, rather than in the bulk. The type of ordering found depends on the growth technique and surface structure, such as crystallographic orientation and surface reconstruction. For example, $CuPt_B$, CuAu-I, chalcopyrite, and an antiphase superlattice based on CuAu-I type ordering have been reported for MBE and MOVPE GaAsSb layers grown on (001) substrates, whereas, CuAu-I ordering has been found in (110) layers. For alloys grown on near (001) orientation substrates, the type of ordering induced in the layers can be selected by choosing a different surface reconstruction during growth. The most commonly observed form of ordering in III-V alloy layers grown on near (001) substrates is a $CuPt_B$-type ordering on {111} planes. The observed $CuPt_B$-type ordering behaviour seems to be best explained at present by a model that proposes a segregation of different-sized atoms at the (2x1) or $\beta2(2x4)$ reconstructed surface of the growing layer associated with the presence of dimer-induced subsurface stresses. The ordering occurs to minimise the strain energy associated with the surface reconstruction and accommodating the different-sized atoms at the reconstructed surface of the growing alloy crystal. Atomic steps on the reconstructed surfaces appear to act as phase-lockers of consecutively ordered (001) monolayers, leading to extended domains of a single variant of ordering. This model can also explain the generation of several other ordered structures in layers grown with different surface reconstructions. The origin of the simultaneous occurrence of CuAu-I and chalcopyrite ordering in (001) MOVPE GaAsSb layers, and CuAu-I ordering in (110) layers, however, remains poorly understood.

ACKNOWLEDGEMENTS

It is a pleasure to acknowledge the valuable contributions to this work of my former colleagues from the Department of Materials, University of Oxford, Prof. G. R. Booker, Dr. I. J. Murgatroyd, and Prof. T.-Y. Seong (now at the Kwangju Institute of Science and Technology, Korea), and from the Interdisciplinary Research Centre for Semiconductor Materials, Imperial College of Science, Technology and Medicine, University of London, Dr. I. T. Ferguson (now at Emcore Corporation, New Jersey, USA.) and Prof. B. A. Joyce. I would also like to thank Prof. S. Mahajan of the University of Arizona USA., Prof. D. E. Laughlin of Carnegie Mellon University, USA., Prof. G. B. Stringfellow of the University of Utah USA., and Prof. T. Suzuki of NEC, Japan, for stimulating discussions. My current colleagues at the National Renewable Energy Laboratory, Dr. J. Olson, Dr. A. Mascarenhas, Dr. M. M. Al-Jassim, and Dr. S. Ferrere, are thanked for valuable discussions and their support while writing this chapter. This research was supported by the U.K. Engineering and Physical Sciences Research Council and by the National Renewable Energy Laboratory (operated by the Midwest Research Institute, Battelle, and Bechtel for the U.S. Department of Energy under Contract No. DE-AC36-99GO10337).

REFERENCES

1 G. B. Stringfellow, J. Electron. Mat. **11,** 903 (1982).
2 M. F. Gratton, R. G. Goodchild, L. Y. Juravel, and J. C. Woolley, J. Electron. Mater. **8,** 25 (1979).
3 K. Ishida, T. Shumiya, T. Nomura, H. Ohtani, and T. Nishizawa, J. of the Less-Common Metals **142,** 135 (1988).
4 K. Ishida, T. Nomura, H. Tokunaga, H. Ohtani, and T. Nishizawa, J. of the Less-Common Metals **155,** 193 (1989).
5 P. Hénoc, A. Izrael, M. Quillec, and H. Launois, Appl. Phys. Lett. **40,** 963 (1982).
6 F. Glas, M. M. J. Treacy, M. Quillec, and H. Launois, J. Phys. (Paris) **43,** C5-11 (1982).
7 S. Mahajan, B. V. Dutt, H. Temkin, R. J. Cava, and W. A. Bonner, J. Cryst. Growth **68,** 589 (1984).
8 A. G. Norman and G. R. Booker, J. Appl. Phys. **57,** 4715 (1985).
9 A. G. Norman and G. R. Booker, in *Microsc. Semicond. Mater. 1985*, edited by A. G. Cullis and D. B. Holt (Inst. of Phys., Bristol, 1985), Inst. Phys. Conf. Ser. **76,** 257 (1985).
10 O. Ueda, S. Isozumi, and S. Komiya, Jpn. J. Appl. Phys. **23,** L241 (1984).
11 T.-Y. Seong, A. G. Norman, G. R. Booker, R. Droopad, R. L. Williams, S. D. Parker, P. D. Wang, and R. A. Stradling, Mater. Res. Soc. Symp. Proc. **163,** 907 (1990).
12 I. T. Ferguson, A. G. Norman, B. A. Joyce, T.-Y. Seong, R. H. Thomas, C. C. Phillips, and R. A. Stradling, Appl. Phys. Lett. **59,** 3324 (1991).
13 A. G. Norman, T.-Y. Seong, I. T. Ferguson, G. R. Booker, and B. A. Joyce, in *Proc. of the 6th Int. Conf. on Narrow Gap Semicond. Southampton 1992*, Semicond. Sci. Technol. **8,** S9 (1993).
14 T.-Y. Seong, A. G. Norman, I. T. Ferguson, and G. R. Booker, J. Appl. Phys. **73,** 8227 (1993).
15 A. G. Norman, D. Phil. Thesis, University of Oxford, 1987.
16 F. Glas, J. Appl. Phys. **62,** 3201 (1987).
17 I. P. Ipatova, V. G. Malyshkin, and V. A. Shchukin, J. Appl. Phys. **74,** 7198 (1993).
18 J. W. Cahn, Acta Metall. **9,** 795 (1961).
19 L. M. Foster and J. F. Woods, J. Electrochem. Soc. **118,** 1175 (1971).
20 C. H. L. Goodman, Nature **179,** 828 (1957).
21 J. L. Shay and J. H. Wernick, *Ternary Chalcopyrites* (Pegamon, Oxford, 1975).
22 W. A. Soffa and D. E. Laughlin, in *Solid-State Phase Transformations Pittsburgh 1982*, edited by H. I. Aaronson, D. E. Laughlin, R. F. Sekerka, and C. M. Wayman (The Metallurgical Society of AIME, Warrendale, 1982), p. 159.
23 V. D. Verner and D. K. Nichugovskii, Sov. Phys. Solid State **16,** 969 (1974).
24 A. G. Khachaturyan, Sov. Phys. JETP **36,** 753 (1973).
25 E. M. Lifshitz, Zh. Eksp. Teor. Fiz. **11,** 255 (1941).
26 N. A. Semikolenova and E. N. Khabarov, Sov. Phys. Semicond. **8,** 1459 (1975).
27 H. Nakayama, A. Nakamura, E. Taguchi, and H. Fujita, in *Abstracts of 45th Fall Meeting of the Japan Society of Applied Physics, Okayama Japan, 1984*, p. 505.
28 H. Nakayama and H. Fujita, in *GaAs and Related Compounds 1985*, (Inst. of Phys., Bristol, 1986), Inst. Phys. Conf. Ser. **79,** 289 (1986).
29 T. S. Kuan, T. F. Kuech, W. I. Wang, and E. L. Wilkie, Phys. Rev. Lett. **54,** 201 (1985).
30 T. S. Kuan, W. I. Wang, and E. L. Wilkie, Appl. Phys. Lett. **51,** 51 (1987).
31 A. Ourmazd and J. C. Bean, Phys. Rev. Lett. **55,** 765 (1985).
32 I. J. Murgatroyd, A. G. Norman, and G. R. Booker, presented at the Inst. Phys. Solid State Physics Conf., Reading U.K., 1985 (unpublished).

33 I. J. Murgatroyd, A. G. Norman, and G. R. Booker, presented at the Mater. Soc. Spring Meeting, Palo Alto, U.S.A., 1986, Abstract B5.2.

34 I. J. Murgatroyd, A. G. Norman, G. R. Booker, and T. M. Kerr, in *Proc. of the XIth Int. Cong. on Electron Microscopy, Kyoto, Japan 1986*, edited by T. Imura, S. Maruse, and T. Suzuki (Jpn. Soc. Electron. Microsc., Tokyo, 1986), J. Electron Microsc. **35**, 1497 (1986).

35 I. J. Murgatroyd, D. Phil. Thesis, University of Oxford, 1987.

36 H. R. Jen, M. J. Cherng, and G. B. Stringfellow, Appl. Phys. Lett. **48**, 1603 (1986).

37 G. P. Srivastava, J. L. Martins, and A. Zunger, Phys. Rev. B **31**, 2561 (1985).

38 P. Boguslawski and A. Baldereschi, Solid State Commun. **66**, 679 (1988).

39 G. P. Srivastava, J. L. Martins, and A. Zunger, Phys. Rev. B **38**, 12694 (1988).

40 P. Boguslawski and A. Baldereschi, Phys. Rev. B **39**, 8055 (1989).

41 J. E. Bernard, L. G. Ferreira, S.-H. Wei, and A. Zunger, Phys. Rev. B **38**, 6338 (1988).

42 J. E. Bernard, R. G. Dandrea, L. G. Ferreira, S. Froyen, S.-H. Wei, and A. Zunger, Appl. Phys. Lett. **56**, 731 (1990).

43 A. Gomyo, K. Kobayashi, S. Kawata, I. Hino, T. Suzuki, and T. Yuasa, J. Cryst. Growth **77**, 367 (1986).

44 A. Gomyo, T. Suzuki, S. Kobayashi, S. Kawata, I. Hino, and T. Yuasa, Appl. Phys. Lett. **50**, 673 (1987).

45 A. Gomyo, T. Suzuki, and S. Iijima, Phys. Rev. Lett. **60**, 2645 (1988).

46 A. Mascarenhas, S. R. Kurtz, A. Kibbler, and J. M. Olson, Phys. Rev. Lett. **63**, 2108 (1989).

47 I. J. Murgatroyd, A. G. Norman, and G. R. Booker, Submitted to Phys. Rev. Lett. (1987).

48 A. G. Norman, R. E. Mallard, I. J. Murgatroyd, G. R. Booker, A. H. Moore, and M. D. Scott, in *Microsc. Semicond. Mater. 1987*, (Inst. of Phys., Bristol, 1987), Inst. Phys. Conf. Ser. **87**, 77 (1987).

49 R. Hull, K. W. Carey, J. E. Fouquet, G. A. Reid, S. J. Rosner, D. Bimberg, and D. Oertel, in *GaAs and Related Compounds 1986*, (Inst. of Phys., Bristol, 1987), Inst. Phys. Conf. Ser. **83**, 209 (1987).

50 M. A. Shahid, S. Mahajan, D. E. Laughlin, and H. M. Cox, Phys. Rev. Lett. **58**, 2567 (1987).

51 Y.-E. Ihm, N. Otsuka, J. Klem, and H. Morkoc, Appl. Phys. Lett. **51**, 2013 (1987).

52 O. Ueda, M. Takikawa, J. Komeno, and I. Umebu, Jpn. J. Appl. Phys. **26**, L1824 (1987).

53 J. P. Goral, M. M. Al-Jassim, J. M. Olson, and A. Kibler, in *Epitaxy of Semiconductor Layered Structures*, (Mater. Res. Soc., Pittsburgh, 1988), Mater. Res. Soc. Symp. Proc. **102**, 583 (1988).

54 S. McKernan, B. C. De Cooman, C. B. Carter, D. P. Bour, and J. R. Shealy, J. Mater. Res. **3**, 406 (1988).

55 S. McKernan, B. C. De Cooman, C. B. Carter, D. P. Bour, and J. R. Shealy, in *Defects in Electronic Materials*, (Mater. Res. Soc., Pittsburgh, 1988), Mater. Res. Soc. Symp. Proc. **104**, 637 (1988).

56 P. Bellon, J. P. Chevalier, G. P. Martin, E. Dupont-Nivet, C. Thiebaut, and J. P. André, Appl. Phys. Lett. **52**, 567 (1988).

57 M. Kondow, H. Kakibayashi, and S. Minagawa, J. Cryst. Growth **88**, 291 (1988).

58 I. J. Murgatroyd, A. G. Norman, and G. R. Booker, J. Appl. Phys. **67**, 2310 (1990).

59 A. G. Norman, in *Proc. of NATO Advanced Research Workshop on the Evaluation of Advanced Semiconductor Materials by Electron Microscopy, Bristol 1988*, edited by D. Cherns (Plenum Press, New York, 1989), NATO ASI Series B **203**, p. 233.

60 A. G. Norman, in *Atomic Ordering in Ternary III-V Semiconductor Alloys Related to Surface Structure During Epitaxial Growth*, SERC visiting fellowship research grant final report, 1991 (unpublished).
61 A. G. Norman, T.-Y. Seong, B. A. Philips, G. R. Booker, and S. Mahajan, in *Microsc. of Semicond. Mater. 1993*, edited by A. G. Cullis, A. E. Staton-Bevan, and J. L. Hutchison (Inst. of Phys., Bristol, 1993), Inst. Phys. Conf. Ser. **134**, 279 (1993).
62 I. T. Ferguson, A. G. Norman, T.-Y. Seong, R. H. Thomas, C. C. Phillips, X. M. Zhang, R. A. Stradling, B. A. Joyce, and G. R. Booker, in *GaAs and Related Compounds 1991*, (Inst. of Phys., Bristol, 1992), Inst. Phys. Conf. Ser. **120**, 395 (1992).
63 T.-Y. Seong, D. Phil. Thesis, University of Oxford, 1991.
64 H. R. Jen, K. Y. Ma, and G. B. Stringfellow, Appl. Phys. Lett. **54**, 1154 (1989).
65 T.-Y. Seong, G. R. Booker, A. G. Norman, and I. T. Ferguson, Appl. Phys. Lett. **64**, 3593 (1994).
66 S. R. Kurtz, L. R. Dawson, R. M. Biefeld, D. M. Follstaedt, and B. L. Doyle, Phys. Rev. B **46**, 1909 (1992).
67 S. R. Kurtz, R. M. Biefeld, L. R. Dawson, and B. L. Doyle, in *GaAs and Related Compounds 1991*, (Inst. of Physics, Bristol, 1992), Inst. Phys. Conf. Ser. **120**, 595 (1992).
68 S.-H. Wei and A. Zunger, Appl. Phys. Lett. **58**, 2684 (1991).
69 D. M. Follstaedt, R. M. Biefeld, S. R. Kurtz, and K. C. Baucom, J. Electron. Mater. **24**, 819 (1995).
70 D. M. Follstaedt, R. M. Biefeld, S. R. Kurtz, L. R. Dawson, and K. C. Baucom, in *Proc. of the 7th Int. Conf. on Narrow Gap Semicond. 1995*, edited by J. L. Reno (Inst. of Physics, Bristol, 1995), Inst. Phys. Conf. Ser. **144**, 224 (1995).
71 T.-Y. Seong, G. R. Booker, A. G. Norman, P. J. F. Harris, and G. B. Stringfellow, in *Microsc. of Semicond. Mater. 1997*, edited by A. G. Cullis and J. L. Hutchison (Inst. of Phys., Bristol, 1997), Inst. Phys. Conf. Ser. **157** 279 (1997).
72 G. B. Stringfellow, J. Cryst. Growth **98**, 108 (1989).
73 J. Shin, Y. Hsu, T. C. Hsu, G. B. Stringfellow, and R. W. Gedridge, J. Electron. Mater. **24**, 1563 (1995).
74 H. R. Jen, M. J. Cherng, M. J. Jou, and G. B. Stringfellow, in *Proc. of 7th Int. Conf. on Ternary and Multinary Compounds, Snowmass, CO 1986*, edited by S. K. Deb and A. Zunger (Mater. Res. Soc., Pittsburgh, 1987), p. 353.
75 H. R. Jen, M. J. Cherng, M. J. Jou, and G. B. Stringfellow, in *GaAs and Related Compounds 1986*, (Inst. of Phys., Bristol, 1987), Inst. Phys. Conf. Ser. **83**, 159 (1987).
76 H. R. Jen, M. J. Jou, Y. T. Cherng, and G. B. Stringfellow, J. Cryst. Growth **85**, 175 (1987).
77 O. Ueda, Y. Nakata, and S. Muto, in *Proc. of the 7th Int. Conf. on InP and Related Materials Hokkaido Japan 1995* (IEEE, New York, 1995), p. 253.
78 O. Ueda, in *Optoelectronic Materials: Ordering, Composition Modulation, and Self-Assembled Structures*, edited by E. D. Jones, A. Mascarenhas, and P. Petroff (Mater. Res. Soc., Pittsburgh, 1996), Mater. Res. Soc. Symp. Proc. **417**, 31 (1996).
79 A. G. Norman, presented at the Mater. Res. Soc.-Japan Meeting, Makuhari Messe Japan, 1996 (unpublished).
80 A. G. Norman, presented at the 40th Electronic Materials Conference, University of Virginia, Charlottesville VA, U.S.A., 1998, Abstract Y2.
81 A. G. Norman, submitted to J. Appl. Phys. (2001).
82 D. J. Chadi, J. Vac. Sci. Technol. A **5**, 834 (1987).
83 T. Hashizume, Q.-K. Xue, A. Ichimiya, and T. Sakurai, Phys. Rev. B **51**, 4200 (1995).
84 H. H. Farrell and C. J. Palmstrøm, J. Vac. Sci. Technol. B **8**, 903 (1990).

85 J. E. Northrup and S. Froyen, Phys. Rev. Lett. **71,** 2276 (1993).

86 P. Moriarty, P. H. Beton, Y.-R. Ma, M. Henini, and D. A. Woolf, Phys. Rev. B **53,** R16148 (1996).

87 W. G. Schmidt and F. Bechstedt, Phys. Rev. B **55,** 13051 (1997).

88 C. Barrett and T. B. Massalaski, *Structure of Metals*, 3rd ed. (Pergamon, Oxford, 1980), Chap. 11.

89 P. B. Hirsch, A. Howie, R. B. Nicholson, D. W. Pashley, and M. J. Whelan, *Electron Microscopy of Thin Crystals* (Krieger, New York, 1977), Chap. 15.

90 Z. Zhong, J. H. Li, J. Kulik, P. C. Chow, A. G. Norman, A. Mascarenhas, J. Bai, T.D. Golding, and S. C. Moss, Phys. Rev. B. **63,** 033314 (2001).

91 P. Gavrilovic, F. P. Dabkowski, K. Meehan, J. E. Williams, W. Stutius, K. C. Hsieh, N. Holonyak, M. A. Shahid, and S. Mahajan, J. Cryst. Growth **93,** 426 (1988).

92 H. Murata, T. C. Hsu, I. H. Ho, L. C. Su, Y. Hosokawa, and G. B. Stringfellow, Appl. Phys. Lett. **68,** 1796 (1996).

93 H. Murata, I. H. Ho, L. C. Su, Y. Hosokawa, and G. B. Stringfellow, J. Appl. Phys. **79,** 6895 (1996).

94 A. Gomyo, K. Makita, I. Hino, and T. Suzuki, Phys. Rev. Lett. **72,** 673 (1994).

95 A. Gomyo, M. Sumino, I. Hino, and T. Suzuki, Jpn. J. Appl. Phys. **34,** L469 (1995).

96 F. K. LeGoues, V. P. Kesan, S. S. Iyer, J. Tersoff, and R. Tromp, Phys. Rev. Lett. **64,** 2038 (1990).

97 J. K. Shurtleff, R. T. Lee, C. M. Fetzer, and G. B. Stringfellow, Appl. Phys. Lett. **75,** 1914 (1999).

98 G. B. Stringfellow, R. T. Lee, C. M. Fetzer, J. K. Shurtleff, Y. Hsu, S. W. Jun, S. Lee, and T.-Y. Seong, J. Electron. Mater. **29,** 134 (2000).

99 R. T. Lee, J. K. Shurtleff, C. M. Fetzer, G. B. Stringfellow, S. Lee, and T.-Y. Seong, J. Appl. Phys. **87,** 3730 (2000).

100 S. W. Jun, C. M. Fetzer, R. T. Lee, J. K. Shurtleff, and G. B. Stringfellow, Appl. Phys. Lett. **76,** 2716 (2000).

101 T. Suzuki, A. Gomyo, and S. Iijima, J. Cryst. Growth **93,** 396 (1988).

102 P. Bellon, J. P. Chevalier, E. Augarde, J. P. André, and G. P. Martin, J. Appl. Phys. **66,** 2388 (1989).

103 G. S. Chen and G. B. Stringfellow, Appl. Phys. Lett. **59,** 324 (1991).

104 A. A. Mbaye, A. Zunger, and D. M. Wood, Appl. Phys. Lett. **49,** 782 (1986).

105 J. C. Mikkelsen and J. B. Boyce, Phys. Rev. Lett. **49,** 1412 (1982).

106 P. Boguslawski, Phys. Rev. B **42,** 3737 (1990).

107 P. Boguslawski, Semicond. Sci. Technol. **6,** 953 (1991).

108 N. Hamada and T. Kurimoto, in *Proc. of 20th Int. Conf. on the Physics of Semiconductors*; *Vol. 1*, edited by E. M. Anastassakis and J. D. Joannopoulos (World Scientific, Singapore, 1990), p. 300.

109 T. Kurimoto and N. Hamada, Superlattices and Microstructures **10,** 379 (1991).

110 S. Froyen and A. Zunger, Phys. Rev. Lett. **66,** 2132 (1991).

111 S. Froyen and A. Zunger, J. Vac. Sci. Technol. **B 9,** 2176 (1991).

112 J. E. Bernard, S. Froyen, and A. Zunger, Phys. Rev. B **44,** 11178 (1991).

113 G. S. Chen, D. H. Jaw, and G. B. Stringfellow, J. Appl. Phys. **69,** 4263 (1991).

114 Q.-K. Xue, T. Hashizume, and T. Sakurai, Appl. Surf. Sci. **141,** 244 (1999).

115 P. H. Fuoss, D. W. Kisker, G. Renaud, K. L. Tokuda, S. Brennan, and J. L. Kahn, Phys. Rev. Lett. **63,** 2389 (1989).

116 D. W. Kisker, P. H. Fuoss, K. L. Tokuda, G. Renaud, S. Brennan, and J. L. Kahn, Appl. Phys. Lett. **56,** 2025 (1990).

117 I. Kamiya, D. E. Aspnes, H. Tanaka, L. T. Florez, J. P. Harbison, and R. Bhat, Phys. Rev. Lett. **68,** 627 (1992).

118 I. Kamiya, H. Tanaka, D. E. Aspnes, L. T. Florez, E. Colas, J. P. Harbison, and R. Bhat, Appl. Phys. Lett. **60,** 1238 (1992).

119 L. Li, B.-K Han, S. Gan, H. Qi, and R. F. Hicks, Surf. Sci. **398,** 386 (1998).

120 B.-K. Han, L. Li, Q. Fu, and R. F. Hicks, Appl. Phys. Lett. **72,** 3347 (1998).

121 L. Li, B.-K. Han, and R. F. Hicks, Appl. Phys. Lett. **73,** 1239 (1998).

122 J. A. Appelbaum and D. R. Hamann, Surf. Sci. **74,** 21 (1978).

123 P. K. Larsen, J. F. Van der Veen, A. Mazur, J. Pollmann, J. H. Neave, and B. A. Joyce, Phys. Rev. B **26,** 3222 (1982).

124 M. D. Pashley, K. W. Haberern, W. Friday, J. M. Woodall, and P. D. Kirchner, Phys. Rev. Lett. **60,** 2176 (1988).

125 J. C. Phillips, *Bands and Bonds in Semiconductors* (Academic Press, New York-London, 1973).

126 T. Suzuki and A. Gomyo, J. Cryst. Growth **111,** 353 (1991).

127 T. Suzuki, A. Gomyo, and S. Iijima, in *Ordering at Surfaces and Interfaces*, edited by A. Yoshimori, T. Shinjo, and H. Watanabe (Springer-Verlag, Berlin, 1992), Springer Series in Materials Science **17,** 363 (1992).

128 T. Suzuki and A. Gomyo, in *Semiconductor Interfaces at the Sub-Nanometer Scale*, edited by H. W. M. Salemink and M. D. Pashley (Kluwer Academic Publishers, Netherlands, 1993), p. 11.

129 B. A. Philips, A. G. Norman, T.-Y. Seong, S. Mahajan, G. R. Booker, M. Skowronski, J. P. Harbison, and V. G. Keramidas, J. Cryst. Growth **140,** 249 (1994).

130 R. Osorió, J. E. Bernard, S. Froyen, and A. Zunger, Phys. Rev. B **45,** 11173 (1992).

131 S. Froyen, J. E. Bernard, R. Osorió, and A. Zunger, Physica Scripta **T45,** 272 (1992).

132 P. C. Kelires and J. Tersoff, Phys. Rev. Lett. **63,** 1164 (1989).

133 A. G. Norman, unpublished work, 1990.

134 D. E. Jesson, S. J. Pennycook, J.-M. Baribeau, and D. C. Houghton, Phys. Rev. Lett. **68,** 2062 (1992).

135 D. E. Jesson, S. J. Pennycook, J. Z. Tischler, J. D. Budai, J.-M. Baribeau, and D. C. Houghton, Phys. Rev. Lett. **70,** 2293 (1993).

136 F. K. LeGoues, R. M. Tromp, V. P. Kesan, and J. Tsang, Phys. Rev. B **47,** 10012 (1993).

137 J. Z. Tischler, J. D. Budai, D. E. Jesson, G. Eres, P. Zschack, J.-M. Baribeau, and D. C. Houghton, Phys. Rev. B **51,** 10947 (1995).

138 F. K. LeGoues, J. Tersoff, and R. M. Tromp, Phys. Rev. Lett. **71,** 3736 (1993).

139 K. L. Whiteaker, I. K. Robinson, J. E. Van Nostrand, and D. G. Cahill, Phys. Rev. B **57,** 12410 (1998).

140 N. Ikarashi, K. Akimoto, T. Tatsumi, and K. Ishida, Phys. Rev. Lett. **72,** 3198 (1994).

141 B. P. Uberuaga, M. Leskovar, A. P. Smith, H. Jónsson, and M. Olmstead, Phys. Rev. Lett. **84,** 2441 (2000).

142 T. Araki, N. Fujimura, and T. Ito, Appl. Phys. Lett. **71,** 1174 (1997).

143 A. G. Norman, T.-Y. Seong, and G. R. Booker, unpublished work, 1990.

144 S. Mahajan and B. A. Philips, unpublished work, 1990.

145 S. Mahajan, in *Proc. of the 5th Brazilian School Semiconductor Physics, Sao Paulo Brazil 1991*, edited by J. R. Leite, A. Fazzio, and A. S. Chaves (World Scientific, Singapore, 1992), p. 79.

146 S. Mahajan and B. A. Philips, in *Ordered Intermetallics-Physical Metallurgy and Mechanical Behaviour*, edited by C. T. Liu, R. W. Cahn, and G. Sauthoff (Kluwer Academic Publishers, Dordrecht, 1991), p. 93.

147 J. E. Bernard, Mater. Sci. Forum **155-156,** 131 (1994).

148 A. Gomyo, T. Suzuki, K. Makita, M. Sumino, and I. Hino, in *Optoelectronic Materials: Ordering, Composition Modulation, and Self-Assembled Structures*, edited by E. D. Jones, A. Mascarenhas, and P. Petroff (Mater. Res. Soc., Pittsburgh U.S.A., 1996), Mater. Res. Soc. Symp. Proc. **417**, 91 (1996).

149 B. A. Philips, Ph.D. Thesis, Carnegie Mellon University, 1994.

150 B. A. Philips, I. Kamiya, K. Hingerl, L. T. Florez, D. E. Aspnes, S. Mahajan, and J. P. Harbison, Phys. Rev. Lett. **74**, 3640 (1995).

151 S. Mahajan, Mater. Sci. Eng. B **30**, 187 (1995).

152 K. Lee, B. A. Philips, R. S. McFadden, and S. Mahajan, Mater. Sci. Eng. B **32**, 231 (1995).

153 M. Sauvage-Simkin, Y. Garreau, R. Pinchaux, M. B. Véron, J. P. Landesman, and J. Nagle, Phys. Rev. Lett. **75**, 3485 (1995).

154 M. Sauvage-Simkin, Y. Garreau, R. Pinchaux, A. Cavanna, M. B. Véron, N. Jedrecy, J. P. Landesman, and J. Nagle, Appl. Surf. Sci. **104/105**, 646 (1996).

155 K. Aid, Y. Garreau, M. Sauvage-Simkin, and R. Pinchaux, Surf. Sci. **425**, 165 (1999).

156 S. Ohkouchi and A. Gomyo, Appl. Surf. Sci. **130-132**, 447 (1998).

157 T. Suzuki, T. Ichihashi, and T. Nakayama, Appl. Phys. Lett. **73**, 2588 (1998).

158 T. Suzuki, T. Ichihashi, and M. Tsuji, in *Self-Organized Processes in Semiconductor Alloys*, edited by A. Mascarenhas, D. M. Follstaedt, T. Suzuki, and B. A. Joyce (Mater. Res. Soc., Warrendale, 2000), Mater. Res. Soc. Proc. **583**, 267 (2000).

159 S. B. Zhang, S. Froyen, and A. Zunger, Appl. Phys. Lett. **67**, 3141 (1995).

160 S. B. Zhang, S. Froyen, and A. Zunger, in *Optoelectronic Materials: Ordering, Composition Modulation, and Self-Assembled Structures*, edited by E. D. Jones, A. Mascarenhas, and P. Petroff (Mater. Res. Soc., Pittsburgh, 1996), Mater. Res. Soc. Symp. Proc. **417**, 43 (1996).

161 S. Froyen and A. Zunger, Phys. Rev. B **53**, 4570 (1996).

162 V. P. Kesan, F. K. LeGoues, and S. S. Iyer, Phys. Rev. B **46**, 1576 (1992).

163 C. M. Fetzer, R. T. Lee, J. K. Shurtleff, G. B. Stringfellow, S. M. Lee, and T.-Y. Seong, Appl. Phys. Lett. **76**, 1440 (2000).

164 T. Ichihashi, K. Kurihara, K. Nishi, and T. Suzuki, Jpn. J. Appl. Phys. **39**, L126 (2000).

165 N. Esser, A. I. Shkrebtii, U. Resch-Esser, C. Springer, W. Richter, W. G. Schmidt, F. Bechstedt, and R. Del Sole, Phys. Rev. Lett. **77**, 4402 (1996).

166 W. G. Schmidt, S. Mirbt, and F. Bechstedt, Phys. Rev. B **62**, 8087 (2000).

167 O. Ueda, Y. Nakata, and S. Muto, J. Cryst. Growth **150**, 523 (1995).

168 O. Ueda, Y. Nakata, and T. Fujii, Appl. Phys. Lett. **58**, 705 (1991).

169 Y. Kangawa, N. Kuwano, K. Oki, and T. Ito, Appl. Surf. Sci. **159-160**, 368 (2000).

Chapter 3

Effects of the Surface on CuPt Ordering During OMVPE Growth

G.B. Stringfellow
College of Engineering, University of Utah
Salt Lake City, Utah 84112

Key words: CuPt Ordering, OMVPE, Surfactants, Surface Structure

Abstract: Ordered structures are formed in semiconductor alloys spontaneously during growth. This paper describes the surface-related driving force for CuPt ordering and the processes occurring at the surface during growth. The effects of several growth parameters on the surface structure and the ordering process are described. The effects of ordering on the optical and electrical properties of these alloys are also described along with a brief discussion of how this phenomenon might be useful in device structures. The bulk of the paper relates to the newly-discovered use of surfactants for control of ordering during the organometallic vapor phase epitaxial growth of GaInP. Several donor and acceptor dopants are known to produce disorder. The mechanisms for the donors Te and Si and the acceptor Zn are discussed. Perhaps the most exciting development in this area is the use of isoelectronic group V elements, for example Sb, to control the surface structure and, hence, the degree of order and bandgap energy of GaInP. At an estimated Sb concentration in the solid of 10^{-4}, the order is eliminated. Surface photo absorption data indicate that the effect is due to a change in the surface reconstruction. Modulation of the TESb flow rate during the growth cycle has been used to produce a heterostructure with a 135 meV bandgap difference between two layers having the same solid composition.

1.   INTRODUCTION

Atomic-scale ordering to produce the CuPt structure frequently occurs in a wide range of zincblende III/V alloys, including essentially all ternary and quaternary alloys composed of combinations of Al, Ga, In, As, P, and Sb [1,2]. Preliminary data indicate that it also occurs in wurtzite GaInN alloys [3]. Similar ordered structures have also been observed in GeSi and

II/VI alloys [1]. The most widely studied system is $Ga_{0.52}In_{0.48}P$, with layers grown by organometallic vapor phase epitaxy (OMVPE) on (001)-oriented GaAs substrates [1,2]. In this case, the Ga and In atoms are spontaneously segregated into alternating {111} monolayers at the surface during growth. This topic has been the topic of several recent review papers [1,4,5].

The first principles calculations of Zunger and co-workers have been used to estimate the relative thermodynamic stabilities of the various ordered structures, including the CuPt structure, in III/V alloys [4]. For the bulk alloys, i.e., ignoring surface effects, the CuPt structure is not found to be stable relative to the disordered alloy. This was initially surprising, since CuPt is nearly the only ordered structure observed experimentally. However, this dilemma is resolved by considering the stabilities of the various ordered structures at the reconstructed surface [6].

For the most commonly observed (2xn), typically (2x4), reconstruction on group V-terminated (001) surfaces, VFF calculations [6] indicate that the B variants of the CuPt structure with ordering on ($\bar{1}11$) and (111) planes are the most stable in the layers just beneath the (001) surface. The [110] rows of [$\bar{1}10$] oriented group V dimers lead to alternating [110] rows of compressive and tensile strain in the 3rd buried layer. For alloys with mixing on the group III sublattice, such as GaInP, this produces the [110] rows of alternating large and small atoms that comprise the CuPt-B variants. These calculations also predict that the surface structure of alloys with mixing on the group V sublattice, such as GaAsP, will also produce the CuPt-B variants, in agreement with experimental observations [7].

Since the CuPt ordering is driven by the [$\bar{1}10$] P dimers on the (001) surface, ordering is found not to occur for growth on other orientations, such as (111) [1,2]. Thus, it is surprising that CuPt-type ordering occurs in wurtzite GaInN [3,8,9] and AlGaN [10] alloys grown on the basal plane, which has a surface configuration similar to that for (111) zincblende surfaces. This may indicate that ordering can be driven by sub-surface stresses at steps on the wurtzite surface during growth [11]. Other factors suggest that surface steps may play an important role in the kinetic processes leading to the formation of the CuPt ordered structure [12].

Clearly, the ordering phenomenon is very sensitive to the surface structure. For this reason, ordering is a perfect vehicle for the study of surface processes during epitaxial growth, in general. The occurrence and mechanism of ordering are fascinating materials science problems that reveal much about the important general features of the surface processes occurring during vapor phase epitaxial growth. This chapter will concentrate on a review of recent advances in our understanding of the effects of surface processes on ordering.

2.   REVIEW OF THE ORDERING PROCESS

The history of ordering in semiconductor alloys began in 1985. Kuan et al. [13] made the surprising discovery that AlGaAs alloys grown on (110) GaAs substrates at 700°C by either OMVPE or molecular beam epitaxy (MBE) spontaneously ordered during the growth process to form the CuAu structure, with ordering on {100} planes. In that same year, Ourmazd and Bean [14] reported the formation of {111} ordered structures composed of alternating Ge and Si atomic pairs in Si-Ge alloys grown on (001) Si substrates by MBE. The next year, Jen et al. [15] observed the CuAu and chalcopyrite structures (with ordering on {211} planes) in GaAsSb alloys grown by OMVPE. These early observations were rapidly followed by reports of mainly CuPt ordering in a number of III/V alloys [1,2]. In fact, CuPt ordering has been observed in virtually all III/V alloys studied when the layers are grown by vapor phase epitaxial processes on (001)-oriented substrates, with the exception of AlGaAs. Ordering is virtually never observed in alloys grown by LPE [1,2,16]. Presumably, the liquid phase passivates the surface during LPE growth, thus removing the need for the surface reconstruction which acts as the major driving force for ordering.

CuPt ordering is found to be strongest in GaInP layers grown by OMVPE. However, even in this case, the degree of order (S) is typically significantly less than unity (a value of S=0 indicates disorder and a value of 1 indicates a superlattice of alternating (111) monolayers of pure GaP and InP). Transmission electron diffraction superspot intensity measurements yield values of the degree of order as large as 0.7 [17]. Recent spin echo nuclear magnetic resonance (NMR) measurements of [71]Ga [18] yield the best independent value to date of 0.6 for a sample grown to maximize the order parameter.

Optical properties are often used to estimate the degree of order. Most commonly, this involves the use of low temperature photoluminescence to determine the bandgap energy and the use of theoretical calculations, described below, to extract the degree of order. This type of analysis leads to a value of 0.56 for the order parameter in highly ordered GaInP grown by OMVPE [19]. A comparison of the values of order parameter obtained from the various techniques shows a high degree of consistency, giving confidence that the maximum degree of order obtained is 0.5 to 0.6. The disorder may be due to two factors: i) regions of the material which are not ordered [20] and ii) a compositional modulation index of less than unity, i.e., the alternating monolayers are not really the pure compounds from which the alloy is formed.

Ordering has extremely important practical consequences. The order-induced property change that elicits the most interest by the device community is the reduction of the bandgap energy. More than 10 years before ordering was discovered an unexplained phenomenon caused the bandgap energies of GaInP layers, all grown lattice matched to GaAs substrates, i.e., all with the same composition, to vary by more than 100 meV from laboratory to laboratory [21]. The reduction of bandgap energy was later found to be caused by CuPt ordering [22]. The first attempts to quantitatively determine the effect of order on the bandgap energy were theoretical [23]. The current best theoretical estimate of the bandgap reduction due to CuPt ordering in perfectly ordered GaInP, ΔE_g, is 0.38 eV [24]. The reduction in bandgap energy has the following dependence on the degree of order, S [25],

$$E_g = E_g (S=0) - \Delta E_g S^2. \tag{1}$$

Ernst et al. [26] examined the bandgap energy vs. order parameter experimentally over a range of bandgap energies, yielding a value of ΔE_g of 471 meV. The largest experimentally measured change in bandgap energy due to CuPt ordering is 160 meV [27].

Bandgap narrowing due to CuPt ordering has also been experimentally observed in GaInAs alloys lattice matched to (001) InP substrates [28]. A maximum bandgap reduction of 65 meV was observed. Experimental determination of the effect of CuPt order on the bandgap of InAsSb alloys reveals a bandgap shrinkage of approximately 45 meV [29].

The large bandgap shrinkage due to order is extremely important for devices, including visible light emitting diodes (LEDs) and injection laser diodes (LDs) in GaInP [30]. To produce the shortest wavelength (most visible) devices, ordering must be avoided. However, in InAsSb alloys, the shrinkage of bandgap energy is potentially beneficial, since it moves the wavelength further into the infrared where an atmospheric window exists between 8 and 12 μm [29]. Thus, ordered InAsSb has the potential to be a useful material for IR detectors, if ordering could be precisely controlled. An additional potential benefit of ordering is use of order/disorder boundaries for the production of heterostructures [31] and quantum wells [32] with absolutely no change in the solid composition.

3.   EFFECTS OF SURFACES

During the development of the OMVPE process, the bonding at the surface during growth was unknown because of the lack of a technique for determination of the surface reconstruction during growth. For the ultra-high vacuum MBE technique, reflection high energy electron diffraction

(RHEED) was used to determine the symmetry of the reconstructed surface during growth. Only with the advent of reliable optical techniques that can be used in the OMVPE environment, such as reflection difference spectroscopy (RDS) [33,34] and surface photo absorption (SPA) [35] has it become possible to determine the reconstruction *in situ* during growth for OMVPE. For growth of the phosphides, including GaInP, using typical OMVPE growth conditions, the surface reconstruction is (2xn), most typically (2x4)-like were the surface is terminated by a single layer of P [110] oriented P dimmers [36,37].

The correspondence between the presence of [$\bar{1}$10] P dimers and CuPt ordering for GaInP layers grown by OMVPE has recently been verified by using the SPA technique for measurement of the surface structure [36,38,39]. This includes the effects due to the loss of [$\bar{1}$10] P dimers at high growth temperatures and low partial pressures of the P precursor as well as the displacement of P at the surface by surfactants such as Sb. These results will be described in more detail below. They make it clear that control of the properties of the solid produced epitaxially depend critically on the reconstruction of the surface during growth.

3.1   Temperature

The effects of temperature during OMVPE growth on the degree of CuPt order have been studied extensively for GaInP. For example, a typical study using the reactants trimethylgallium, trimethylindium, and PH_3 [40] shows a clear maximum in the degree of order at temperatures near 620°C, with a continuous decrease in order with increasing temperature until at 720°C the material is essentially disordered. Similarly, a decrease in growth temperature to 520°C produces material with only a small degree of order. Nearly identical results were obtained by Murata et al. [39] for the OMVPE growth of GaInP using tertiarybutylphosphine (TBP) rather than phosphine, as seen in *Figure 1*.

Determination of the concentration of [$\bar{1}$10] P dimers on the surface using SPA reveals that the major cause for the effect of temperature on ordering observed in GaInP grown by OMVPE is loss of the (2x4)-like surface reconstruction that provides the thermodynamic driving force for ordering at the surface during growth. This is clearly seen in Fig. 1, where

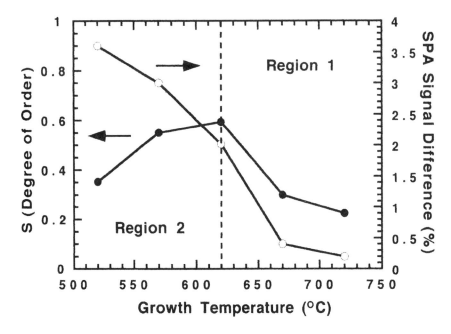

Figure 1. The effect of growth temperature on the degree of order and SPA difference signal at 400 nm for GaInP layers by OMVPE (after Murata et al. [39]).

the SPA signal intensity at 400 nm, due to the $[\overline{1}10]$ P dimers, decreases sharply as the temperature increases from 620°C. At 720°C the signal has nearly disappeared.

However, the data presented in Fig. 1 make it very clear that the loss of order at low temperatures must be related to another factor. The SPA spectra for samples grown at low temperatures show a feature at approximately 480 nm that is attributed to an "excess phosphorus" phase: A second layer of P accumulates on the surface at low temperatures and high P partial pressures [41]. This may be the reason for the loss of CuPt order at low growth temperatures.

3.2   Partial Pressure of the P Precursor

The flow rate of the group V precursor is also found to have a significant effect on the ordering process. Again, the degree of order has been closely correlated with the surface reconstruction [42]. In Fig. 2 the degree of order is plotted versus the TBP partial pressure during growth at 620 and 670°C.

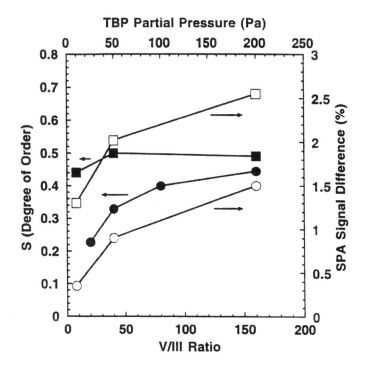

Figure 2. The effect of input TBP partial pressure on the degree of order and the 400nm SPA difference signal for GaInP layers grown by OMVPE. The open and filled data points are for T=670°C and T=620°C, respectively (after Murata et al. [42]).

The SPA intensity at 400 nm, due to the [$\overline{1}$10] P dimers, is also plotted. Clearly, the loss of CuPt ordering at low V/III ratios correlates closely with the loss of the (2x4)-like reconstruction.

The data from the studies of changing temperature and TBP partial pressure are combined for the plot of the degree of order versus the SPA signal in Fig. 3. A one-to-one relationship between the degree of order and the concentration of [$\overline{1}$10] P dimers on the surface is observed for changes in both temperature (620-720°C) and TBP partial pressure (<200 Pa).

Other factors can also affect the degree of order for GaInP layers grown by OMVPE. For example, as mentioned above, variation of the misorientation of the substrate [12] reveals two factors: 1) An increasing misorientation angle leads to a reduction in the concentration of P dimers at the surface, which leads to the growth of less ordered material. 2) [$\overline{1}$10] steps have an additional disordering effect and, conversely, [110] steps are found to enhance the ordering process. This latter effect is presumably kinetic in nature.

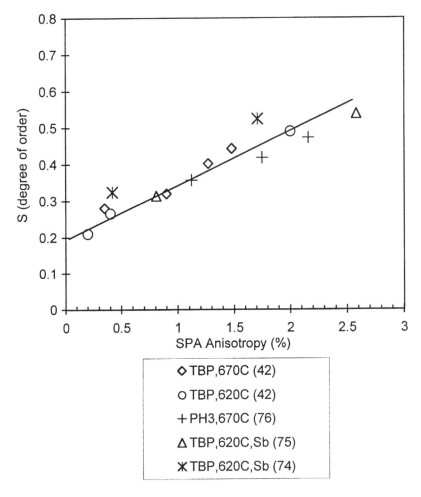

Figure 3. Degree of CuPt order plotted versus SPA difference signal at 400 nm for GaInP grown byOMVPE on singular GaAs substrates (after Stringfellow et al. [75]).

The results of growth rate studies also strongly suggest that kinetic factors may be significant in the ordering process. Considering only thermodynamic factors, the growth rate should not affect the ordering in GaInP for OMVPE growth at high V/III ratios. Changing the partial pressures of the Ga and In precursors will change the growth rate, but will not affect the P partial pressure at the interface [43]. Thus, the P coverage of the surface and the surface reconstruction should be independent of growth rate. In recent experiments at a temperature of 670°C the group III flow rates were changed while holding the TBP partial pressure constant at 1.5 Torr. Changing the growth rate from 0.25 to 2 μm/hr was found to have no detectable effect on the degree of order [44]. This clearly demonstrates the lack of a kinetic factor in the ordering process under these conditions. However, at higher growth rates, a kinetic factor becomes clearly evident. Cao et al. [45] studied the effect of growth rate on ordering in GaInP grown by OMVPE at rates from 4 to 12 µm/hr at a temperature of 680°C. They found a marked decrease in the degree of order at the higher growth rates. The ordering is virtually eliminated at a growth rate of 12 µm/hr (about 10 monolayers/sec). The reduction in order parameter with increasing growth rate seen for rates above 4 µm/hr gives a rough measure of the rate of the ordering process occurring on the surface during growth. The data indicate that the time constant is approximately 0.25 sec. Cao et al. attributed this to surface diffusion.

As discussed in more detail below, the addition of Te to the system during GaInP growth results in a marked increase in the $[\bar{1}10]$ step velocity. This is also thought to result in a dramatic decrease in the degree of order.

4.   EFFECTS OF SURFACTANTS

The discussion in the last section indicates that the surface structure can be controlled during growth by the use of temperature and the partial pressure of the P precursor. More recent experiments indicate that this can also be accomplished via the use of surfactants during the OMVPE growth process.

The original use of surfactants during the epitaxial growth (mainly MBE) of semiconductors was for control of the structure and morphology of highly strained layers in both elemental [46,47] and III/V semiconductors [48,49]. The addition of dopants during MBE growth was shown to affect both adatom attachment at step edges [50,51] and surface reconstruction [52]. It probably also affects the specific surface energy [53] and the diffusion of adatoms on the surface [54]. The use of isoelectronic surfactants was practically unknown until the recent work described below, although the isoelectronic surfactant As has been shown to modify the surface reconstruction of cubic GaN grown by MBE [19].

Several studies in GaInP have demonstrated a connection between ordering and n-type [56-60] or p-type [61-67] dopant concentration. The results show that a drastic decrease in ordering is caused by introducing a high concentration of dopants during OMVPE growth. The effect for some impurities has been attributed to diffusion in the bulk [63,66-70]. However, recent data for Te added during OMVPE growth of GaInP indicate that the disordering is due to a surfactant effect, namely an increase in the $[\bar{1}10]$ step velocity, which dramatically decreases the degree of order of the GaInP [58-60]. The effect would be similar to that of an increased growth rate, as described above. This is one of the early examples of surfactant effects during the OMVPE growth of semiconductor materials. More detailed results of the effects of dopants on the surface structure (both reconstruction and step structure) and ordering are briefly reviewed in the following sub-sections.

5.   TELLURIUM

The reduction in order parameter and the change in step structure associated with the addition of Te (as DETe) during OMVPE growth has been documented [58-60]. As seen in Fig. 4, for growth on a vicinal substrate, the reduction in order parameter associated with the addition of Te begins at a concentration of approximately 10^{17} cm^{-3} and the material is completely disordered for concentrations exceeding $5x10^{17}$ cm^{-3} [58]. The disordering effect of Te was verified by transmission electron diffraction results [58]. The addition of Te was found to have no effect on the surface reconstruction [59]. The disordering effect was attributed to the marked change in group III adatom attachment kinetics at the step edge, due to a change in the step edge reconstruction and bond configuration.

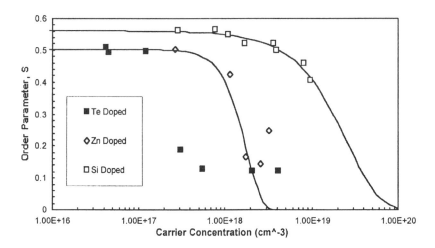

Figure 4. Degree of CuPt order versus doping level, both n- and p-type, for Te, Si, and Zn for GaInP layers grown by OMVPE (after Fetzer et al. [69]).

This is supported by data for another donor, Si. As Si was added to the GaInP at levels exceeding 10^{18} cm^{-3}, the PL peak energy was observed to shift to higher energy, as seen in Fig. 4. The Si doping levels required to increase the PL peak energy were found to be much higher than for Te. In fact, at a Si doping level of 10^{19} cm^{-3} the material still had an order parameter of 0.4, as judged from the PL peak energy. These results were verified by the TED patterns obtained for samples with various Si doping levels [68]. Samples with electron concentrations of approximately 10^{19} cm^{-3} were still somewhat ordered. This contrasts sharply with the results for the Te doped samples, where a Te doping level of 10^{18} cm^{-3} was found to yield completely disordered material. The line through the Si data points in Fig. 4 was calculated for disordering due to the increase in Ga and In diffusion coefficients due to the shift of the Fermi level with increasing doping level [69]. The good fit to the data suggests that the disordering mechanism for Si is simply diffusion in the bulk during the growth cycle.

Data showing the degree of order versus p-type doping level for Zn doped GaInP is also included in Figure 4. The disordering induced by Zn was verified by TEM results [70]. As for the Si data, the line for Zn in Fig. 2 indicates the calculated disordering for a bulk disordering process [69].

By changing the Te flow rate during the growth process, order on disorder (O/D) and disorder on order (D/O) heterostructures were produced. However, the adsorption/desorption kinetics dictate that the growth cycle be interrupted for several minutes in order to produce abrupt interfaces. An example of such a structure is shown in Fig. 5, the TEM image of a D/O heterostructure produced by adding Te to the system during the growth cycle [68]. The bottom layer was grown without Te, followed by a 10 min interruption during which the group III precursors were removed from the system and DETe was added. After the 10 minute interruption, growth of Te doped GaInP was initiated. The TEM image shows an abrupt decrease in the order parameter at the point where Te was added to the system. This also produced a marked change in the APB spacing and the propagation angle of the APBs. D/O heterostructures were also produced, with PL consisting of two peaks, due to the undoped (highly ordered) and Te doped (less ordered) layers, with an energy separation of 60 meV [71]. This was the first report of heterostructures with the change in bandgap energy produced by a modulation of the concentration of a surfactant. Similar techniques have also been used to produce quantum well structures with widths as small as 100 Å [71].

6. ANTIMONY

Perhaps the most interesting surfactant, shown to strongly affect the ordering in GaInP, is the isoelectronic dopant Sb. Since it has the same number of valence electrons as P, no significant change in the Fermi level

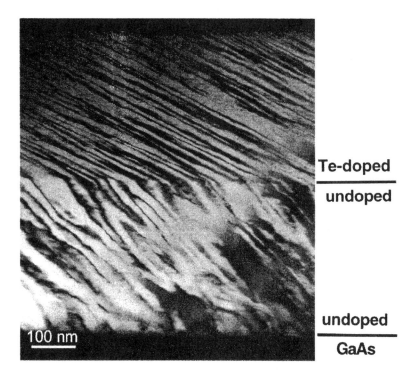

Figure 5. [110] TEM cross section of a heterojunction produced by using Te doping to disorder the top layer. A clear change in the APB spacing and angle is seen at the interface (after Lee et al. [68]).

position at the growth temperature is expected. Thus, it should have essentially no effect on the Ga and In diffusion coefficients. Because it is much larger than P, the solubility is small [72,73]. Thus, it is likely to accumulate at the surface during growth. The amount of TESb added to the system was insufficient to produce a measurable amount of Sb incorporation into the solid by x-ray diffraction. From previous studies [73], the Sb concentration in the solid is expected to be approximately 10^{-4}. However, the addition of this tiny amount of Sb produces a marked increase in the low temperature PL peak energy [69,74,75]. The results indicate that the layers are highly ordered without Sb and essentially disordered when grown with a small amount of Sb present, as shown in Fig. 6.

Figure 6 includes the SPA anisotropy intensity at 400 nm versus TESb flow rate during growth. The results indicate a dramatic change in the surface reconstruction induced by the addition of Sb to the system. As discussed above, the intensity of the peak near 400 nm in the SPA anisotropy spectra directly correlates to the concentration of $[\bar{1}10]$ P dimers on the surface. Comparison of the spectra for the undoped $Ga_{0.52}In_{0.48}P$ layers and layers with Sb clearly indicates that the P dimer concentration on the surface of the samples grown with Sb present has been significantly reduced [74,75]. This is thought to be the first experimental evidence that an isoelectronic dopant, such as Sb, can act as a surfactant to change the surface reconstruction of a III/V semiconductor layer during OMVPE growth.

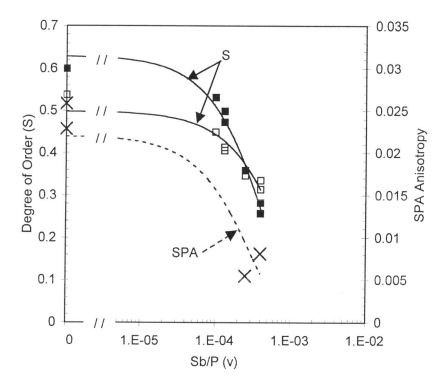

Figure 6. The effect of TESb flow rate on the degree of order and SPA difference signal at 400 nm (after Stringfellow et al. [75]).

Since the $[\bar{1}10]$ P dimers produce the surface thermodynamic driving force for formation of the CuPt structure during growth, it is not surprising that the removal of $[\bar{1}10]$ P dimers by the addition of Sb, described above, also eliminates ordering. These results confirm the observation that the degree of order in $Ga_{0.52}In_{0.48}P$ layers is directly related to the P dimer concentration. In fact, the degree of order versus 400 nm SPA anisotropy signal for the layers grown using Sb fall very near the data discussed above for the effects of temperature and flow rate of the P precursor, as indicated in Fig. 3.

The ability to control the ordering and, thus, the bandgap energy of GaInP layers by the addition of Sb during growth suggests the possibility of producing heterostructures and QWs. To grow a D/O heterostructure, the surface of the undoped, ordered $Ga_{0.52}In_{0.48}P$ layer grown first was exposed to TBP and TESb for five minutes, without growth, to allow Sb to accumulate. The disordered layer, grown with Sb present, followed. Figure 7 shows the PL spectrum for an O/D heterostructures grown by this procedure. A remarkable difference in the bandgap energy between the two layers of 135 meV was observed. This is the first demonstration of a potentially powerful method for the production of atomically engineered structures for advanced electronic and photonic devices. The ability to independently modulate the bandgap and the Fermi level position is the key to the potential usefulness of this technique. The 135 meV bandgap discontinuity is more than 5 kT at room temperature, which should be sufficient for many devices.

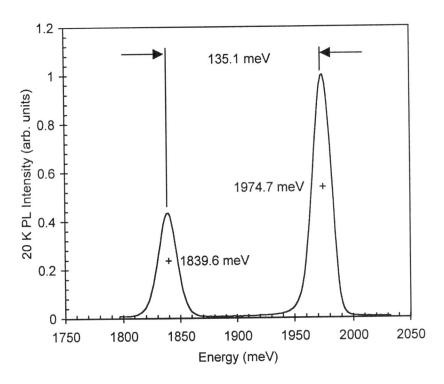

Figure 7. 20-K PL spectrum for a GaInP order/disorder heterostructure grown by OMVPE, with TESb present during growth of the top layer. A 10-second interrupt was included at the point when Sb was added to the system (after Shurtleff et al. [74]).

These results suggest a new and powerful concept: determination of the surface reconstruction during growth, using a surfactant, as a method of controlling the properties of the resultant epitaxial layer. This has been clearly demonstrated for the control of the bandgap energy in GaInP. However, the implications may be much broader. Surface reconstruction will certainly affect morphology, and hence the ability to grow low dimensional structures. It would not be surprising to find that the surface reconstruction during growth also controls other important processes such as spinodal decomposition, a process known to occur at the surface during growth [4,5]. It seems likely that the surface reconstruction during growth will also affect the formation of native defects as well as dopant incorporation. Thus, surfactants, such as Sb, are likely to affect the conductivities, mobilities, and minority carrier lifetimes of layers grown by vapor phase processes.

7.   SUMMARY

The phenomenon of atomic scale ordering during vapor phase epitaxial growth to produce natural monolayer superlattices was discovered in III/V alloys a little over a decade ago. During the intervening years, the CuPt ordered structure, with ordering on {111} planes, has been found to occur in virtually all III/V semiconductors for growth on the (001) plane. Both theory and experimental results indicate that the CuPt structure is not thermodynamically stable in the bulk alloys. It is stabilized by the formation of [110] rows of [$\overline{1}$10] group V dimers on the surface during growth. For GaInP grown by OMVPE, a direct link between formation of the CuPt structure and the concentration of [$\overline{1}$10] P dimers, for changes in either the temperature or the partial pressure of the P precursor, has now been firmly established using SPA to measure the dimer concentration. However, other factors also affect ordering. Evidence of the role of surface steps indicates the enhancement of ordering by [110] steps and diminution of ordering for [$\overline{1}$10] steps. This strongly indicates that the step structure also plays a role in the ordering process, as it ultimately must since this is where all adatoms are incorporated into the lattice. Examination of the effect of growth rate gives some idea of the kinetics of the ordering process. At rates above 4 μm/hr the ordering weakens in GaInP grown by OMVPE and almost disappears at a growth rate of 12 μm/hr. This indicates that the kinetics of the ordering process are relatively rapid. At this time, we have no compelling model to explain the effects of steps on the ordering process.

One reason that ordering has received so much attention is because the bandgap energy is strongly dependent on the CuPt order parameter. Theoretical and experimental studies indicate that the bandgap energy of completely ordered GaInP will be 300-500 meV less than for disordered material. For this reason, order must be strictly controlled for materials used for both electronic and photonic devices. We are now beginning to explore the use of ordering to improve the performance of devices.

Dopants are known to result in the destruction of ordering for GaInP. The effects of Si and Zn are due to an increased Ga/In interdiffusion. However, Te is found to act as a surfactant, causing disorder by changing the group III adatom attachment at steps during growth.

Perhaps the most dramatic surfactant effect for GaInP grown by OMVPE is due to the isoelectronic element Sb, which is expected to accumulate at the surface during growth. A very small concentration of TESb in the vapor, resulting in an Sb concentration in the solid estimated to be approximately 10^{-4}, results in the growth of disordered material. The Sb is isoelectronic with P, so it does not change the Fermi level position. Thus, it is expected to produce no increase in the Ga/In bulk diffusion. The step structure is not changed appreciably. However, SPA results indicate that the GaInP surface reconstruction is markedly changed by the addition of Sb. The $[1\overline{1}0]$ P dimers, responsible for the thermodynamic driving force for CuPt ordering, are eliminated by Sb at concentrations sufficient to produce the growth of completely disordered GaInP. The results clearly show that the introduction of TESb during growth can be used to control the GaInP bandgap energy. A single heterostructure was produced by modulating the TESb flow rate during growth. PL peaks from the two layers, grown with and without Sb present, were found to differ by 135 meV. This is the first example of the use of an isoelectronic surfactant to significantly modulate the bandgap energy of a material grown by OMVPE. It is anticipated that surfactant Sb can also be used to produce more complex structures, such as QWs.

ACKNOWLEDGMENTS

The author wishes to acknowledge the work of his students, K. Shurtleff, R.T. Lee, S.H. Lee, C.M. Fetzer, and Yu Hsu. The work reported here could not have been performed without the financial support of the Department of Energy and the National Science Foundation.

REFERENCES

1. G. B. Stringfellow, in *Thin Films: Heteroepitaxial Systems*, ed. M. Santos and W.K. Liu (World Scientific Publishing, 1998) pp. 64-116.

2. G. B. Stringfellow, in *Common Themes and Mechanisms of Epitaxial Growth*, edited by P. Fuoss, J. Tsao, D. K. Kisker, A. Zangwill, and T. Kuech (Materials Research Society, Pittsburgh, 1993), pp. 35-46.

3. D. Doppalapudi, S.N. Basu, K.F. Ludwig, and T.D. Moustakas, J. Appl. Phys. 84, 1390 (1009).

4. A. Zunger and S. Mahajan, in *Handbook on Semiconductor*, edited by T. S. Moss (Elsevier Science B.V., Amsterdam, 1994), p. 1399.

5. See several papers in "Compositional Modulation and Ordering in Semiconductors," ed. T.P. Pearsall and G.B. Stringfellow, MRS Bulletin 22 #7, 1997.

6. S.B. Zhang, S. Froyen, and A. Zunger, Appl. Phys. Lett. 67, 3141 (1995).

7. G.S. Chen, D.H. Jaw, and G.B. Stringfellow, J. Appl. Phys. 69, 4263 (1991).

8. M. Shimotomai and A. Youshikawa, Appl. Phys. Lett. 73, 3256 (1998).

9. M.K. Behbehani, E.I. Piner, S.X. Liu, N.A. ElMastry, and S.M. Gedair, Paper presented at Electronic Materials Conference, July 1999, Santa Barbara, CA.

10. D. Korakakis, K.F. Ludwig, and T.D. Moustakas, Appl. Phys. Lett. 71, 72 (1997).

11. A. Zunger (private communication).

12. H. Murata, I.H. Ho, L.C. Su, Y. Hosokawa, and G.B. Stringfellow, J. Appl. Phys. 79, 6895 (1996).

13. T.S. Kuan, T.F. Kuech, W.I. Wang, and E.L. Wilkie, Phys. Rev. Lett. 54, 201 (1985).

14. A. Ourmazd and J.C. Bean, *Phys. Rev. Lett.* 55, 765 (1985).

15. H.R. Jen, M.J. Cherng, and G.B. Stringfellow, Appl. Phys. Lett. 48, 1603 (1986).

16. O. Ueda, M. Hoshino, M. Takechi, M. Ozeki, T. Kato, and T. Matsumoto, *J. Appl. Phys.* 68, 4268 (1990).

17. L.C. Su, I.H. Ho, and G.B. Stringfellow, J. Appl. Phys. 75, 5135(1994).

18. D. Mao, P.C. Taylor, S.R. Kurtz, M.C. Wu, and W.A. Harrison, Phys. Rev. Lett. 76, 4769 (1996).

19. S.H. Wei and A. Zunger, Phys. Rev. B 49, 14337 (1994).

20. T.Y. Seong, G.R. Booker, A.G. Norman, P.J.F. Harris, and A.G. Cullis, Inst. Phys. Conf. Ser. 146, 241 (1995); T.Y. Seong, A.G. Norman, G.R. Booker, and A.J. Cullis, J. Appl. Phys. 75, 7852 (1994).

21. G.B. Stringfellow, P.F. Lindquist, and R.A. Burmeister, J. Electron. Mater. 1, 437 (1972).

22. A. Gomyo, K. Iobgayashi, S. Kawata, I. Hino, T. Suzuki, and T. Yuasa, J. Crystal Growth 77, 367 (1987).

23. S.H. Wei and A. Zunger, Phys. Rev. B 39 (1989) 3279.

24. S.H. Wei, A. Franceschetti, and A. Zunger, Mat. Res. Symp. Proc. 147, 3 (1996).

25. D.B. Laks, S.H. Wei, and A. Zunger, Phys. Rev. Lett. 69, 3766 (1992).

26. P. Ernst, C. Geng, F. Scholz, H. Schweizer, Y. Zhang, and A. Mascarenhas, Appl. Phys. Lett. 67, 2347 (1997).

27. L. C. Su, I. H. Ho, N. Kobyashi, and G. B. Stringfellow, *J. Crystal Growth* 145, 140 (1994).

28. D.J. Arent, M. Bode, K.A. Bode, K.A. Bertness, S.R. Kurtz, and J.M. Olson, Appl. Phys. Lett. 62, 1806 (1993).

29. S.R. Kurtz, L.R. Dawson, R.M. Biefeld, D.M. Follstaedt, and B.L. Doyle, Phys. Rev. B 46, 1909 (1992).

30. G.B. Stringfellow, in *High Brightness Light Emitting Diodes*, edited by G.B. Stringfellow and M.G. Craford (Academic Press, Boston, 1997).

31. L.C. Su, I.H. Ho, N. Kobayashi, and G.B. Stringfellow, J. Cryst, Growth 145, 140 (1994).

32. R.P. Schneider, E. D. Jones, and D. M. Follstaedt, *Appl. Phys. Lett.* 65, 587 (1994).

33. I. Kamiya, L. Mantese, D.E. Aspnes, D.W. Kisker, P.H. Fuoff, G.B. Stephenson, and S. Brennan, J. Crystal Growth 173, 67 (1996).

34. D.E. Aspnes et al., J. Crystal Growth 120, 71 (1992).

35. Y. Kobayashi and N. Kobayashi, J. Electron. Mater. 25, 691 (1996).

36. H. Murata, I.H. Ho, T.C. Hsu, and G.B. Stringfellow, Appl. Phys. Lett. 67, 3747 (1995).

37. N. Kobayashi, Y. Kobayashi, and K. Uwai, J. Crystal Growth 170, 225 (1997).

38. H. Murata, T.C. Hsu, I.H. Ho, L.C. Su, Y. Hosokawa, and G.B. Stringfellow, Appl. Phys. Lett. 68, 1796 (1996).

39. H. Murata, I.H. Ho, L.C. Su, Y. Hosokawa, and G.B. Stringfellow, J. Appl. Phys. 79, 6895 (1996).

40. L.C. Su, I.H. Ho, and G.B. Stringfellow, J. Crystal Growth, 146, 558 (1995).

41. H. Murata, I.H. Ho, and G.B. Stringfellow, J. Crystal Growth 170, 219 (1997).

42. H. Murata, S.H. Lee, I.H. Ho, and G.B. Stringfellow, J. Vac. Sci. Technol. B14, 3013 (1996).

43. G.B. Stringfellow, *Organometallic Vapor Phase Epitaxy: Theory and Practice, 2nd Edition* (Academic Press, Boston, 1999), Chapter 2.

44. Y.S. Chun, S.H. Lee, I.H. Ho, and G.B. Stringfellow, J. Appl. Phys. 81, 646 (1997).

45. D.S. Cao, E.H. Reihlen, G.S. Chen, A.W. Kimball, and G.B. Stringfellow, J. Crystal Growth 109, 279 (1991).

46. E. Tournie and K. H. Ploog, Thin Solid Films, 231, 43 (1993).

47. D. Reinking, M. Kammier, M. Horn-von Hoegen, and K.R. Hofmann, Appl. Phys. Lett. 71, 924 (1997).

48. J.E. Cunningham, K.W. Goossen, W. Jan, and M.D. Williams, J. Vac. Sci. Technol. B 13, 646 (1994).

49. B.R.A. Neves, M.S. Andrade, W.N. Rodriques, G.A.M. Safar, M.V.B. Moreira, and A.G. de Oliveira, Appl. Phys. Lett., 72, 1712 (1998).

50. D. Kandel and E. Kaxiras, Phys. Rev. Lett. 75, 2742 (1995).

51. C.W. Oh, E. Kim, and Y.H. Lee, Phys. Rev. Lett. 76, 776 (1996).

52. H. Oigawa, M. Wassermeier, J. Behrend, L. Daweritz, and K.H. Ploog, Surf. Sci. 376, 185 (1997).

53. S.J. Jenkins and G.P. Srivastava, Surf. Sci. 398, L308 (1998).

54. B. Voigtlander, A. Zinner, T. Weber, H.P. Bonzel, Phys. Rev. B 51, 7583, (1995).

55. H. Okumura, H. Hamaguchi, G. Feuillet, Y. Ishida, and S. Yoshida, Appl. Phys. Lett, 72, 3056 (1998).

56. A. Gomyo, H. Hotta, I. Hino, S. Kawata, K. Kobayashi, and T. Suzuki, Jpn. J. Appl. Phys. 28, L1330 (1989).

57. J. P. Goral, S. R. Kurtz, J. M. Olson, and A. Kibbler, J. of Electronic Materials 19, 95, (1990).

58. S. H. Lee, C. M. Fetzer, and G. B. Stringfellow, D. H. Lee, and T. Y. Seong, J. Appl. Phys. 85, 3590 (1999).

59. S. H. Lee. T. C. Hsu, and G. B. Stringfellow, J.Appl. Phys, 84, 2618 (1998).

60. S. H. Lee, C. M. Fetzer, and G. B. Stringfellow, J. Cryst, Growth, 195, 13 (1998).

61. C. H. Wu, M. S. Feng, and C. C. Wu, Mat. Res. Soc. Symp. Proc. 300, 477 (1993).

62. A. Gomyo, T. Suzuki, K. Kobayashi, S. Kawata, I. Hino, T. Yuasa, Appl. Phys. Lett. 50, 673 (1987).

63. S. R. Kurtz, J. M. Olson, D. J. Friedman, A. E. Kibbler, and S. Asher, J. Electronic Materials, 23, 431 (1994).

64. E. Morita, M. Ikeda, O. Kumagai, and K. Kaneko, Appl. Phys. Lett. 53, 2164 (1988).

65. F. P. Dabkowski, P. Gavrilovic, K. Meehan, W. Stutius, J. E. Williams, M. A. Shahid, and S. Mahajan, Appl. Phys. Lett. 52, 2142 (1998).

66. T. Suzuki, A. Gomyo, I. Hino, K. Kobayashim, S. Kawata, and S. Iijima, Jpn. J. Appl. Phys. 27, L1549 (1988).

67. M.K. Lee, R.H. Horng, and L.C. Haung, Appl. Phys. Lett. 59, 3261, (1991).

68. R.T. Lee, C.M. Fetzer, K. Shurtleff, Y. Hsu, G.B. Stringfellow, S. Lee, and T.Y. Seong, J. Electron. Mater., 29,134 (2000).

69. C.M. Fetzer and G.B. Stringfellow (unpublished results).

70. S.H. Lee, C.M. Fetzer, G.B. Stringfellow, C.J. Choi, and T.Y. Seong, J. Appl. Phys. (to be published).

71. Y. Hsu, C.M. Fetzer, and G.B. Stringfellow, J. Appl. Phys. (submitted).

72. G.B. Stringfellow, J. Crystal Trowth 27, 21 (1974).

73. M.J. Jou, D.H. Jaw, Z.M. Fang, and G.B. Stringfellow, J. Crystal Growth, 190, 208 (1990).

74. J.K. Shurtleff, R.T. Lee, C.M. Fetzer, and G.B. Stringfellow, Appl. Phys. Lett., 75, 1914 (1999).

75. G.B. Stringfellow, J.K. Shurtleff, R.T. Lee, C.M. Fetzer, and S. W. Sun, "Surface processes in OMVPE — the Frontiers," J. Crystal Growth, 221, 1 (2000), Figs. 1 and 3 used with permission of Elsevier Science.

76. T.C. Hsu, Y. Hsu, and G.B. Stringfellow, J. Crystal Growth 193, 18 (1998).

Chapter 4

X-Ray Diffraction Analysis of Ordering in Epitaxial III-V Alloys

Rebecca L. Forrest
University of California-Los Angeles, Department of Materials Science

Key words: X-ray diffraction, long-range order, short-range order, diffuse scattering, order parameter, lateral composition modulation, reciprocal space map

Abstract: The X-ray diffraction investigation of atomic ordering and lateral composition modulation is discussed. An overview of X-ray diffraction theory is given so that the reader can appreciate the results, potential, and challenges of the technique. Quantitative analysis of long- and short-range order parameters, lateral composition modulation, and ordered domain structure, including experimental results, are discussed.

1.   INTRODUCTION

X-ray diffraction is an important technique for materials characterization. It is used non-destructively to study crystalline structure, interatomic spacings, alloy phase constitution, crystalline perfection, and order/disorder in a wide variety of materials including metals, organic and polymeric solids, liquids, biological compounds, and electronic materials. It is used on epitaxial layers to determine lattice mismatch, composition, layer thickness, strain, and interfacial roughness. It was first applied to atomic ordering in metals in 1938, and lately has been used to study ordering in ternary epitaxial semiconductor alloys.

This chapter will discuss x-ray diffraction analysis of ordering in ternary epitaxial semiconductor alloys; several examples of recent work will be presented, along with some background in x-ray diffraction theory. This is intended to allow readers to follow existing x-ray research, and to motivate those interested in doing their own x-ray diffraction experiments. The later are strongly advised to read the more detailed references cited throughout the chapter. There are several good references on x-ray diffraction analysis of ordered structures, including the books by Warren,[1] Guinier,[2] Schweika,[3,4] and Krivoglaz,[5] to which the reader can refer for further detail. Also recommended are the books on scattering and diffraction from thin films, surfaces, interfaces, and multilayers by Yang et al.[6] and Holý et al.[7] This chapter will generally follow the introductory treatment of Warren, and apply it to ordering in III-V ternary alloys.

Radiation and matter can interact in three basic ways; the atomic electrons can 1) absorb the radiation and emit an electron (*Auger effect*) or a photon (*fluorescence*) through electronic excitation, 2) scatter the radiation incoherently (inelastic *Compton scattering*), or 3) scatter the radiation coherently (elastic scattering). Radiation elastically scattered from different atoms can interfere; constructive interference of radiation coherently scattered by an arrangement of atoms is referred to as *diffraction*.

When radiation and matter interact, the arrangement of the atoms and the incident energy determines the distribution of diffracted radiation. Diffraction can occur if the wavelength of the incident radiation is of the same order of magnitude as the spacing between the scattering atoms. X-rays are particularly well suited to probe the structure of crystalline solids, since their wavelengths (approximately $0.1 - 5$ Å) are of the same order of magnitude as the atomic spacings within most solids. A periodic arrangement of atoms, as in a crystal, will result in a periodic distribution of diffracted x-rays. The amplitude of the x-rays in this diffraction pattern is essentially given by the Fourier transform of the arrangement of the atoms (i.e. of their electron density). The square of this complex amplitude yields the measured intensity.

Since an x-ray diffraction pattern is determined by the atomic arrangement of the scattering material and the associated electron density, it can be used to study directly the material's structure. This obviously extends to the study of order and disorder in crystals. The following sections will briefly describe x-ray diffraction theory and discuss the application of x-ray diffraction to different topics in the ordering of semiconductor materials, including long-range order, short-range order, and compositional modulation.

2.   OVERVIEW OF X-RAY DIFFRACTION THEORY

2.1   Kinematical Diffraction Theory

In order to discuss the application of x-ray diffraction to topics in ordering we first present some basics of the theory of diffraction. What follows is a brief overview; a more detailed treatment can be found in the references.[1-5]

According to the kinematical theory of x-ray diffraction, a plane wave of radiation incident on a small crystal is scattered by the electron distribution associated with the atoms in the crystal. (In the kinematical theory, the incident intensity is not reduced in the scattering process; it is attenuated only by photoelectric absorption. See section 2.2.) The amplitude of the radiation scattered elastically by a single atom is given by the normalized Fourier transform of the electron distribution, and is called the atomic form factor, f. The intensity distribution scattered by a small crystal is obtained by summing the intensity scattered by all the atoms in the crystal,

$$I = I_e \sum_{m,n} f_m f_n e^{\frac{2\pi i}{\lambda}(\mathbf{S}-\mathbf{S}_o)\bullet(\mathbf{R}_m-\mathbf{R}_n)}, \tag{1}$$

where λ is the x-ray wavelength, $\mathbf{S}_o$ and $\mathbf{S}$ are unit vectors in the directions of the incident and scattered waves, and $\mathbf{R}_m$ and $\mathbf{R}_n$ are vectors from any origin to the atoms m and n, and the summation is over all such atom pairs. I_e is the Thompson scattering equation for the intensity scattered per electron for a polarized incident beam given by

$$I_e = I_o \frac{e^4}{m^2 c^4 R^2} \left(\frac{1+|\cos 2\theta_m|\cos^2 2\theta}{1+|\cos 2\theta_m|} \right)$$

where $2\theta_m$ is the (perfect crystal) monochromator reflection angle (see below), e is the charge of an electron, m is the mass of an electron, R is the distance to the detector, c is the speed of light, and I_o is the intensity of the main beam. For a plane wave incident on a small crystal, the scattered intensity reduces to[1]

$$I = I_e |F_T|^2 \times \frac{\sin^2[\frac{\pi}{\lambda}(\mathbf{S}-\mathbf{S}_o)\cdot N_1\mathbf{a}_1]}{\sin^2[\frac{\pi}{\lambda}(\mathbf{S}-\mathbf{S}_o)\cdot\mathbf{a}_1]} \times \frac{\sin^2[\frac{\pi}{\lambda}(\mathbf{S}-\mathbf{S}_o)\cdot N_2\mathbf{a}_2]}{\sin^2[\frac{\pi}{\lambda}(\mathbf{S}-\mathbf{S}_o)\cdot\mathbf{a}_2]}$$
$$\times \frac{\sin^2[\frac{\pi}{\lambda}(\mathbf{S}-\mathbf{S}_o)\cdot N_3\mathbf{a}_3]}{\sin^2[\frac{\pi}{\lambda}(\mathbf{S}-\mathbf{S}_o)\cdot\mathbf{a}_3]}. \tag{2}$$

In Eqn. 2, F_T is the geometrical structure factor of the crystal, including a thermal (Debye-Waller) factor for each atom, given by a summation over all of the atoms in a unit cell of the crystal (see Section 2.3). The last three terms in Eqn. 2 effectively restrict the diffracted intensity to Bragg peak positions. In these three terms, the vectors $\mathbf{a}_i$ are the unit cell crystal axes, N_i is the number of unit cells along the $\mathbf{a}_i$ direction, and $\mathbf{S}_o$ and $\mathbf{S}$ are unit vectors in the directions of the incident and scattered x-ray beams, respectively. 2θ is the angle between the incident and scattered waves, $\mathbf{S}_o$ and $\mathbf{S}$. The Bragg condition described by these three terms results when in all three $(\mathbf{S} - \mathbf{S}_o) \cdot \mathbf{a}_i = n\lambda$, where n is an integer. This can be written as $\mathbf{S} - \mathbf{S}_o = \lambda \, (h\mathbf{b}_1 + k\mathbf{b}_2 + \ell\mathbf{b}_3) = \lambda \, \mathbf{H}_{hk\ell}$, or as the more common Bragg Law, $\lambda = 2 \, d_{hk\ell} \sin \theta$. Here h, k, and ℓ are the familiar integer Miller indices, $\mathbf{b}_1$, $\mathbf{b}_2$, and $\mathbf{b}_3$ are the reciprocal lattice basis vectors, and a reciprocal lattice vector $\mathbf{H}_{hk\ell}$ is perpendicular to the planes (h k ℓ) with a length equal to the reciprocal of the planar spacing, $1 / d_{hk\ell} = | \mathbf{H}_{hk\ell} |$. (Hereafter (h k ℓ) will refer to a position in reciprocal space, defined by the reciprocal lattice vector $\mathbf{H}_{hk\ell}$.)

If the Bragg condition is satisfied, the diffracted intensity reduces to

$$I = I_e \left| F_T \right|^2 N_1^2 N_2^2 N_3^2 .$$

Diffracted intensity will only be found in the immediate vicinity of a Bragg peak; an infinite perfect crystal would produce diffracted intensity located in very sharp functions at Bragg points. For a real, finite crystal, on a real diffractometer (with a finite resolution), the kinematical Bragg intensity will be spread out over a somewhat larger width. (On a high-resolution diffractometer, a high-quality Si crystal typically has a Bragg peak full-width at half maximum (fwhm) of 10 arc sec.)

If a crystal is rocked about a Bragg angle, the total Bragg intensity can be collected by a fixed, open detector. This is referred to as the integrated intensity. Mathematically it is obtained by integrating the intensity over the small volume in reciprocal space surrounding the peak. According to the kinematical theory of x-ray scattering, the integrated intensity for an extended face imperfect crystal is given by[1]

$$E = \frac{P_o}{\omega} \left(\frac{e^4}{m^2c^4} \right) \frac{\lambda^3 \left| F_T \right|^2}{2\mu V_a^2} \left(\frac{1 + \left| \cos 2\theta_m \right| \cos^2 2\theta}{\sin 2\theta (1 + \left| \cos 2\theta_m \right|)} \right) \left(1 - e^{-2\mu L / \sin \theta} \right), \qquad (3)$$

where P_0 is the incident x-ray power, ω is the angular scan rate, μ is the linear absorption coefficient of the crystal, V_a is the volume of a unit cell in the crystal, L is the sample thickness, and $L/\sin \theta$ is the x-ray path length inside the sample. The term

$$\left(\frac{1+|\cos 2\theta_m|\cos^2 2\theta}{\sin 2\theta (1+|\cos 2\theta_m|)} \right)$$

is the combined Lorentz and polarization factor for x-rays polarized by a perfect monochromator. The last term in Eqn. 3 accounts for absorption of the x-rays by the sample when the sample thickness is less than or comparable to $1/\mu$.

The integrated intensity (E) is measured by rocking the sample through a range of incident angles (ω). The integrated intensity is given by the area under the curve in a plot of intensity versus ω. It is assumed that all of the Bragg intensity scattered by the sample at a given ω is detected. In order to verify this, the integrated intensity should be measured with several different detector slit sizes and plotted versus slit size. E should increase with slit size until the slits are large enough to accept all intensity at a given ω, after which E should be constant (aside from any diffuse background that may surround the Bragg reflection, i.e., thermal diffuse scattering or fluorescence). The smallest slit that measures the entire integrated intensity should be used; larger slits only decrease the instrumental resolution and increase the aforementioned measured background intensity.

2.2 Dynamical Diffraction Theory

The above kinematical treatment of the diffracted intensity ignores the possibility of multiple reflections of the x-rays within the crystal. This assumption is valid if multiple reflections are eliminated by imperfections in the crystal. The kinematical theory is therefore appropriate for imperfect, mosaic crystals. Since it is possible to obtain very high quality III-V crystals, it may be necessary to use a theory of diffraction that takes into account the interaction between multiple reflections, called dynamical diffraction theory.

In a perfect crystal, the x-ray beam is transmitted and reflected by many planes at a given (h k ℓ) Bragg angle. These transmitted and reflected waves interact strongly, leading to phase cancellation of the transmitted wave and thus attenuation which is much greater than that due to photoelectric absorption alone. For this reason, the integrated intensity from a perfect crystal will be less than the integrated intensity from a mosaic crystal; essentially less volume is measured in a perfect crystal. In fact, the stronger the reflection, the greater the cancellation and the less volume is measured. The intensity distribution is typically modeled based on the approach of Takagi[8] and Taupin.[9] This formalism leads to somewhat complicated expressions that are nevertheless useful in computer simulations, but the

basic result may be found, for example, in Warren's book.[1] There are several
good references for the interested reader. [6,7,10-15] For our study of the order
in epitaxial layers we may need a dynamical expression for the integrated
intensity, which turns out to be much simpler than that for the intensity
distribution. The integrated intensity of a reflection from a perfect crystal in
which absorption can be neglected is given by dynamical theory as[1]

$$E = \frac{8}{3\pi}\frac{P_o}{\omega}\left(\frac{e^2}{mc^2}\right)\frac{\lambda^2|F_T|}{2V_a}\left(\frac{1+|\cos 2\theta_m||\cos 2\theta|}{\sin 2\theta(1+|\cos 2\theta_m|)}\right) \tag{4}$$

where all of the terms have the same definitions as in the kinematical theory.
Note that for a perfect crystal, the integrated intensity is proportional to $|F_T|$,
and not $|F_T|^2$ as for an imperfect crystal.

If the integrated intensity is calculated using dynamical theory assuming
a reflection with weak reflecting power compared to absorption, i.e., small
$|F_T|$, one obtains Eqn. 3, the same expression given by kinematical theory.
Thus, Eqn. 3 describes the integrated intensity of any reflection from a
mosaic crystal, and also very weak reflections from a perfect crystal. A
strong reflection from a high quality, "perfect" crystal has an integrated
intensity given by Eqn. 4. Real crystals are usually neither ideally perfect
nor ideally imperfect, in which case a peak's integrated intensity would lie
somewhere between the values given by Eqns. 3 and 4. For this reason, a
crystallographer will often either artificially "damage" a crystal, or remove
strong reflections from the structural analysis, in order to use Eqn. 3.

The perfection of a crystal can be determined from the breadth of its
Bragg peaks in rocking (ω) scans. The more perfect a crystal, the more
narrow its peaks in ω scans, i.e., the smaller their ω fwhm value. The
perfection of an epilayer can be judged by comparing its ω-fwhm to that of a
crystal known to be of high quality (i.e., essentially perfect, such as Si or
Ge). A substrate peak can usually be used as a reference, since most
substrates are high-quality crystals. Based on this measurement, one can
judge whether Eqn. 3 or Eqn. 4 most appropriately models the integrated
intensity of the diffracted peaks. If the peaks can not be reasonably modeled
as either ideally perfect or imperfect, Eqns. 3 and 4 will at least give upper
and lower bounds on the measured integrated intensity. In this case,
approximations for the integrated intensity from non-ideal crystals can be
used, as, for example, in the paper by Stevenson.[14]

2.3   Structure Factor

The geometrical structure factor of a crystal, F_T, at a Bragg reflection is given by

$$F_T = \sum_n f_n e^{-M_n} e^{2\pi i \left(hx_n + ky_n + \ell z_n \right)} \tag{5}$$

where (x_n, y_n, z_n) are the fractional coordinates of atom n in the unit cell, $(h\ k\ \ell)$ are the Miller indices (integers if the true unit cell is used), f_n is the atomic scattering factor, or form factor, for atom n, and the sum is over all atoms in the unit cell. The form factor, f_n, is a function of the scattering angle; at zero scattering angles f_n equals the number of electrons in atom n. Values of f_n can be found in the International Tables for X-ray Crystallography.[16] The term e^{-Mn} is the conventional Debye-Waller term that accounts for the thermal vibrations of the atoms.[1] The structure factor can be calculated for known structures (i.e., known x_n, y_n, z_n), and the integrated intensities at specific $(h\ k\ \ell)$ reflections can be evaluated. Conversely, the positions of the diffraction peaks reveal information about the periodicity of the crystal, and the size and shape of an average unit cell in the crystal.

We will now calculate the structure factor for a III_A-III_B-V alloy with the zinc-blende crystal structure. Zinc-blende, the structure of most III-V alloys, consists of two interpenetrating face-centered cubic sublattices offset from each other by one quarter of a body diagonal. The two group-III constituents share one sublattice and the group-V atoms occupy the other. A III-V zinc-blende unit cell contains four group-III atoms at (x, y, z) positions (0, 0, 0), (0.5, 0.5, 0), (0.5, 0, 0.5) and (0, 0.5, 0.5) and four group-V atoms at positions (0.25, 0.25, 0.25), (0.75, 0.75, 0.25), (0.75, 0.25, 0.75), and (0.25, 0.75, 0.75). In a random, III_A-III_B-V alloy, i.e., $A_xB_{1-x}C$, each group-III site has a probability of x of being occupied by an A atom, and a probability of 1-x of being occupied by a B atom. Ignoring for the moment the thermal factors, the zinc-blende structure factor is given by

$$F = \sum_{III} \left(xf_A + (1-x)f_B \right) e^{2\pi i \left(hx_{III} + ky_{III} + \ell z_{III} \right)} + \sum_V f_C e^{2\pi i \left(hx_V + ky_V + \ell z_V \right)},$$

where the first summation is over the four group-III atoms, and the second is over the four group-V atoms. This F will be complex, so we will give the result of the complex square of the structure factor, $|F|^2$. For an $A_xB_{1-x}C$ zinc-blende alloy,

$|F|^2 = 0$, if h, k, and ℓ are mixed odd and even integers,

$$= 16[|f_C|^2 + |xf_A + (1-x)f_B|^2], \text{ if h, k, and } \ell \text{ are all odd,}$$

$$= 16[|f_C|^2 + |xf_A + (1-x)f_B|^2 + 2f_C(xf_A + (1-x)f_B)], \qquad (6)$$

$$\text{if } h + k + \ell = 4n,$$

$$= 16[|f_C|^2 + |xf_A + (1-x)f_B|^2 - 2f_C(xf_A + (1-x)f_B)],$$

$$\text{if } h + k + \ell = 4n + 2.$$

To include the Debye-Waller thermal factors, we use

$$f_A \Rightarrow f_A e^{-M_A}, \; f_B \Rightarrow f_B e^{-M_B}, \text{ and } f_C \Rightarrow f_C e^{-M_C},$$

where M_A, M_B, and M_C are determined for the alloy of interest. If these values are not available, a reasonable approximation for reflections at lower (h k ℓ) indices would be to use the Debye-Waller factors for the III-V binary constituents,

$$f_A \Rightarrow f_A e^{-M_{AC}}, \; f_B \Rightarrow f_B e^{-M_{BC}}, \text{ and } f_C \Rightarrow \tfrac{1}{2} f_C (e^{-M_{AC}} + e^{-M_{BC}}).$$

3.   LONG-RANGE ORDER

3.1   Long-Range Order Parameter

Ordering in a crystal refers to an additional periodicity imposed on the initial structure that is not present under all conditions. If this ordering is coherent across many unit cells it is called long-range order (LRO) and it will lead to additional diffraction peaks, termed superstructure peaks. (In principle, long-range order implies an ordering over an entire mosaic block of the crystal; often, however, this is interrupted by various defects. Short-range order, by contrast, is usually confined to the first several atomic neighbors.) The structure factor for an ordered crystal can be calculated, and will predict the intensity of the superstructure reflections whose locations are determined by the new period. Information about the ordering in the crystal is obtained by comparing the superstructure reflections with the fundamental reflections, which do not depend on order.

LRO has been detected, by observation of superstructure peaks in diffraction patterns, in most epitaxial III-III-V and III-V-V alloys. Ordered structures observed to date include CuPt, CuAu, and chalcopyrite ordering, named for the alloys in which they were first observed, and triple-period ordering, which is a three-layer-period version of the two-layer-period CuPt structure. Much can be learned about the ordering in III-V alloys by studying the superstructure peaks. For example, the degree of order in a sample, quantified by the long-range order parameter, can be measured. This parameter is of interest because the electronic properties of an ordered alloy depend directly on this parameter.

In order to study long-range order we define the degree of order using the modified Bragg-Williams order parameter.[1] Consider an ordered III_A-III_B-V alloy, and let x_A and x_B be the fractions of type III atoms that are A and B atoms, respectively (the fractional compositions). In a perfectly ordered crystal, the sites occupied by A atoms are defined as α sites, and those occupied by B atoms are β sites. Let y_α and y_β be the fractions of type III sites that are α and β sites, respectively. Now consider an imperfectly ordered alloy; some A atoms will "rightly" occupy α sites while some A atoms will "wrongly" occupy β sites, similarly for the B atoms. The fraction of α sites that are "rightly" occupied is defined as r_α, the fraction of α sites that are "wrongly" occupied is defined as w_α. The same will be true of the β sites, with fractions r_β and w_β. Some relations that follow from these definitions are $x_A + x_B = 1$, $y_\alpha + y_\beta = 1$, $r_\alpha + w_\alpha = 1$, $r_\beta + w_\beta = 1$, $y_\beta r_\beta + y_\alpha w_\alpha = x_B$, and $y_\alpha r_\alpha + y_\beta w_\beta = x_A$. The modified Bragg-Williams order parameter is defined so that it is linearly proportional to ($r_\alpha + r_\beta$), and ranges from 0 for a random crystal to 1 for a perfectly ordered crystal. Thus the order parameter, S, is given by[1]

$$S = \frac{(r_\alpha - x_A)}{y_\beta} = \frac{(r_\beta - x_B)}{y_\alpha}. \tag{7}$$

For the completely random crystal $r_\alpha = x_A$ and $S = 0$. For the perfectly ordered structure $r_\alpha = 1$, $w_\alpha = 0$, and $S = 1$. According to this definition, only a stoichiometric alloy ($x_A = y_\alpha$) can result in $S = 1$, i.e., perfect order.

3.2 Structure Factor of an Ordered Crystal

We can now use these definitions to calculate the structure factor for an ordered crystal. We will calculate the structure factor for a CuPt-ordered III_A-III_B-V zinc-blende crystal. In this structure, the two group-III constituents are preferentially located in alternating (111) planes of their sublattice. Theoretically, the CuPt ordering direction could be any of the four $\{111\}$ variants (namely [1 1 1], [$\bar{1}$ 1 1], [1 $\bar{1}$ 1], and [1 1 $\bar{1}$]).

A CuPt ordered unit cell is twice as large in each of the x, y, and z directions as that of the random structure. (Note that there are thus half as many unit cells in each direction, N_i.) In a perfect $A_{0.5}B_{0.5}C$ ordered structure there are 16 III_A atomic sites, 16 III_B atomic sites, and 32 C (group-V) atomic sites in the unit cell. The fractional coordinates of the sites in a perfect CuPt [111] ordered unit cell are given in Table I. The coordinates are given in terms of the random unit cell dimensions and therefore range between 0 and 2. In this structure, $y_\alpha = y_\beta = 0.5$, and thus $S = 2(r_\alpha - x_A) = 2(r_\beta - x_B)$.

Table 1. (x, y, z) coordinates of III_A α sites, III_B β sites, and V sites in a CuPt [111] ordered zinc-blende unit cell, in terms of the random unit cell dimensions.

	\multicolumn{3}{c}{α}			\multicolumn{3}{c}{β}			\multicolumn{3}{c}{V}		
	x	y	z	x	y	z	X	y	z
1	0.0	0.0	0.0	0.5	0.5	0.0	0.25	0.25	0.25
2	1.0	1.0	0.0	0.5	0.0	0.5	0.25	0.75	0.75
3	0.0	1.0	1.0	0.0	0.5	0.5	0.75	0.25	0.75
4	1.0	0.0	1.0	1.5	1.5	0.0	0.75	0.75	0.25
5	1.5	0.5	0.0	1.0	1.5	0.5	1.25	0.25	0.25
6	0.5	1.5	0.0	1.5	1.0	0.5	1.25	0.75	0.75
7	1.5	0.0	0.5	0.5	1.5	1.0	1.75	0.25	0.75
8	0.0	1.5	0.5	0.0	1.5	1.5	1.75	0.75	0.25
9	0.0	0.5	1.5	0.5	1.0	1.5	0.25	1.25	0.25
10	0.5	0.0	1.5	1.5	0.5	1.0	0.25	1.75	0.75
11	1.0	0.5	0.5	1.0	0.5	1.5	0.75	1.25	0.75
12	0.5	1.0	0.5	1.5	0.0	1.5	0.75	1.75	0.25
13	0.5	0.5	1.0	1.0	0.0	0.0	0.25	0.25	1.25
14	1.5	1.5	1.0	0.0	1.0	0.0	0.25	0.75	1.75
15	1.0	1.5	1.5	0.0	0.0	1.0	0.75	0.25	1.75
16	1.5	1.0	1.5	1.0	1.0	1.0	0.75	0.75	1.25
17							1.25	1.25	0.25
18							1.25	1.75	0.75
19							1.75	1.25	0.75
20							1.75	1.75	0.25
21							1.25	0.25	1.25
22							1.25	0.75	1.75
23							1.75	0.25	1.75
24							1.75	0.75	1.25
25							0.25	1.25	1.25
26							0.25	1.75	1.75
27							0.75	1.25	1.75
28							0.75	1.75	1.25
29							1.25	1.25	1.25
30							1.25	1.75	1.75
31							1.75	1.25	1.75
32							1.75	1.75	1.25

In an imperfectly ordered CuPt III-V alloy, the average form factor for an α site is $(r_\alpha f_A + w_\alpha f_B)$ and that for a β site is $(r_\beta f_B + w_\beta f_A)$. Ignoring for the moment the thermal factors, the CuPt ordered zinc-blende structure factor is given by

$$F = \sum_{III\alpha} \left(r_\alpha f_A + w_\beta f_B \right) e^{2\pi i \left(hx_{III\alpha} + ky_{III\alpha} + \ell z_{III\alpha} \right)}$$
$$+ \sum_{III\beta} \left(r_\beta f_B + w_\alpha f_A \right) e^{2\pi i \left(hx_{III\beta} + ky_{III\beta} + \ell z_{III\beta} \right)} + \sum_V f_C e^{2\pi i \left(hx_V + ky_V + \ell z_V \right)}$$

where the sums are over the group-III α and β, and group-V atomic sites.

After simplification, and using the relationships between $y_{\alpha(\beta)}$, $r_{\alpha(\beta)}$, $w_{\alpha(\beta)}$, and $x_{A(B)}$, the complex square of the CuPt ordered zinc-blende structure factor is given by

$$|F|^2 = 0, \text{ if h, k, and } \ell \text{ are mixed odd and even integers,}$$
$$= 32^2 [|f_C|^2 + |x_A f_A + x_B f_B|^2], \text{ if h, k, and } \ell \text{ are all odd,}$$
$$= 32^2 [|f_C|^2 + |x_A f_A + x_B f_B|^2 + 2 f_C (x_A f_A + x_B f_B)], \quad (8)$$
$$\text{if h} + \text{k} + \ell = 4n,$$
$$= 32^2 [|f_C|^2 + |x_A f_A + x_B f_B|^2 - 2 f_C (x_A f_A + x_B f_B)],$$
$$\text{if h} + \text{k} + \ell = 4n+2.$$

These solutions of $|F|^2$, for integer values of (h k ℓ), correspond to the fundamental zinc-blende reflections given by Eqn. 6. Although the factors of 16 in Eqns. 6 have been replaced by 32^2, the magnitude of the intensities for these (h k ℓ) reflections will match those for random zinc-blende. This is because there are two times fewer unit cells along x, y, and z in the ordered structure, and thus

$$I \propto N_1^2 N_2^2 N_3^2 |F|^2 \propto \frac{1}{8^2} \times 32^2 = 16.$$

Since CuPt ordering doubles the zinc-blende unit cell, certain half-integer (h k ℓ) superstructure reflections, indexed on the basis of the random unit cell size, are allowed. Since the four different CuPt variants correspond to different atomic coordinates, their superstructure reflections will be allowed at different (h k ℓ) positions. The complex square of the structure factor at a CuPt superstructure reflection is

$$|F|^2 = 16^2 S^2 |f_A - f_B|^2, \tag{9}$$

where S is the order parameter. If the ordered planes are along the [111] direction, this structure factor corresponds to Miller indices such that h, k, and ℓ are all $(4n + 1)/2$ or all $(4n + 3)/2$. If the ordering is along the [$\bar{1}\bar{1}1$] direction, the same $|F|^2$ value results if h and k = $(4n + 1)/2$, while $\ell = (4n + 3)/2$, or if h and k = $(4n + 3)/2$ while $\ell = (4n + 1)/2$. If the ordered planes are in the [$\bar{1}11$] direction, then the above value of $|F|^2$ is obtained if k and $\ell = (4n + 1)/2$ while h = $(4n + 3)/2$, or if k and $\ell = (4n + 3)/2$ while h = $(4n + 1)/2$. If the fourth variant, [$1\bar{1}1$], of CuPt ordering exists, then the above superstructure value of $|F|^2$ results if h and $\ell = (4n + 1)/2$ while k = $(4n + 3)/2$, or if h and $\ell = (4n + 3)/2$ while k = $(4n + 1)/2$. Since each of the four CuPt variants results in unique (h k ℓ) reflections, it is a simple matter to distinguish among them with single crystal diffraction techniques, once the indexing scheme is referred to the substrate lattice. (Cross-sectional transmission electron diffraction (TED) is often useful for this as it provides an overview of reciprocal space and reveals the relevant variants.)

The presence of CuPt ordering is thus characterized by the presence of superstructure peaks at certain half-integer (h k ℓ) positions. The intensity of these peaks depends on the difference between the form factors of the ordering atoms, $|f_A - f_B|$, and on the perfection of the ordering, given by the order parameter S.

Note that in order to measure superstructure reflections at half-integer (h k ℓ) positions, one must ensure that no x-rays with wavelength $\lambda/2$ are present in the main beam. A $\frac{1}{2}$(h k ℓ) peak measured with λ x-rays will have the same Bragg angle as the (h k ℓ) peak measured with $\lambda/2$ x-rays. This can be accomplished either by energy discrimination of the detected beam, or by using a monochromator that does not diffract $\lambda/2$, such as Ge or Si (111). If one is measuring an ordered structure with peaks at other non-integer positions, care must be taken to eliminate the pertinent wavelength contamination from the incident beam.

3.3   Measurement of the Long-Range Order Parameter

The integrated intensity of a reflection is proportional to either $|F_T|^2$ (kinematical diffraction theory) or to $|F_T|$ (dynamical diffraction theory). $|F_T|$ for a superstructure reflection is proportional to the order parameter, S. Therefore, if one takes the ratio of the integrated intensities of a superstructure reflection and a fundamental reflection, one can determine S, since all other terms cancel out or can be determined. If one chooses superstructure and fundamental reflections accessed in similar geometries, additional geometrical corrections also cancel out. For example, for an imperfect $A_xB_{1-x}C$ zinc-blende crystal with CuPt ordering, the ratio of the integrated intensities of the (1/2 1/2 3/2) superstructure and (113) fundamental peaks is, using Eqn. 2,

$$
\frac{E_{\frac{1\,1\,3}{2\,2\,2}}}{E_{113}} = \frac{\left|F_T\right|^2\left(\dfrac{1+\left|\cos 2\theta_m\right|\cos^2 2\theta}{\sin 2\theta}\right)\left(1-e^{-2\mu L/\sin\theta}\right)\Big|_{\frac{1\,1\,3}{2\,2\,2}}}{\left|F_T\right|^2\left(\dfrac{1+\left|\cos 2\theta_m\right|\cos^2 2\theta}{\sin 2\theta}\right)\left(1-e^{-2\mu L/\sin\theta}\right)\Big|_{113}}.
$$

(If the crystal were of very high crystalline quality, one would use Eqn. 3 for the integrated intensity of the strong fundamental peak, though not for the weak superstructure peak). Substituting in for $|F_T|^2$ from Eqns. 5 and 6 and solving for S, we have

$$
S = \left\{\frac{E_{\frac{1\,1\,3}{2\,2\,2}}}{E_{113}}\frac{4\left[\left[\left|f_C\right|^2+\left|xf_A+(1-x)f_B\right|^2\right]\left(\dfrac{1+\left|\cos 2\theta_m\right|\cos^2 2\theta}{\sin 2\theta}\right)\left(1-e^{-2\mu L/\sin\theta}\right)\right]_{113}}{\left[\left|f_A-f_B\right|^2\left(\dfrac{1+\left|\cos 2\theta_m\right|\cos^2 2\theta}{\sin 2\theta}\right)\left(1-e^{-2\mu L/\sin\theta}\right)\right]_{\frac{1\,1\,3}{2\,2\,2}}}\right\}^{1/2}
\tag{10}
$$

The form factors in Eqn. 10 should include the Debye-Waller thermal factors. This results in a straightforward experimental technique for quantitatively measuring the order parameter for each CuPt order variant, since each variant results in a unique set of reflections.

These ideas were used by Forrest et al.[17] to measure the long-range order parameter in CuPt ordered GaInP. They measured the integrated intensities of the {113} fundamental and {½ ½ ³⁄₂} superstructure peaks from six partially ordered GaInP samples, and calculated the order parameters using Eqn. 10. They also measured the band gap reduction (ΔE_g) and valence band splitting (ΔE_{vbs}) caused by the ordering, using excitation photoluminescence. Their results for the dependence of ΔE_{vbs} and ΔE_g on S are plotted in Figure 1.

More recently, Zhong et al.[18] studied long-range quadruple-period ordering in $GaAs_{0.87}Sb_{0.13}$ grown on (001) GaAs, at beamline X-14A at the National Synchrotron Light Source (NSLS), Brookhaven National Laboratory. The basic ordering in this epitaxial film is of the CuAu type, i.e., alternating (001) planes of V_A and V_B species, on which a four-fold [110]

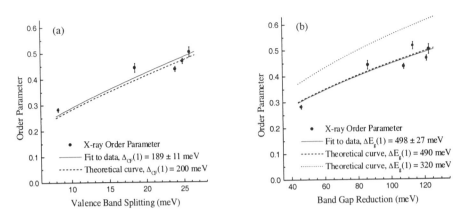

Figure 1. Long-range order parameter versus a) $\Delta E_{VBS}(S)$ and b) $\Delta E_g(S)$ for CuPt-ordered GaInP. Forrest et al.[17]

periodicity is imposed due to an $\alpha(2 \times 4)$ surface reconstruction during growth. In this non-stoichiometric case the nominally Sb sites are occupied by 74% As and 26% Sb. Figure 2 shows their measurement of the superstructure peaks between the (004) and (224) fundamental Bragg peaks; strong superstructure peaks are visible at (h $\pm$ 0.75, k $\pm$ 0.75, ℓ) and weak peaks at (h $\pm$ 0.25, k $\pm$ 0.25, ℓ). Zhong et al. modeled the resulting superstructure peaks using kinematical diffraction theory and Eqn. 2. The intensity model was fair using the basic structure, but was rendered excellent through the inclusion of atomic displacements calculated by minimizing a Keating potential in which a valence-force-field model was employed. In this calculation the Ga atoms suffer the largest displacements. The displacements, when introduced into the structure factor, produced the fit shown in Figure 2. The inset shows the fit without displacements.

4.   RECIPROCAL SPACE MAPPING OF ORDERED DOMAINS

In typical x-ray diffraction experiments, a one-dimensional line scan of a peak is performed, normally a $\theta/2\theta$ (radial) or ω (rocking) scan. The shape of the peak in these scans provides information about the average shape of the contributing crystal domains in real-space, since the diffracted intensity distribution is essentially a Fourier transform of the real-space crystal structure. A one-dimensional scan provides one-dimensional information

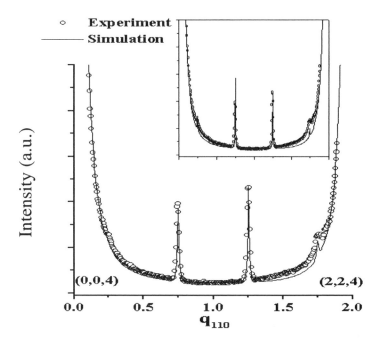

Figure 2. A [110] scan using synchrotron radiation from a $GaAs_{0.87}Sb_{0.13}$ film in which quadruple-period [110] LRO is observed. The simulation includes displacements based on a VFF calculation (see text) while the inset shows the fit without displacements. Zhong et al.[18]

about the crystal. A two or three-dimensional map of the diffracted intensity would provide two or three-dimensional information about the crystal. Such a map is referred to as a reciprocal space map (RSM). A two-dimensional RSM is made by combining a series of line scans within a small region of reciprocal space. This technique can be extended to three dimensions by mapping orthogonal planes in reciprocal space, yielding a three-dimensional representation of the intensity. For epitaxial layers, RSM's can determine the relative orientation of the layers, as well as the shape and orientation of the crystalline domains within the layers. When applied to ordered epitaxial alloys, the characteristics of the ordered domains are revealed.

The shape of a crystalline domain directly determines the shape of its diffraction peaks in reciprocal space. The average correlation length (L) of the crystalline domains along an arbitrary direction can be estimated from the width (Δq) of a resulting peak by $L = 2\pi/\Delta q$, where q is the scattering vector $q \equiv (2\pi/\lambda)(S-S_o)$, and Δq is measured in the direction of interest. The correlation length perpendicular to the diffracting planes can be estimated from a θ/2θ scan using this equation, which in diffractometer (angular) units is called the Scherrer equation[1]

$$L \cong \lambda \big/ (2\theta \; fwhm) \cos\theta \; .$$

Using this relationship it is possible to measure the average size and orientation of the domains in a sample from an RSM.

X-ray RSM's can provide different and complementary information compared to transmission electron microscopy and diffraction (TEM and TED). First of all, TEM images focus on small regions of a sample, while x-ray measurements average over the sampled volume, with a typical beam cross-section of 1 mm x 5 mm. Secondly, x-ray RSM's have higher resolution, and are more easily quantified than TED patterns, mainly because TED tends to symmetrize patterns through multiple scattering. On the other hand, TED can more easily detect weak features and measure large areas of reciprocal space. Both techniques have advantages and provide important details about crystal structure. If x-ray RSM's and TEM measurements of a sample can be combined, a thorough picture of the structure can be obtained.

Reciprocal space mapping has been applied to ordered III-V alloys.[19-22] Hess et al. used RSM's to determine the three-dimensional shape of ordered domains in CuPt-ordered $Ga_{1-x}In_xP$ grown by metal organic vapor phase epitaxy (MOVPE) on GaAs. Their comparison of a TED pattern and x-ray RSM of the same sample, shown in Figure 3, nicely illustrates their complementary nature.

5.   SHORT-RANGE ORDER

5.1   Short-Range Order Diffuse Intensity and Order Parameters

Short-range order (SRO) refers to a non-random distribution of the first several atomic neighbors of an atom in a crystal, rather than over a large number of unit cells as in LRO. Most alloys, even those not in a state of long-range order, tend to show a preference for either like or unlike nearest neighbors. Because of the short-range nature of this phenomenon, it results in diffuse diffracted intensity: intensity distributed in broad, weak features. A great deal of theoretical and experimental attention has been focused on SRO because pair interaction energies (interatomic potentials) can be derived from the short-range order parameters. These interaction potentials

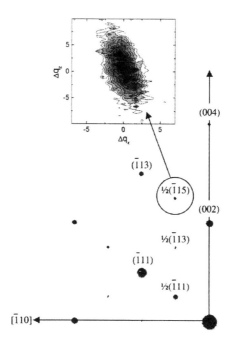

Figure 3. TED pattern and x-ray RSM (inset) of double-variant CuPt-ordered GaInP grown on GaAs miscut 2° toward [100] (Hess et al.)[19]

can then be used to calculate the energy of the structure using an Ising Hamiltonian, which, together with techniques such as Monte Carlo methods[3], permits a determination of the structural phase diagram. Many good references exist on the theory and experimental details of SRO, some of which are listed in the references.[1-5,23-30] We will describe the main concepts of the theory of SRO and discuss its application to epitaxial III-V alloys.

Eqn. 1 in section 2.1 gives the total coherent scattered intensity from a sample. For simplicity, we will express the intensity in atomic units as $I_{eu} = I/I_e$. Consider a binary A-B alloy with fractional compositions x_A and x_B. (We can apply this approach to III_A-III_B-V or III-V_A-V_B alloys since the interactions of interest are between the two similar atoms on their shared sub-lattice, analogous to a binary alloy.) We assume for now that there is no long-range order present, so that the occupancy of two sites m and n which are not near neighbors is uncorrelated.

We are interested in obtaining the diffuse intensity apart from the intensity due to the sharp fundamental Bragg peaks, $I_{Diff} = I_{eu} - I_{Fund}$. Two slightly different derivations of I_{Diff} are given by Warren[1] and Sparks and Borie[25], and Moss[23] gives a review of the formalism due to Krivoglaz[5]. The basic idea is that the fundamental intensity is given by the average occupation $(x_A f_A + x_B f_B)^2$, while the diffuse intensity takes into account correlations between near neighbor sites m and n. The result for the diffuse intensity is given by a summation over all m and n sites[1,25]

$$I_{Diff} = \sum_{m,n} x_A x_B \left(f_A - f_B\right)^2 \alpha_{mn} e^{iq\bullet(R_m - R_n)}, \tag{8}$$

where α_{mn} are the Warren-Cowley short-range order parameters

$$\alpha_{mn} = 1 - \frac{p_{mn}^{BA}}{x_B}, \tag{9}$$

and p_{mn}^{BA} is the probability of a B atom at site m, given an A atom at site n. For convenience, we have defined the scattering vector $q \equiv (2\pi/\lambda)(S-S_o)$. If there is no correlation between sites m and n, $p_{mn}^{BA} = x_B$, and $\alpha_{mn} = 0$. Values of $\alpha_{mn} > 0$ correspond to a preference for like neighbors, called clustering, and $\alpha_{mn} < 0$ correspond to a preference for unlike neighbors, referred to as short-range order. If the crystal contains no LRO, α_{mn} tends to zero for distant neighbors, i.e. (m − n) large. The case in which LRO and SRO coexist will be discussed in section 5.2.

We now include the possibility of atomic displacements from the ideal lattice sites by both thermal (dynamic) and static displacements. This is done by replacing the vector R_m with a vector to the average atomic position, r_m, plus a displacement vector, u_m. Thus $(R_m - R_n)$ becomes $(r_m - r_n) + (u_m - u_n)$, where $u_m = \delta_m + s_m$, a sum of a static and dynamic term, respectively. With this, the exponential in Eqn. 8 becomes

$$e^{iq\bullet(R_m - R_n)} = e^{iq\bullet(r_m - r_n)} e^{iq\bullet(\delta_m - \delta_n)} e^{iq\bullet(s_m - s_n)}.$$

We make the assumption that both types of displacements are small and that s_m is independent of occupation, and take the thermal average. We assume that the static displacements are small enough that they can be expanded to second order.

$$e^{iq\bullet(\delta_m - \delta_n)} \cong 1 + \left[iq \bullet (\delta_m - \delta_n)\right] + \tfrac{1}{2}\left[iq \bullet (\delta_m - \delta_n)\right]^2. \tag{10}$$

The average of the dynamic term is approximated by

$$\left\langle e^{iq\bullet(s_m - s_n)} \right\rangle \cong e^{-\tfrac{1}{2}\left\langle [iq\bullet(s_m - s_n)]^2 \right\rangle} \cong e^{-2M\phi(mn)}, \tag{11}$$

where e^{-2M} is the standard Debye-Waller thermal factor and $\phi(mn)$ is a coupling-factor dependent on the neighbor distance. $\phi(mn)$ accounts for the fact that near neighbors have smaller *relative* displacements than do distant neighbors (see Warren,[1] p.238). Inserting Eqns. 10 and 11 into Eqn. 8 we obtain for the diffuse intensity (following Reinhard et al.[26])

$$I_{Diff} = I_{SRO} + I_{SE} + I_2 + I_{TDS}. \tag{12}$$

In Eqn. 9, I_{SRO} is the short-range order diffuse intensity we have been seeking,

$$I_{SRO} = \sum_{m,n} x_A x_B \left(f_A - f_B \right)^2 \alpha_{mn} e^{-2M\phi(mn)} e^{iq\bullet(r_m - r_n)}. \tag{13}$$

I_{SRO} describes the diffuse intensity due to near neighbor non-random chemical occupation. For a totally random crystal, all of the SRO parameters α_{mn} are zero, except for the self-term α_{00} which is 1. In this case, $I_{SRO} = N x_A x_B (f_A - f_B)^2$, which is the Laue monotonic diffuse scattering (N = number of sites). This term decreases monotonically with increasing scattering angle. It has been found, however, that most alloys have non-zero α_{mn}'s. In this case, Eqn. 13 modulates the Laue diffuse scattering. If clustering is present, the diffuse intensity will be greater near the Bragg peaks. If short-range ordering is present, the diffuse intensity will peak in the vicinity of the superstructure positions.

The term I_{SE} in Eqn. 12 is the size-effect term, or local atomic displacement term, which represents the cross term of SRO and local displacements,

$$I_{SE} = \sum_{m,n} x_A x_B \left(f_A - f_B \right)^2 e^{-2M\phi(mn)}$$

$$\times \left[\left(\frac{x_A}{x_B} + \alpha_{mn} \right) f_A iq \bullet \delta_{mn}^{AA} - \left(\frac{x_B}{x_A} + \alpha_{mn} \right) f_B iq \bullet \delta_{mn}^{BB} \right] \frac{e^{iq\bullet(r_m - r_n)}}{\left(f_A - f_B \right)}.$$

I_{SE} assumes no restrictions on how the atoms are displaced from their average positions, accept that they be small. It was derived to account for the fact that because different atoms in an alloy have different sizes, all atoms will be displaced from the average lattice sites. The effect is quite pronounced, even for very small size differences, as in Fe-Cr.[26] Because I_{SE} is linear in q, it increases at increasing scattering angles, contrary to I_{Fund} and I_{SRO}. I_{SE} is typically straightforward to measure, since it is an asymmetric modulation of the diffuse scattering whereas I_{SRO} is symmetric about its maxima, and it typically displaces the diffuse maxima from their symmetrical positions.

The term I_2 in Eqn. 12 is commonly called Huang diffuse scattering and it is largest in the vicinity of the fundamental Bragg peaks. It reflects the effects of the asymptotic static displacements at large interatomic separations that are correlated with occupational order, and is obtained by keeping second order terms in the expansion of the displacement exponential. I_2 is quadratic in the static displacements, δ_m, and the scattering vector, $\mathbf{q}$, and therefore increases strongly with increasing scattering angle. For further discussion see Borie and Sparks[24], Reinhard et al.[26], and Moss[23].

I_{TDS} is the thermal diffuse scattering term. It does not depend on the SRO, and can therefore be subtracted from the SRO diffuse intensity. For more details, see the work by Schweika,[4] Reinhard et al.,[26] and Borie and Sparks.[24] (It should also be noted that I_{TDS} can be removed using *elastic neutron scattering.*[4,23])

As illustrated by Eqn. 12, a sample's diffuse diffracted intensity holds a great deal of information about the interactions of the atoms within the crystal. In particular, the equilibrium SRO parameters, obtained from I_{SRO}, can be used to derive the atomic interaction energies for the alloy. The pair interaction energies, V_{mn}, are typically expressed using an Ising Hamiltonian. The SRO parameters can be used to derive the interaction energies using the Krivoglaz–Clapp–Moss (KCM) approximation,[5,26,27,30] which is a mean-field expression for the conditions above the ordering critical temperature,

$$\alpha(\mathbf{q}) = \frac{D}{1 + \dfrac{2x_A x_B}{k_B T} V(\mathbf{q})},$$

where $\alpha(\mathbf{q})$ and $V(\mathbf{q})$ are the Fourier transforms of the SRO-parameters and effective pair interactions, V_{mn}, respectively, k_B is Boltzman's constant, and D is usually set to one.[30] The interaction energies are used for theoretical calculations of the phase diagrams of the structure and predictions of the stability of ordered configurations.[3,31]

Measurements of SRO in III-V alloys are clearly very interesting, as are the V_{mn} values that can be extracted if the short-range correlations are in thermal equilibrium. Long-range ordering in these materials, however, is predicted to be energetically preferred due to the surface reconstruction present during growth[32-34]. In a real sample the realization of the favored configuration can be limited by growth kinetics. Thus the preferred LRO may not be realized; it could be retarded or in fact eliminated. Growth kinetics should not, however, have as strong an effect on SRO. Therefore,

I_{SRO} should be a good indicator of the interaction energies in epitaxial III-V alloys, even in the case where LRO exists.

5.2   Coexistence of Long- and Short-Range Order

LRO and SRO can coexist in a crystal provided the LRO is imperfect, i.e., $0 < S < 1$. In epitaxial III-V alloys, this could occur if the LRO were limited by kinetics, or if the sample were non-stoichiometric ($x_A \neq y_A$). In such samples the long and short-range order parameters are related.

For a sample with perfect LRO, all of the intensity due to the ordering is concentrated into the superstructure peaks. In an imperfectly ordered sample, some of that intensity is distributed elsewhere, and is associated with the "wrongly" occupied sites. In a crystal with no LRO, α_{mn} tends to zero for large (m − n) and is determined by the SRO for small (m − n). In a sample with imperfect LRO, all values of α_{mn} will depend on whether the "wrongly" occupied sites are distributed randomly or not. In particular, in the case of imperfect LRO, the limiting value of α_{mn} for large (m − n) can be expressed in terms of the LRO parameter S. Following Schwartz and Cowley[28,35], if the "wrongly" occupied sites are randomly distributed, the limit of the Warren-Cowley SRO parameters for distant neighbors is

$$\alpha_{\ell mn} = \pm S^2/4x_A x_B$$

where + and − correspond to (ℓ, m, n) even and odd, respectively. (The neighbor indices (m, n) have been replaced by the actual atomic coordinates (ℓ, m, n) in units of the real space lattice parameters normalized so that ℓ, m, and n are integers. For example, on the group-III fcc sublattice of zinc-blende one of the twelve first nearest neighbors of the (0, 0, 0) atom is located at the fractional unit cell position (0.5, 0.5, 0), for which (ℓ, m, n,) = (1, 1, 0).[1,28]) If the "wrongly" occupied sites are not randomly distributed, the ordering intensity can be represented by two series, one with coefficients $\pm S^2/4x_A x_B$ associated with the sharp superlattice reflections, and one with coefficients $\psi_{\ell mn} = (\alpha_{\ell mn} \pm S^2/4x_A x_B)$ associated with broad diffuse scattering surrounding the superlattice reflections.

5.3   Measurement of SRO in Epitaxial III-V Alloys

Diffuse x-ray scattering measurements are well suited to a quantitative investigation of local order. One is typically is interested in the Warren-Cowley SRO parameters, α_{mn}, and thereby the atomic interaction energies. In order to measure the α_{mn} values, one must measure I_{SRO} throughout an entire repeating volume in reciprocal space and calculate the α_{mn} values as

Fourier coefficients of the I_{SRO} Fourier series. This measurement is usually reduced, by the high symmetry of most alloys, to a small fraction of the entire Brillouin zone. Since the diffuse intensity is by nature weak, and I_{SRO} must be separated from the other contributions to I_{Diff}, such measurements are not simple. However, many "tricks of the trade" have been published, and good measurements have been obtained.[1-5,25,26]

Yasuami et al. measured the diffuse intensity and calculated the Warren–Cowley SRO parameters for two MOVPE grown III-V samples with imperfect LRO; an InGaP sample with [1 $\bar{1}$ 1] and [$\bar{1}$ 1 1] CuPt ordering, and an InAlP sample with [1 $\bar{1}$ 0] CuAu ordering.[36] Their α_{mn} values show a tendency towards SRO in both samples, with larger $|\alpha_{mn}|$ values for the CuPt-ordered sample due to its larger degree of LRO. They interpreted their data as indicating the mechanism leading to ordering in the early stages of growth. They state that both ordered structures begin with chains of like-atoms along the [110] direction, followed by a stage of strong like-atom chains in the [001] direction, leading to [1 $\bar{1}$ 0] ordering. Yasuami et al. propose a possible third stage in which the [001] like-atom correlation is replaced by a preference for unlike atoms, resulting in [1 $\bar{1}$ 1] ordering.[36]

Forrest et al.[20] and Kulik et al.[21] measured diffuse intensity due to both [1 1 1] and [$\bar{1}$ $\bar{1}$ 1] CuPt-A SRO as well as [1 1 1] and [$\bar{1}$ $\bar{1}$ 1] triple-period-A SRO in AlInAs grown by molecular beam epitaxy (MBE) on [001] InP. The work was originally done on a laboratory x-ray diffractometer with modest resolution[20], and was repeated at the NSLS, beamline X-14A.[21] Figure 4 shows their reciprocal space map of the diffuse intensity due to SRO in the vicinity of the (111), (222), and (331) fundamental peaks. (The strong ridge of intensity running from (331) to (222) is associated with finely-spaced twin lamellae in the sample and obscures the SRO diffuse intensity there.) They also observed size-effect modulations of the diffuse intensity. Figure 5 shows their line scan between the (111) and (333) fundamental peaks. Diffuse intensity maxima due to CuPt-A SRO are located at 3/2 and 5/2 x (111), diffuse intensity maxima due to triple-period-A SRO are located at 4/3, 5/3, 7/3, and 8/3 x (111). The observed *increase* in diffuse intensity with scattering angle is typical of atomic displacement effects.

Much more work needs to be done in this area. It would be interesting to compare the SRO parameters for samples with single versus double variant ordering, samples grown under differing thermodynamic (surface reconstruction) conditions, and samples composed of different III-V species. Diffuse scattering in samples with lateral compositional modulation, discussed below, would also be interesting to pursue with the high intensity synchrotron radiation sources, and such work is in progress.[37] Such experimental measurements of SRO diffuse scattering could be used to assess and refine the theoretical predictions of ordering in epitaxial III-V alloys.

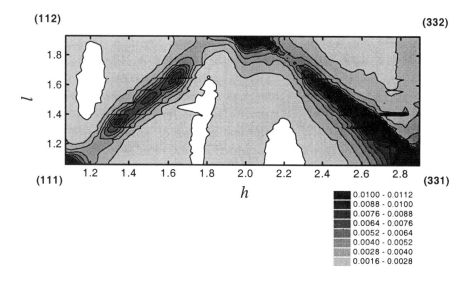

Figure 4. Reciprocal space map of the intensity from an AlInAs sample grown by MBE on [001] InP, in the rectangle defined by (111), (331), (112), and (332). The (111), (222), and (331) fundamental peaks as well as SRO diffuse intensity are visible (Kulik et al.).[21]

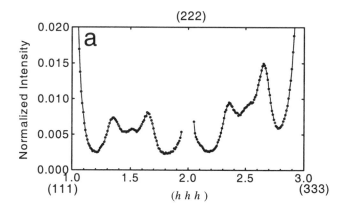

Figure 5. X-ray scan of AlInAs grown by MBE on [001] InP. The scan is along [1 1 1], between the (1 1 1) and (3 3 3) fundamental peaks. Diffuse intensity due to triple-period-A and CuPt-A SRO are both present, modulated by the size-effect (Kulik et al.).[21]

6.   LATERAL COMPOSITIONAL MODULATION

Lateral compositional modulation (LCM) refers to a regular variation in the composition of an $A_xB_{1-x}C$ III-V alloy in a direction parallel to the substrate. This has been observed to occur spontaneously in GaInP, GaInAs, and AlInAs. Highly regular modulation has been seen in samples grown as nominally latticed-matched short-period superlattices. In addition to the intended (man-made) vertical superlattice, a spontaneous lateral composition modulation (lateral superlattice) occurs. Samples exhibiting LCM can have composition variations of more than $\pm$ 0.1 about the average value, with lateral wavelengths of 10 – 40 nm.[22] If the lateral modulation is sufficiently periodic, then the structure is essentially a lateral superlattice. Lateral compositional modulation is included in this chapter because, like ordering, it is a spontaneous occurrence in epitaxial III-V alloys, most probably induced by strain effects, consisting of a preferential positioning of the atoms.[38] One could describe LCM as very long-range one-dimensional ordering, whose diffraction spots are called superlattice satellite peaks as opposed to superstructure peaks.

When sufficiently periodic, LCM results in lateral superlattice peaks in a diffraction pattern, distributed about the fundamental peak in the direction of the modulation. These satellite peaks are completely analogous to superlattice peaks in vertical superlattice structures. In typical vertical superlattices, x-ray diffraction is used to determine the wavelength of the structure, the thickness of each layer in the repeating unit, the compositions of each layer, the dispersion in the period, and the character of the interface between layers (i.e., roughness and grading). Some of these characteristics can only be determined for highly regular superlattices, resulting in many observable superlattice peaks. The presence of only one superlattice peak, however, is nonetheless sufficient to identify a modulation and its period. The wavelength of the superlattice, Λ, is given by the sum of the thicknesses of the constituent layers (D_i), $\Lambda = D_1 + D_2$. The average composition (x_{ave}) of the superlattice is given in terms of the compositions of the constituent layers (x_i) as $x_{ave} = (D_1x_1 + D_2x_2) / \Lambda$. The superlattice peaks are distributed about the point in reciprocal space corresponding to the average fundamental peak, and they are oriented along the direction of the modulation. Λ can be determined from the spacing of two superlattice peaks, of orders i and j, by $\Lambda = 2\pi(i - j)/(|\mathbf{q}_i| - |\mathbf{q}_j|)$, where $\mathbf{q}$ is the scattering vector. For vertical superlattices this can be expressed in angular terms as $\Lambda = (i - j)\lambda / 2(\sin \theta_i - \sin \theta_j)$, where λ is the x-ray wavelength, and θ_i is the angle of the superlattice peak of order i. (For a detailed discussion of diffraction by superlattices see Bowen and Tanner[13] and Holý et al.[7], and the references therein.)

X-ray diffraction has been used to characterize LCM in III-V alloys. Lee et al. used reciprocal space mapping to characterize LCM in AlAs-InAs superlattices, and give a good description of the experimental issues involved.[22,38,39]   Figure 6 shows their RSM of the (224) peak of an AlAs/InAs superlattice grown on an InP [001] substrate (lattice constant = 5.8687 Å) and an $Al_{0.48}In_{0.52}As$ lattice-matched buffer.   The superlattice consists of 100 periods of 1.95 monolayers AlAs and 1.98 monolayers InAs, with a resulting average InAs composition of 0.505 (lattice constant = 5.8614 Å).   The resulting LCM peaks, shown in Figure 6, are centered about the superlattice peak, slightly above the InP substrate peak, and indicate a [110] LCM wavelength of 20 nm.   (The periodic peaks in Figure 6 lying along [001], with a spacing of approximately 0.02 Å^{-1}, were identified as interference effects between the finite-thickness buffer and superlattice epilayers. [40])

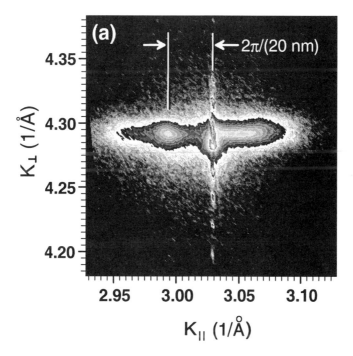

Figure 6. Reciprocal space map of AlAs-InAs superlattice (224) peak showing satellite peaks due to lateral composition modulation. $K_\perp$ and $K_\parallel$ lie along [001] and [110], respectively (Lee et al.).[22]

7.   SUMMARY

Many aspects of III-V ordering remain to be investigated by x-ray diffraction. Refinement of both atomic and electronic models remains for those materials and ordered structures that have been studied so far. Some ordered III-V alloys and structures have yet to be investigated by x-ray diffraction. Both LRO and SRO in these materials are of interest, and therefore work exists for both laboratory and high intensity synchrotron sources.

This chapter was intended to be an introduction to the x-ray diffraction analysis of ordering in ternary epitaxial III-V alloys. The x-ray experimentalist who seeks further detail is urged to read the many references that are given herein. This is an active field of research and our review cannot be completely inclusive, especially as new research will undoubtedly be published subsequent to this writing. Readers are therefore also referred to the body of published articles on the subject.

ACKNOWLEDGEMENTS

I would like to thank my former teacher, Professor Simon C. Moss, for his thoughtful reading of this review. Research performed at the University of Houston was supported by the National Science Foundation on grants DMR 92-08450 and DMR 97-29297, and by the R. A. Welch Foundation of Texas. Research performed at the University of California, Los Angeles was supported by the National Science Foundation on grant DMR 95-02117 and by the California Energy Institute.

REFERENCES

1   B. E. Warren, X-ray Diffraction (Dover Publications, Inc., New York, 1990).
2   A. Guinier, X-ray Diffraction In Crystals, Imperfect Crystals, and Amorphous Bodies (Dover Publications, Inc., New York, 1994).
3   W. Schweika, Disordered Alloys: Diffuse Scattering and Monte Carlo Simulation, Vol. 147 (Springer Verlag, New York, 1998).
4   W. Schweika, in Statics and Dynamics of Alloy Phase Transformations; Vol. 319, edited by A. G. P. Turchi (Plenum Press, New York, 1994), p. 103-126.
5   M. A. Krivoglaz, Diffuse Scattering of X-rays and Neutrons by Fluctuations (Springer-Verlag, Berlin, 1996).
6   H.-N. Yang, G.-C. Wang, and T.-M. Lu, Diffraction from Rough Surfaces and Dynamic Growth Fronts (World Scientific Pub., River Edge, New Jersey, 1993).
7   V. Holý, U. Pietsch, and T. Baumbach, High Resolution X-ray Scattering from Thin Films and Multilayers, Vol. 149 (Springer, New York, 1999).

8   S. Takagi, J. Phys. Soc. Japan 26, 1239 - 1253 (1969).
9   D. Taupin, Bull. Soc. Fr. Mineral. Cristallogr. 87, 469 (1964).
10   R. Zaus, J. Appl. Cryst. 26, 801 - 811 (1993).
11   W. Zachariasen, Acta Cryst. 23, 558 - 564 (1967).
12   W. J. Bartels, J. Hornstra, and D. J. W. Lobeek, Acta Cryst. A42, 539 - 545 (1986).
13   D. K. Bowen and B. K. Tanner, High Resolution X-ray Diffraction and Topography
     (Taylor and Francis Inc., Bristol, PA, 1998).
14   A. W. Stevenson, Acta Cryst. A49, 174 - 183 (1993).
15   L. Tapfer and K. Ploog, Phys. Rev. B 40, 9802 - 9810 (1989).
16   International Tables for X-ray Crystallography; Vol. III, edited by C. H. MacGillavry, G.
     D. Rieck, and K. Lonsdale (Kynoch Press, Birmingham, England, 1962).
17   R. L. Forrest, T. D. Golding, S. C. Moss, Z. Zhang, J. F. Geisz, J. M. Olson, A.
     Mascarenhas, P. Ernst, and C. Geng, Phys. Rev. B 58, 15355 - 15358 (1998).
18   Z. Zhong, J. H. Li, J. Kulik, P. C. Chow, A. G. Norman, A. Mascarenhas, J. Bai, T. D.
     Golding, and S. C. Moss, Phys. Rev. B, 63, 033314 (2001).
19   R. R. Hess, C. D. Moore, R. T. Nielsen, and M. S. Goorsky, in 9th Biennial Workshop on
     Organometallic Vapor Phase Epitaxy, 1999.
20   R. L. Forrest, J. Kulik, T. D. Golding, and S. C. Moss, J. Mat. Res. 15, 45 - 55 (2000).
21   J. Kulik, R. Forrest, J. Li, T. Golding, S. C. Moss, and J. Bai, Mat. Res. Soc. Symp. Proc.
     (1999).
22   S. R. Lee, J. M. Millunchick, R. D. Twesten, D. M. Follstaedt, J. L. Reno, S. P. Ahrenkiel,
     and A. G. Norman, J. Mat. Sci.: Materials in Electronics 10, 191 - 197 (1999) Kluwer
     Academic Publishers.
23   S. C. Moss, Mat. Res. Soc. Symp. Proc. Vol. 376, 675 - 687 (1995).
24   B. Borie and C. J. Sparks, Acta Cryst. A 27, 198 - 201 (1971).
25   C. J. Sparks and B. Borie, in Local Atomic Arrangements Studied by X-ray Diffraction;
     Vol. 36, edited by J. B. Cohen and J. E. Hilliard (Gordon and Breach Science Publishers,
     New York, 1965), p. 5 - 50.
26   L. Reinhard, J. L. Robertson, S. C. Moss, G. E. Ice, P. Zschack, and C. J. Sparks, Phys.
     Rev. B 45, 2662 - 2676 (1992).
27   P. C. Clapp and S. C. Moss, Phys. Rev. 142, 418 - 427 (1966).
28   J. M. Cowley, Phys. Rev. 77, 669 - 675 (1949).
29   J. M. Cowley, J. Appl. Phys. 21, 24 - 30 (1949).
30   I. V. Masanskii, V. I. Tokar, and T. A. Grishchenko, Phys. Rev. B 44, 4647 - 4649 (1991).
31   F. Ducastelle, Order and Phase Stability in Alloys, Vol. III (North-Holland, New York,
     1991).
32   B. A. Philips, A. G. Norman, T. Y. Seong, S. Mahajan, G. R. Booker, M. Skowronski, J.
     P. Harbison, and V. G. Keramidas, J. Cryst. Growth 140, 249 (1994).
33   S. B. Zhang, S. Froyen, and A. Zunger, in Theory of Surface Dimerization-Induced
     Ordering in GaInP Alloys, Boston, MA, 1995 (Materials Research Society), p. 43 - 48.
34   A. Zunger and S. Mahajan, in Handbook on Semiconductors; Vol. 3, edited by T. S. Moss
     and S. Mahajan (Elsevier Science, Amsterdam, 1994), p. 1399 - 1514.
35   L. H. Schwartz, in Local Atomic Arrangements Studied by X-ray Diffraction; Vol. 36,
     edited by J. B. Cohen and J. E. Hilliard (Gordon and Breach Science Publishers, New
     York, 1965), p. 123 - 158
36. S. Yasuami, K. Koga, K. Oshima, S. Sasaki, and M. Ando, "Diffuse X-ray Scattering
     Study of Sublattice Ordering Among Group III Atoms in $In_{0.5}Ga_{0.5}P$ and $In_{0.5}Al_{0.5}P$," J.
     Appl. Cryst. 25, 514-518 (1992).
37   S. C. Moss, (private communication, 2000).

38 R. D. Twesten, D. M. Follstaedt, S. R. Lee, E. D. Jones, J. L. Reno, J. M. Millunchick, A. G. Norman, S. P. Ahrenkiel, and A. Mascarenhas, Phys. Rev. B 60, 13619 - 13635 (1999).
39 D. M. Follstaedt, R. D. Twesten, J. M. Millunchick, S. R. Lee, E. D. Jones, S. P. Ahrenkiel, Y. Zhang, and A. Mascarenhas, Physica E 2, 325 - 329 (1998).
40 S. R. Lee, (private communication, 2000).

Chapter 5

Surface Morphology and Formation of Antiphase Boundaries in Ordered (GaIn)P — A TEM Study

Torsten Sass* and Ines Pietzonka
Dept. of Solid State Physics, Lund University, Box 118, S-221 00 Lund, Sweden

Key words: surface morphology; superlattices; metal-organic vapor-phase epitaxy
              (MOVPE); transmission electron microscopy (TEM)

Abstract: The surface of metal-organic vapor-phase epitaxy (MOVPE) grown $CuPt_B$ -
             type ordered (GaIn)P layers and their interface to the substrate have been
             investigated using high-resolution transmission electron microscopy (TEM).
             In accordance with previous atomic force microscopy investigations, we find
             the (GaIn)P surface to consist of bunched supersteps, (001) terraces, and
             vicinal regions. The dependence of the height and the density of supersteps,
             the portion of monolayer steps contained in supersteps as well as the density
             of antiphase boundaries on the growth temperature and the substrate
             misorientation is investigated. The density of the supersteps on the surface of
             (GaIn)P and at its interface to GaAs has been compared to the density of the
             antiphase boundaries. We find that the supersteps at the interface are the
             preferential sites for the nucleation of antiphase boundaries in (GaIn)P.
             However, the existence of vicinal regions and the effect of the width of
             supersteps and (001) terraces disturb this correlation. The results are utilized
             to reduce the number of antiphase boundaries. This has been achieved by the
             growth of ordered (GaIn)P on the tailor-made surface of an AlAs/GaAs
             superlattice containing less supersteps than a simple GaAs buffer layer.

1.   INTRODUCTION

The spontaneous formation of $CuPt_B$ - type ordering is well known to occur in $Ga_xIn_{1-x}P$ [(GaIn)P hereafter] grown on (001) GaAs by metal-organic vapor-phase epitaxy (MOVPE) or molecular beam epitaxy (MBE).[1] This ordered structure consists of Ga- and In-rich lattice planes alternating in the $[1\bar{1}1]$- or $[\bar{1}11]$-direction [so-called (111)B-planes], which doubles the unit cell in this direction. This causes extra spots in electron diffraction patterns at $\frac{1}{2}\overline{\frac{1}{2}}\frac{1}{2}$ and $\overline{\frac{1}{2}}\frac{1}{2}\frac{1}{2}$ that are often used to detect the superstructure. The initially cubic symmetry of the random alloy (space group $F\overline{4}3m$) is reduced to the trigonal space group $R3m$.[2] The formation of the CuPt$_B$ structure can be attributed to the existence of P-dimer rows lying parallel to [110] as the consequence of a (2×4)-surface reconstruction.[3] On the other hand, surface steps play an important role, too. The ordering is supported by monoatomic steps running parallel to [110], whereas it is hindered by [1$\overline{1}$0]-oriented steps.[4] Recently, it was suggested that a change in the step structure causes a change of the degree of order.[5]

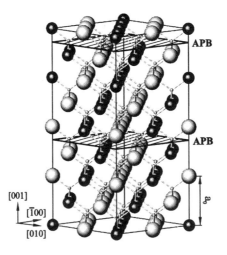

Figure 1. Structure model of CuPt$_B$ - type ordered (GaIn)P. The ordering direction is [1$\overline{1}$1]. Antiphase boundaries (APB) lying parallel to the [001] direction subdivide the structure into domains. In order of decreasing size: In, Ga, P.

Epitaxial layers of ordered (GaIn)P always contain a two-dimensional defect, so-called antiphase boundaries (APBs),[6] which subdivide the material into domains (see *Figure 1*). The distances between APBs, being equivalent to the widths of the domains, are strongly dependent on growth temperature,[7] substrate misorientation, and growth rate.[8] A diminution of domain size to about 1.5 nm further reduces the symmetry of ordered (GaIn)P to the orthorhombic space group $Pmn2_1$, which again alters the electronic band structure.[9] Moreover, APBs deteriorate the electronic properties, e.g., they cause a reduction of the carrier mobilities.[10] Thus, a low density of APBs is desirable for device applications. According to the model of Philips *et al.*,[3] an antiphase boundary is assumed to form at a monoatomic surface step with odd terrace width lying among steps with even terrace width. The term odd or even terrace width means that the terrace contains an odd or even number of incorporation sites for adatoms. Steps with odd terrace width shift the sequence of the ordered {111} planes from Ga/In to In/Ga. Such a shift represents an antiphase boundary (see *Figure 1*). Therefore, the distribution of such steps should be the determinant for the domain width.

The process of step bunching is known to influence the distribution of monolayer (ML) steps on the surface, as observed in several III/V alloys, e.g., GaAs,[11] (GaIn)P,[12] and (AlGaIn)P.[13] The monolayer steps, caused by the substrate misorientation, bunch into supersteps of multiatomic height, which are separated by large (001) terraces. Due to this concentration of monoatomic steps in supersteps the latter might provide preferential sites for the formation of APBs. Based on atomic force microscopy (AFM) measurements, Su and Stringfellow[14] found a 1:1 correlation between the APBs and bunched supersteps on the sample surface. The present work will investigate the correlation between the accumulation of monolayer steps and the width of the ordered domains by comparing the densities of supersteps and APBs, determined by high-resolution transmission electron microscopy (HRTEM) and dark-field (DF) TEM, respectively.

2. TRANSMISSION ELECTRON MICROSCOPY

For TEM the specimen is irradiated with electrons of energies usually between 100 and 1000 keV. A part of these electrons is diffracted within the crystal. The transmitted electron wave is then transformed into the diffraction pattern in the back focal plane of the objective lens, which mathematically corresponds to a Fourier analysis of the crystal potential. Afterwards, the diffraction pattern is transformed into the image of the sample (Fourier synthesis). Due to the small wavelength and small diffraction angles there are always several diffracted beams visible in electron diffraction patterns. Depending on the number of beams that are used for the imaging, there are different imaging modes. For example, a DF image is made with only one diffracted beam, whereas HRTEM imaging requires the interference of several beams with the direct beam. This phase-contrast image can be thought of as the superposition of the lattice fringes corresponding to the used diffracted beams. One should note that such a "lattice image" represents only in the most straightforward case a two-dimensional projection of the structure, depending on the resolution of the microscope (at present 1-3 Å - limited by lens aberrations and not the wavelength of the electrons) and the complexity of the structure. For instance, in *Figure 2* each dot represents two atoms at a time - the light and dark ones are GaP and InP, respectively. Thus, the ordered CuPt$_B$ structure can be easily recognized in HRTEM images by the light/dark contrast of the alternating (1 $\bar{1}$ 1) or ($\bar{1}$ 11) planes. For more information concerning TEM see the comprehensive overview of Williams and Carter[15] or the treatise of Spence.[16]

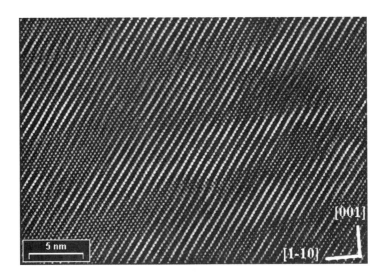

Figure 2. Digitally Bragg-filtered [110] cross-sectional HRTEM image of (GaIn)P/GaAs ordered along [1 $\bar{1}$ 1] ($T_g = 650°C$, $\theta_m = 2°[1 \bar{1} 0]$). The ordered domains can be recognized by the light/dark contrast of the (1 $\bar{1}$ 1) planes. Light and dark dots represent atomic columns of GaP and InP, respectively. The areas, where adjacent (1 $\bar{1}$ 1) planes have the same brightness, are the APBs. They are tilted by approximately 15° towards [1 $\bar{1}$ 0] against the [001] direction. The APBs do not appear atomically sharp, since they are additionally inclined by 15° towards the [110] direction. The antiphase boundaries can be seen best if the image is viewed along the ordered lattice planes.

In comparison to AFM measurements,[14,17,18] the HRTEM has the advantage of being capable of directly resolving even the monoatomic steps that are contained in the supersteps. On the other hand, HRTEM is not well suited to investigate large areas, which limits the counting statistics. We have analyzed at least 1.7 μm of the surface of each sample. Since the TEM samples are only 20 nm thick, this is approximately comparable to one line scan in an AFM image of 2 × 2 μm.

3.   EXPERIMENTAL

Epitaxial (GaIn)P layers were grown lattice matched on (001) GaAs by low-pressure MOVPE in a commercial AIX 200 reactor. To achieve different surface topologies and domain widths, growth temperatures of 650° and 720°C were chosen and the nominally (001) GaAs substrates were misoriented by angles of 0°, 2°, and 6° towards the azimuthal [$\bar{1}$ 10] direction. A single GaAs buffer layer without (GaIn)P was grown at $T_g = 720$°C on (001) GaAs misoriented 2° towards [$\bar{1}$ 10] to enable a comparison of the surface structures of both the GaAs buffer layer and the (GaIn)P layer. Trimethylgallium and trimethylindium were used as group-III precursors and the hydrides AsH_3 and PH_3 as group-V sources. The V/III ratio and the growth rate were kept at 340 and 1.2 μmh^{-1}, respectively. The solid composition of the epitaxial layers has been calculated from X-ray diffraction measurements assuming the validity of Vegard's law. The layer thickness was determined by TEM to be 25 nm for all (GaIn)P layers and 130 nm for the GaAs buffer layer. The [110] cross-section specimens for TEM were prepared in the standard manner by mechanical thinning and polishing followed by Ar-ion beam etching until electron transparency was reached (10-20 nm). Cross-sectional dark-field TEM and HRTEM have been carried out using a Philips CM200 TEM operated at 200 kV to investigate the domain width and the surface morphology of (GaIn)P and GaAs as well as the interface structure between (GaIn)P and GaAs.

4.   RESULTS AND DISCUSSION

The surface of the (GaIn)P and GaAs layers was characterized by HRTEM using [110] cross-sections. We find the (GaIn)P surface to consist of bunched supersteps, which are not only separated by (001) terraces, but also by vicinal regions [*Figure 3*(a)], similar to the findings of Su and Stringfellow.[14] In those vicinal regions the terrace width t between monoatomic steps corresponds to the value given by the substrate misorientation angle θ_m, that can be calculated by

$$t = \frac{h_{ML}}{\tan \theta_m},$$

where h_{ML} is the height of a monoatomic step. On the surface of the GaAs buffer layer no such vicinal regions were observed, but only a very regular array of supersteps and [001] terraces [*Figure 3*(b)]. The terrace width between 2 monolayer steps within a superstep varies from 0.6 to 6 nm for both materials. The terrace widths between the supersteps decrease at both growth temperatures with increasing substrate misorientation.

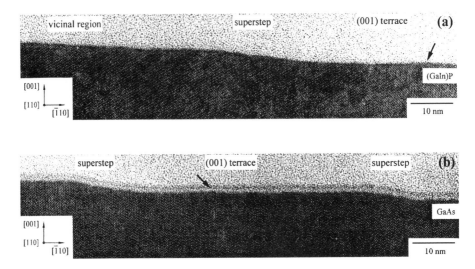

Figure 3. [110] cross-sectional HRTEM image of (a) (GaIn)P and (b) GaAs, grown at $T_g = 720°C$ and misoriented 2° towards [$\bar{1}$ 10]. In addition to supersteps and (001) terraces, vicinal regions appear on the (GaIn)P surface. The arrows mark islands of monoatomic height on the large terraces.

Small islands form on large (001) terraces of (GaIn)P as well as GaAs. Island growth takes place also on the exactly oriented (GaIn)P layer. The edges of these 1-3-ML-high islands consist mainly of single monoatomic steps. There are only a few supersteps of not more than 2 ML height.

The height of the supersteps is plotted versus the misorientation angle in *Figure 4* and it increases with increasing substrate misorientation. From the error bars in this Figure, it is also apparent that the variation of the superstep height is distinctly more pronounced on (GaIn)P than on GaAs. On (GaIn)P we observed very small supersteps of only 0.56 nm (2 ML) height as well as very large supersteps with a height of 7.9 nm (28 ML). The AFM measurements of Stringfellow and Su[17] yielded larger superstep heights for 1-µm-thick (GaIn)P layers grown at atmospheric pressure using a V/III ratio of 160. All other growth parameters were comparable to those of this work. As is known, the superstep height becomes larger with increasing layer thickness[14] and decreasing V/III ratio.[18] Thus, our distinctly smaller superstep heights are mainly attributed to the small layer thickness of 25 nm and the higher V/III ratio. A possible effect of the different reactor pressures remains unknown, because no literature data are available.

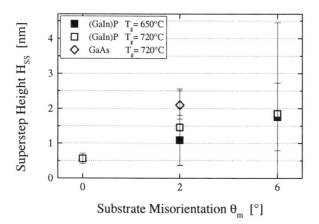

Figure 4. Dependence of the height of supersteps H on growth temperature and substrate misorientation for (GaIn)P (■,□) and GaAs (◊) grown at $T_g = 650°C$ (filled symbols) and 720°C (open symbols). The error bars are the standard deviations of the mean values.

Since the monoatomic steps can be located in supersteps or vicinal regions, a value R is introduced to quantify this distribution.[17] This value represents the fraction of monolayer steps that are contained in supersteps. It is derived as follows. A certain number ML_{SM} of monoatomic steps is necessary to accommodate a substrate misorientation θ_m over a length L (see inset in *Figure 5*). As the monoatomic steps form either supersteps and/or vicinal regions, we write $ML_{SM} = ML_{SS} + ML_{VR}$. Here, ML_{SS} and ML_{VR} are the numbers of monolayer steps contained in supersteps and vicinal regions, respectively. However, it is difficult to reveal and, thus, to count precisely every monolayer step over the distance L (at least 1.7 µm) by means of HRTEM. Therefore, we consider only the monoatomic steps in supersteps and assume all others are located in vicinal regions. Then, the value R is given by

$$R = \frac{ML_{SS}}{ML_{SM}}.$$

If $R = 1$, all monoatomic steps are accumulated in the supersteps, the surface consists only of supersteps and (001) terraces. The difference $1-R$ corresponds to the number of monoatomic steps in the vicinal regions. Thus, the value R represents the ratio between supersteps and vicinal regions on the surface.

Figure 5 shows the dependence of the value R on the substrate misorientation θ_m. R decreases with increasing growth temperature and substrate misorientation. At a temperature of 720°C and a misorientation angle of 2° only 20 % of the (GaIn)P surface consist of vicinal regions, whereas at 6° misorientation it is already more than 80%. In terms of a thermodynamic description of step bunching the supersteps vanish above a critical temperature and a vicinal surface is stable due to the predominance of the entropy.[19] As calculations for other material systems prove, the surface energy of a (001) plane misoriented by 2° and consisting only of monoatomic steps is smaller than that of a 6° misoriented (001) plane.[20,21] Hence, at the same growth conditions a 2° misoriented surface should contain more vicinal regions than a 6° misoriented surface and should have, therefore, a lower R value. However, *Figure 5* shows an opposite dependence, which means that the thermodynamic description cannot explain the dependence of R on the substrate misorientation.

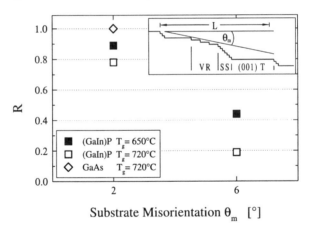

Substrate Misorientation θ_m [°]

Figure 5. Fraction R of monolayer steps contained in supersteps as a function of growth temperature and substrate misorientation for (GaIn)P (■,□) and GaAs (◊) grown at $T_g = 650°C$ (filled symbols) and 720°C (open symbols). The inset shows a schematic drawing of the surface to explain the parameters used for the derivation of R. The terms VR, SS, and (001) T represent vicinal regions, supersteps, and (001) terraces, respectively. L is the characterized overall length of the surface and θ_m the substrate misorientation.

Thus, kinetic models such as that of Schwoebel[22] may apply. The model is based on two ideas: (i) The attachment of adatoms at ascending and descending monolayer steps has different probabilities. (ii) The initial terrace width shows a certain variation so that wider and smaller terraces exist simultaneously. Naturally, the wider terraces capture more incoming adatoms than the smaller ones. If incoming adatoms are preferably attached at descending steps, wider terraces become even wider and step bunching occurs. If the adatoms are mainly captured at ascending steps, terrace widths become more regular - a vicinal surface evolves.

Within this description, the different attachment energies at ascending and descending steps are less important at higher growth temperatures, which explains the reduction of R with increasing growth temperature, (see *Figure 5*). The significant decrease of the R value with increasing substrate misorientation θ_m is caused by the variation of terrace widths. On a 6° misoriented vicinal surface, the terrace width of a monolayer step is 2.7 nm, whereas it is 8.1 nm for a misorientation angle of 2°. Therefore, the same percentage variation of the terrace width leads to a significantly smaller difference in the number of adatoms captured on wider and smaller terraces on a 6° misoriented vicinal surface than on a 2° misoriented one. This limits the superstep formation on higher misoriented surfaces.

That our values of R are remarkably larger than those observed by Stringfellow and Su[17] is probably attributed to the different investigation methods of HRTEM (this work) and AFM[17] and the above mentioned different growth conditions. HRTEM is capable of resolving the monoatomic steps even in supersteps. Therefore, we considered every surface region as a superstep, which consisted of more than one monoatomic step and formed an angle to the (001) plane of at least 2° more than the misorientation angle. Thus, even small supersteps of 2 or 3 ML height, lying close together due to their small terrace widths, contribute to our R value. Such steps are very difficult to resolve with an AFM even if using very fine tips. Hence, the R value resulting from AFM measurements tends to be lower. Moreover, in AFM investigations the value R is determined from the average height of the supersteps and not, as in this work, from the exact number of monoatomic steps contained in the supersteps. Effects of the different growth conditions are also possible, however, they are unknown yet.

The decay of the superstep density with increasing growth temperature can be seen in *Figure 6*. The cause is obviously the reduced R value, i.e., the higher number of vicinal regions. The dependence of the density of supersteps on the substrate misorientation is more complicated. For $T_g = 650°C$, the superstep density increases with increasing substrate misorientation, since even at 6° misorientation R is still almost 0.5 (see *Figure 5*). That means the additional monolayer steps caused by the higher substrate misorientation angle contribute mainly to the formation of additional or larger supersteps. However, at $T_g = 720°C$, the predominant part of the additional monolayer steps forms vicinal regions - R is very small. Therefore, at high growth temperatures the superstep density diminishes with increasing substrate misorientation. Thus, the opposite dependence of the superstep density on the substrate misorientation observed at high and low growth temperatures is attributed to the different decrease of R with increasing substrate misorientation found for these two cases. The very low density of supersteps for exactly oriented (001) (GaIn)P results from island growth. There are only very few supersteps 2 ML in height.

The effect of the surface morphology on the formation of APBs is discussed below. The density of APBs, measured in dark-field TEM micrographs, is plotted versus the misorientation angle in *Figure 6*. In agreement with data in the literature,[23,24] it distinctly decreases with increasing growth temperature, substrate misorientation, and layer thickness.

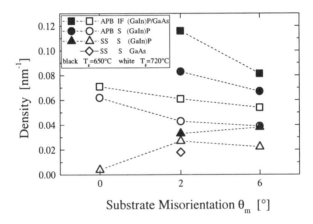

Figure 6. Density of APBs at the (GaIn)P surface (●,○) and the interface (■,□) and density of supersteps on the (GaIn)P surface (▲,△) vs. the substrate misorientation angle. Filled and open symbols represent growth temperatures of 650° and 720°C, respectively. The dashed lines are only a guide to the eye.

Obviously, we find no 1:1 correlation between the APB density and the density of supersteps, as revealed by AFM investigations elsewhere.[14] The different correlation might be again caused by the significantly different layer thicknesses since the density of APBs diminishes with increasing layer thickness (see *Figure 6* and cf. Ref. 8). Hence, for the 1-µm-thick layers of Ref. 14 a 1:1 correlation possibly exists. However, according to *Figure 6*, this is not the case for the very thin layers investigated here, where the density of APBs is always higher than that of supersteps. This indicates, that antiphase boundaries are formed not only at every superstep, but additionally in other surface areas, e.g. at monolayer steps in vicinal regions. The reasons for this behavior can be explained as follows. The data in *Figure 6* are values averaged over the whole layer. *Figure 7*(a) shows the distribution of spacings between APBs, measured by DF-TEM, and between supersteps, determined using HRTEM, on the surface of a (GaIn)P layer grown at $T_g = 720°C$ and misoriented by 2° towards [$\bar{1}$ 10]. Apparently, the maximum values of the frequencies for APBs and supersteps occur at almost the same spacing. However, the mean value of the superstep spacing (A_{SS} in *Figure 7*) is considerably higher due to the existence of a few very large values. Such large superstep spacings are caused by large supersteps and terraces as well as additional vicinal regions between supersteps. This implies, that the number of APBs is always higher than that of the supersteps for the following reasons: (i) The formation of more than one APB can occur at large supersteps since the likelihood is increased that the superstep contains more than one monolayer step with odd terrace width. It is emphasized again, that monolayer steps with odd terrace width and not the supersteps themselves are responsible for the formation of APBs.[3] (ii) Antiphase boundaries are also formed in vicinal regions consisting of a high number of monolayer steps (see *Figure 5*). (iii) Due to the formation of small islands on large (001) terraces (see *Figure 3*), similar to the growth process on exactly oriented (001) substrates, APBs may nucleate even there. The island edges itself are formed by ML steps. If their terrace widths are odd, an APB starts there.

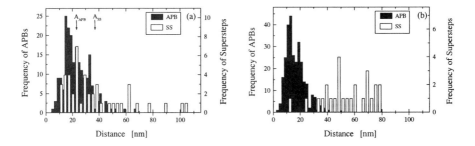

Figure 7. Frequency of the spacings between APBs (black columns) and between supersteps (white columns) at (a) the surface of (GaIn)P and (b) the interface between (GaIn)P and GaAs for a (GaIn)P layer grown at $T_g = 720°C$ and misoriented 2° towards [$\bar{1}$ 10]. A_{SS} and A_{APB} are the mean values of the spacings between the supersteps and the antiphase boundaries, respectively. The superstep spacings in (b) have been measured on the surface of a GaAs layer grown at the same conditions as the heterostructure. Note, that the APB spacings were determined by DF-TEM, the superstep spacings using HRTEM.

As can be seen in *Figure 7*(b), the situation at the interface between (GaIn)P and GaAs is different compared to that of the free surface of the layer. There is no correlation between the distances of APBs and GaAs supersteps. The spacings between the APBs in (GaIn)P are smaller, whereas the superstep spacings of the GaAs are remarkably larger than that of the (GaIn)P surface. The latter are caused exclusively by the large (001) terraces [see *Figure 3*(b)]. Due to the already mentioned island growth on these large terraces, a high number of APBs is nucleated. Therefore, the APBs are closer spaced than the supersteps on the GaAs. This seems to be the main process for the formation of APBs at the very beginning of the (GaIn)P growth on GaAs since still no vicinal regions or giant supersteps exist.

These interpretations are proved by *Figure 8*. The DF micrograph in *Figure 8*(a) is an example for the ideal case of a 1:1 correlation between APBs and supersteps at the surface of (GaIn)P. Only one antiphase boundary intersects the surface at each superstep. Hence, according to the model of Philips *et al.*,[3] there is only one monoatomic step with odd terrace width within each superstep. The case where more than one APB belongs to one superstep is displayed in *Figure 8*(b). Two APBs (1,2) meet the (GaIn)P surface within a large superstep, which then, following Ref. 3, contains two single, not adjacent monoatomic steps with odd terrace width. All other monoatomic steps in that superstep have to be of even terrace width. Three other APBs (3-5) intersect the surface within the large (001) terrace. Therefore, monolayer islands should exist on this terrace. Such islands are not resolvable in DF-TEM but could indeed be observed on these layers by HRTEM (see *Figure 3*). That means, in this example five APBs are associated with just one superstep. Thus, the ratio between the APB density and that of supersteps is always larger than unity, as seen in *Figure 6*.

Another conclusion can be inferred from *Figure 5* and *Figure 6*. A 2° misoriented (GaIn)P surface consists mainly of supersteps and (001) terraces, whereas a 6° misoriented surface is covered more than 50% by vicinal regions. Although at $T_g = 650°C$ the superstep density increases with increasing substrate misorientation, the APB density diminishes at both growth temperatures with increasing substrate misorientation. Hence, less APBs are formed in vicinal regions than on large (001) terraces. This indicates that most of the monolayer steps in vicinal regions have even terrace widths. However, the preferential sites for the APB formation are the supersteps due to the high local concentration of monoatomic steps.

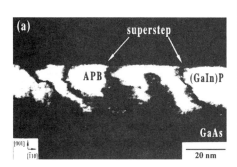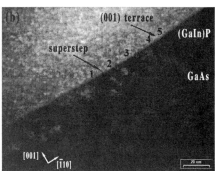

Figure 8. [110] Dark-field TEM micrographs of (GaIn)P misoriented 6° towards [$\bar{1}$ 10] grown at (a) $T_g = 650°C$ and (b) $T_g = 720°C$. (a) Only one APB intersects the surface at each of the marked superstep edges. (b) Two APBs (1,2) end in the superstep and three (3-5) in the (001) terrace. The superstep in (b) can be seen best if the micrograph is viewed along the layer surface.

These results can be used to reduce the number of APBs in ordered (GaIn)P by preparing well-defined surfaces for the growth. They should favorably consist of only monolayer steps ($R = 0$) with defined even terrace widths to impede or even prevent the nucleation of APBs. This cannot be achieved by growing a single GaAs buffer layer since supersteps and large (001) terraces are always present on it. However, surfaces consisting of only vicinal regions can be prepared by the growth of an AlAs/GaAs superlattice.[25] Since the surface diffusion coefficient of Al on GaAs is smaller than that of Ga,[26] the velocity of the step bunching of AlAs and GaAs is different. This leads to a smoothing effect due to the compensation of the superstep formation of the both materials. To investigate the influence of the superlattice on the ordering 20 ML (5.6 nm) AlAs and 3 ML (0.84 nm) GaAs have been stacked 25 times at $T_g = 750°C$ on (001) GaAs misoriented by 6° towards [1 $\bar{1}$ 0]. The superlattice was completed by another 20 ML AlAs. The following 370-nm-thick (GaIn)P layer was grown at $T_g = 650°C$.

The domain thickness of this (GaIn)P layer is indeed larger than that of (GaIn)P grown on a simple GaAs buffer layer (see *Figure 9*). Thus, the superlattice increases the number of vicinal regions and, since the APBs preferentially form at supersteps, decreases the APB density.

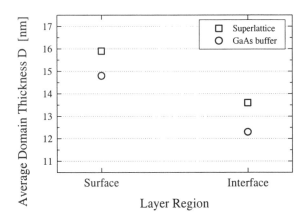

Figure 9. Average thickness D of domains ordered along the [$\bar{1}$11] direction in (GaIn)P ($T_g = 650°C$, $\theta_m = 6°$[$\bar{1}$10]), that was grown on an AlAs/GaAs superlattice (□) and on a simple GaAs buffer layer (O). Since the (GaIn)P layers are 370 and 25 nm thick, the term "surface" means a height of 25 nm above the substrate interface.

However, the difference in the domain thickness is small because the uppermost AlAs layer still contains supersteps (*Figure 10*). This is due to the quality of the superlattice that has not been optimized. As HRTEM investigations revealed, the AlAs/GaAs interfaces are not abrupt and the layer thicknesses vary locally over 1-3 ML. With a better quality of the superlattice we expect an enhanced impact on the domain thickness. This would also enable the analysis of the possible effect of the terrace width of the monolayer steps on the domain thickness.

Figure 10. [110] cross-sectional HRTEM image of the interface between the ordered (GaIn)P layer and the last AlAs layer of the AlAs/GaAs superlattice. The AlAs layer still contains supersteps (marked by arrows).

Such efforts to influence the domain thickness by the preparation of well-defined surface morphologies are limited to a layer thickness not exceeding 500 nm by the growth mechanism of the (GaIn)P itself. With increasing layer thickness the step bunching process of the (GaIn)P affects essentially the density of APBs. Above a certain layer thickness (500-1000 nm) the domain thickness has adjusted itself back to the values that are typical for the given growth conditions. This holds true also for the success of experiments to increase the domain thickness by the introduction of a "temperature step" during growth.[27] However, both methods offer the opportunity to reduce the number of APBs without influencing the degree of order.

5.   SUMMARY

The influence of the surface morphology on the formation of antiphase boundaries has been determined for $CuPt_B$ - type ordered (GaIn)P. The surfaces and interfaces of (GaIn)P and GaAs epitaxial layers were characterized by HRTEM with regard to their dependence on growth temperature and substrate misorientation. The (GaIn)P surface consists of supersteps, (001) terraces, and vicinal regions. The latter were not observed on GaAs. The superstep height was found to increase with substrate misorientation. A significant diminution of the fraction R of monolayer steps contained in supersteps has been observed with increasing misorientation angle and growth temperature. This dependence is explained in terms of the Schwoebel model, because thermodynamic descriptions fail. Our values of R, ranging from 89% at $T_g = 650°C$, 2° [$\bar{1}$ 10] to 19 % for $T_g = 720°C$, 6° [$\bar{1}$ 10], are remarkably larger than those found in the literature. This is probably caused by the different investigation techniques, HRTEM and AFM, and different growth conditions. The superstep density increases with decreasing growth temperature. The opposite dependence of this density on the misorientation angle at different growth temperatures is explained by the different R values. Dark-field TEM micrographs were used to determine the density of APBs of ordered (GaIn)P layers. Higher growth temperatures, misorientation angles, and layer thicknesses cause the APB density to drop. By comparing the APB density to the density of supersteps, a correlation between both parameters has been found. The supersteps are therefore the preferential formation site of APBs. However, the correlation is blurred by the additional APB formation in vicinal regions and on large (001) terraces as well as by the formation of more than one APB per superstep. This explains why the density of APBs is always higher than the superstep density. In the initial stage of layer growth the substrate surface morphology determines the APB density. With increasing layer thickness the step bunching process of the (GaIn)P affects essentially the density of APBs. All these findings have been used to reduce the number of APBs in ordered (GaIn)P by growing it on an AlAs/GaAs superlattice that contains less supersteps than a simple GaAs buffer layer.

ACKNOWLEDGEMENTS

The authors are grateful to G. Wagner, V. Gottschalch, H. Schmidt, and M. Schubert for useful discussions. This work was carried out at the University of Leipzig (Germany) and was supported by the Deutsche Forschungsgemeinschaft and the Bundesministerium für Bildung und Forschung.

REFERENCES

* e-mail: torsten.sass@ftf.lth.se

[1] A. Zunger and S. Mahajan, in *Handbook of Semiconductors,* 2nd ed., edited by S. Mahajan (Elsevier, Amsterdam, 1995) Vol. 3, 1399.

[2] Thus, the description of the ordered structure with the centrosymmetrical CuPt structure type (space group $R\bar{3}m$) is actually inaccurate, since it neglects the existence of the P atoms. The correct description for the structure would be the $CrCuS_2$ structure type.

[3] B.A. Philips, A.G. Norman, T.-Y. Seong, S. Mahajan, G.R. Booker, M. Skowronski, J.P. Harbison, and V.G. Keramidas, J. Cryst. Growth 140 (1994) 249.

[4] H. Murata, I.H. Ho, Y. Hosokawa and G.B. Stringfellow, Appl. Phys. Lett. 68 (1996) 2237.

[5] S.H. Lee, C.Y. Fetzer, G.B. Stringfellow, D.H. Lee, and T.Y. Seong, J. Appl. Phys. 85 (1999) 3590.

[6] O. Ueda, M. Takikawa, M. Takechi, J. Komeno, and I. Umebu, J. Cryst. Growth 93 (1988) 418.

[7] P. Ernst, C. Geng, G. Hahn, F. Scholz, H. Schweizer, F. Phillipp and A. Mascarenhas, J. Appl. Phys. 79 (1996) 2633.

[8] L.C. Su, I.H. Ho, and G.B. Stringfellow, J. Appl. Phys. 75 (1994) 5135.

[9] T. Saß, I. Pietzonka, and H. Schmidt, J. Appl. Phys. 85, 3561 (1999). The monoclinic space group *Pm* given in this paper is incorrect. The structure has indeed the space group *Pmn*2_1.

[10] S.P. Najda, A. Kean, and G. Duggan, J. Appl. Phys. 82 (1997) 4408.

[11] M. Kasu and T. Fukui, Jpn. J. Appl. Phys. 31 (1992) L864.

[12] D.J. Friedman, J.G. Zhu, A.E. Kibbler, J.M. Olson, and J. Moreland, Appl. Phys. Lett. 63 (1993) 1774.

[13] A. Gomyo, H. Hotta, F. Miyasaka, K. Tada, H. Fujii, K. Fukagai, K. Kobayashi, and I. Hino, J. Cryst. Growth 145 (1994) 126.

[14] L.C. Su and G.B. Stringfellow, J. Appl. Phys. 78 (1995) 6775.

15 D.B. Williams and C.B. Carter, *Transmission Electron Microscopy*, Plenum Press, New York (1996).
16 J.C.H. Spence, *Experimental High-Resolution Electron Microscopy*, 2nd ed., Oxford University Press, Oxford (1988).
17 G.B. Stringfellow and L.C. Su, J. Cryst. Growth 163 (1996) 128.
18 I. Pietzonka, T. Sass, R. Franzheld, G. Wagner, and V. Gottschalch, J. Cryst. Growth. 195 (1998) 21.
19 E.D. Williams and C. Bartelt, Ultramicroscopy 31 (1989) 36.
20 Z.-J. Tian and T.S. Rahman, Phys. Rev. B 47 (1993) 9751.
21 T.W. Poon and S. Yip, Phys. Rev. Lett. 65 (1990) 2161.
22 R.L. Schwoebel and E.J. Shipsey, J. Appl. Phys. 37 (1966) 3682; R.L. Schwoebel, J. Appl. Phys. 40 (1969) 614.
23 L.C. Su, I.H. Ho, and G.B. Stringfellow, J. Appl. Phys. 76 (1993) 3520.
24 S.R. Kurtz, J.M. Olson, D.J. Arent, A.E. Kibbler, and K.A. Bertness, in *Common Themes and Mechanisms of Epitaxial Growth*, edited by P. Fuoss, J. Tsao, D.W. Kisker, A. Zangwill and T. Kuech (Materials Research Society, Pittsburgh, PA, 1993), 83.
25 V. Gottschalch, R. Schwabe, G. Wagner, S. Kriegel, R. Franzheld, I. Pietzonka, F. Pietag and J. Kovac, Workshop Booklet of the 7th European Workshop on Metal-Organic Vapour-Phase Epitaxy and Related Growth Techniques, June 8-11, 1997, Berlin, Germany, E4.
26 M. Kasu and N. Kobayashi, Appl. Phys. Lett. 67 (1995) 2842.
27 C. Geng, A. Moritz, S. Heppel, A. Mühe, J. Kuhn, P. Ernst, H. Schweizer, F. Phillipp, A. Hangleitner, and F. Scholz, J. Cryst. Growth 170 (1997) 418.

Chapter 6

X-Ray Characterization of CuPt Ordered III-V Ternary Alloys

Jianhua Li
Physics Department, University of Houston

Key Words: X-ray diffraction, order parameter, reciprocal area map, anti-phase boundary

Abstract: A quantitative model, based on x-ray diffraction, is proposed to analyze the CuPt ordered III-V ternary semiconductor alloy films. The model takes into account the size distribution of the two different laminae-shaped variants, the random distribution of the anti-phase domain boundaries, and the atomic displacements due to the bond length difference between the two constitutive binary materials. The model enables us to extract quantitatively the structural information of the ordered films from the x-ray diffraction data.

1.   INTRODUCTION

Recent development in x-ray diffraction techniques, including the use of advanced synchrotron sources and sophisticated instruments, as well as newer experimental methods, such as grazing incidence diffraction, truncation rod analysis, reflectometry, and quantitative reciprocal space mapping, etc., have made x-ray diffraction, a traditionally bulk material characterization method, a powerful tool in thin film analysis. Along with its advantages of nondestructiveness and straightforward sample preparation, x-ray diffraction can be applied to most of the known materials (metallic, ceramic, semiconductor, polymer, biological, etc.), and is highly sensitive to crystallinity, strain, texture, surface morphologies, interfaces, and defects in the films. These advanced x-ray diffraction techniques also allow one to perform depth dependent analyses - from the top atomic layer to micrometers deep. The normal weak interaction of x-ray photons with the analyzed materials make interpretation of the diffraction data simpler, which in general can be done within the framework of Fourier transform of correlation functions. A recent book by Holy et.al. presents an excellent review of these techniques.[1]

Atomic ordering in epitaxial semiconductor alloy films is a substrate induced phenomenon.[2-4] In particular, CuPt-type ordering has been observed in a large variety of III-V ternary and quaternary alloy films grown on (001)-oriented substrates.[3-5] It leads to a reduction of the film structure from original $F\bar{4}3m$ symmetry to a rhombohedral R3m symmetry. This consequently has a strong impact on the optical and electronic properties of the film, and thus requires full understanding. Although transmission electron diffraction (TED) and microscopy (TEM) analyses are able to show most of the qualitative features of the CuPt ordered phases and the complex morphologies within the ordered film over a large range of length scales,[5] quantitative analysis of the TED data of a CuPt ordered film is difficult.[6,7] X-ray diffraction, on the other hand, has long been used for quantitative analysis of atomic ordering in bulk materials. The basic principles of x-ray diffraction measurements and determination of ordered structure and order parameters of a uniformly ordered crystal, particularly metallic alloys, are well understood.[8] However, such approaches are difficult to apply to ordered III-V compounds for a number of reasons, as follows:

(1) The epitaxial films are usually grown in a well-defined crystalline direction, subject to the orientation of the substrates that are often chosen so that the lattice constants of the films and substrates closely match each other. Therefore, in conventional co-planar diffraction geometry, only a few ordered reflections are accessible.

(2) The ordering is often not uniform in the films. Coexistence of two ordering variants is frequently observed in the CuPt ordered films.

(3) The ordering can be both short-range or long-range, depending on the growth conditions of the films.

(4) Unlike metallic alloys, the atoms in a semiconductor alloy are covalently bonded. Extended x-ray absorption fine structure (EXAFS) analyses have indicated that the first-nearest-neighbour distances in a ternary or quaternary alloy deviate significantly from those of an ideal crystal.[9,10]

This chapter will describe a diffraction method, that combines a non-coplanar diffraction geometry, achieved by using a four-circle diffractometer, and a proper structural model that considers not only the symmetry in the CuPt ordered III-V ternary compounds of the form of $A_xB_{1-x}C$, but also the size distributions of the two coexisting variants and the anti-phase domain boundaries, as well as the atomic displacement due to the bond length difference between the constitutive binary materials, to analyze the CuPt ordered films. To permit the reader, who is mainly interested in the diffraction analysis presented in this chapter, to see all necessary information and thus obtain a better understanding of the application of diffraction technique to materials of this class, the chapter is organized as follows. A brief review of the TED and TEM results is given in section 2, to give the readers a basic overview of the CuPt-type ordering. A non-coplanar diffraction geometry which is more flexible than a co-planar diffraction geometry will be introduced in section 3. A structural model of CuPt ordered films and the corresponding kinematical diffraction theory are described in section 4. The effect of the atomic displacements due to the bond length difference on the diffraction data is discussed in section 5. Examples of experimental data analyses are given in section 6. We finally discuss the determination of the order parameter in section 7. Section 8 presents a brief summary and suggestions to further work currently underway.

2.   RESULTS FROM TED AND TEM STUDIES

A typical TED pattern from a CuPt ordered $A_xB_{1-x}C$ film with double variants is shown in Fig. 1. The ordering reflections are elongated along directions ~10° away from the [001], forming a characteristic wavy pattern. The cause of this unusual diffraction phenomenon is believed to be the complex structure of the ordered film which allows the two variants to interlock with the anti-phase domain boundaries to form plate-like domains

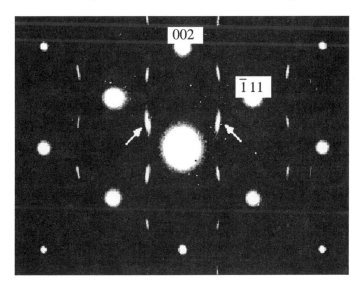

Figure 1. A typical [110]-zone TED pattern taken from a CuPt ordered $Ga_{0.5}In_{0.5}P$ alloy. In addition to the fundamental reflection spots, tilted and streaky ordering reflection spots are also seen at (h/2,k/2,l/2) positions.

slightly inclined with respect to the [001] direction.[5,11,12] In Fig.2, atomic images taken from a single phase, double-variant domain and a single phase, single-variant domain are shown. In the case of double variants, a laminar structure with alternating variants along [001] is clearly observed. A detailed description of the laminar structure of the two variants and their complex interlock with the anti-phase domain boundaries can be found in Ref. 5 and the references therein.

(a)   **(b)**

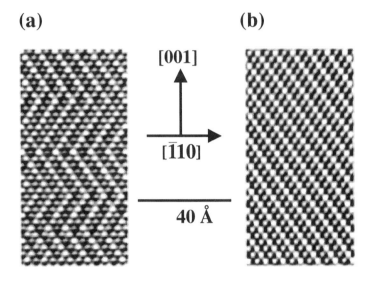

Figure 2. High resolution images taken from (a) a double-variant, and (b) a single-variant ordered domain by TEM.

3.   X-RAY DIFFRACTION MEASUREMENTS

For a (001)-oriented III-V compound thin film, conventional coplanar diffraction geometry, where the diffraction plane lies perpendicular to the sample surface, does not have access to most of the CuPt ordering reflections. We therefore introduce a non-coplanar diffraction geometry by using a four-circle diffractometer. Since the ordering reflection is caused primarily by the difference between the x-ray scattering factors of the two group III elements, and is thus weak, a synchrotron x-ray source is preferred and is actually essential as long as the counting time is concerned. The measurements shown in this chapter were performed on beamline X14A of the National Synchrotron Light Source (NSLS) at the Brookhaven National Laboratory (BNL) with a x-ray energy of 8.0478keV. The diffraction geometry is schematically shown in Fig. 3, where a χ angle on a standard four-circle diffractometer is used to maximise the accessible region in reciprocal space. Most of the measurements discussed in this chapter were done in a reciprocal space region shown in Fig. 4. Both line scans and area scans can be realised in this geometry.

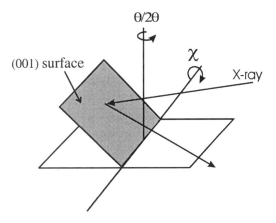

Figure 3. Schematic diagram showing the diffraction geometry for x-ray measurements.

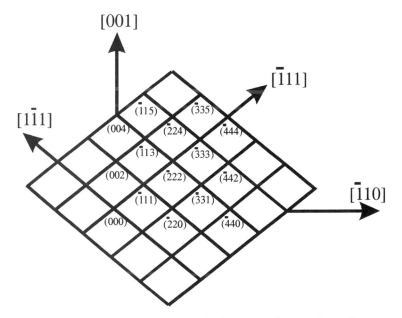

Figure 4. The reciprocal space area within which most of the experimental measurements were made in this chapter.

To minimise the instrumental errors, at the beamline we used, a Si (111) monocromator set slightly away from the Bragg position and a pulse height analyser with its window set to accept only the energy in interest were used to remove the higher order harmonics ($\lambda_n = \lambda/n$). This consideration is particularly important for our experiments because the second order harmonic would appear exactly at the positions where the CuPt ordering reflections would occur (although in principle, there is no Si (222) reflection, it actually exists due to a loss in inversion symmetry of the binding electrons). A proportional counter with a reasonable energy resolution was used to record the diffracted x-rays. The diffractometer set-up used for taking the data shown in this chapter gives a resolution of better than 0.005 in dimensionless reciprocal lattice units.

4.   STRUCTURAL MODEL AND DIFFRACTION THEORY

4.1   Structural model

The two possible CuPt ordering variants are schematically shown in Fig. 5a and b, respectively. In the text of the paper, we will call the ($\overline{1}$ 11) variant and the (1 $\overline{1}$ 1) variant as variant I and II, respectively. In the diagram of variant I, an anti-phase domain boundary perpendicular to the [$\overline{1}$ 10] direction is also shown. Such boundaries will inevitably exist in the film because the initial occupation of a group III lattice site by atoms A or B is equally possible and the growth actually occurs simultaneously everywhere over the substrate surface. In Figs. 5c and 5d, two examples of the laminar structure with alternating variants are shown. The difference between Figs. 5c and 5d lies in the thickness of the variant II lamina and the effect this has on the relationship between the neighbouring variant I laminae. When the two variant I laminae are separated by a lamina of variant II containing an even number of group III atomic layers (as in Fig. 5c) the variant I laminae are in phase with each other. Conversely, if the intervening variant II lamina is an odd number of group III atomic layers in thickness (e.g., Fig. 5d) the variant I laminae will then be out of phase. In a real case, the number of group III atomic layers in a lamina of either variant may be random.

Based on these considerations, a realistic structural model must consist of statistically distributed lateral anti-phase domains and a varying phase relationship in the growth direction, as shown in Fig. 6, where the anti-phase domains in the two variants are considered to be independent from each other. The number of atomic layers in each lamina is random.

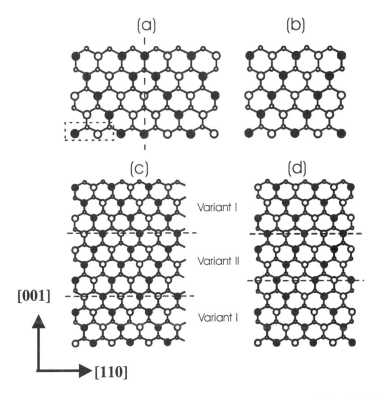

Figure 5. Schematic diagram showing the CuPt ordered structure. (a) Variant I with an anti-phase domain boundary; (b) Variant II. (c) and (d) are laminar structures with alternating ariants. Dashed rectangle marks the unit cell used in our model calculation.

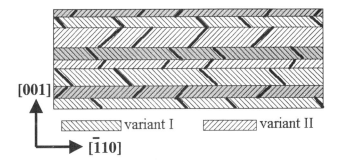

Figure 6. Structural model of a double-variant film containing randomly distributed anti-phase domain boundaries (represented by the thick bars) and random number of atomicayers in each lamina. Shaded and unshaded laminae of the same variant are out-of-phase.

4.2   Diffraction theory

To describe the diffraction theory in steps, let us first neglect the anti-phase domain boundaries in the $[\bar{1}\,10]$ direction, the atomic layers in the $[\bar{1}\,10]$ direction are assumed to be infinite. The basic structural units that constitute the ordered phase are a layer of group III atoms and a layer of group V atoms (see Fig. 5). In order to obtain a structure of variant I, we have to shift the next group III atomic layer with respect to the previous one by a vector $(\bar{a}/2,0,a/2)$ in the crystallographic coordinate system. The phase shift in the structure factor of this layer caused by such an operation is:

$$\alpha = e^{\pi i(q_1+q_3)},\tag{1}$$

where (q_1,q_2,q_3) are the coordinates of the scattering vector $\mathbf{Q}$:

$$\mathbf{Q} = \frac{2\pi}{a}(q_1,q_2,q_3).$$

Similarly, the phase shift between the structure factors of neighbouring atomic layers in variant II is:

$$\beta = e^{\pi i(-q_1+q_3)}\tag{2}$$

Let us assume a vertical sequence of N pairs of laminae of variant I and variant II. The j-th pair of laminae consists of n_{Aj} and n_{Bj} atomic layers of group III atoms in variants I and II, respectively. These numbers are random. If we denote $F_{0A,B}(\mathbf{Q})$ the structure factors of infinite layers in variants I and II, the structure factor of the whole stack of laminae is then given by:

$$F(\mathbf{Q}) = \sum_{j=1}^{N}\left[F_{0A}(\mathbf{Q})\frac{\alpha^{n_{Aj}}-1}{\alpha-1} + F_{0B}(\mathbf{Q})\beta\alpha^{n_{Aj}-1}\frac{\beta^{n_{Bj}-1}}{\beta-1}\right]$$
$$\cdot(\alpha\beta)^{j-1}\prod_{k=1}^{j-1}\alpha^{n_{Ak}-1}\beta^{n_{Bk}-1},\tag{3}$$

The x-ray intensity scattered by such a structure can thus be written in the form:

$$I(\mathbf{Q}) \propto \left\langle |F(\mathbf{Q})|^2 \right\rangle = \sum_{j=1}^{N} \left\langle \left| F_{0A}(\mathbf{Q}) \frac{\alpha^{n_{Aj}} - 1}{\alpha - 1} + F_{0B}(\mathbf{Q}) \beta \alpha^{n_{Aj}-1} \frac{\beta^{n_{Bj}} - 1}{\beta - 1} \right|^2 \right\rangle$$

$$+ 2\,\text{Re} \left[\sum_{j=2}^{N} \sum_{r=1}^{j-r} (\alpha\beta)^{j-r} \left\langle \left(F_{0A}(\mathbf{Q}) \frac{\alpha^{n_{Aj}} - 1}{\alpha - 1} + F_{0B}(\mathbf{Q}) \beta \alpha^{n_{Aj}-1} \frac{\beta^{n_{Bj}} - 1}{\beta - 1} \right) \right. \right. \quad (4)$$

$$\left. \left. \times \left(F_{0A}(\mathbf{Q}) \frac{\alpha^{n_{Ar}} - 1}{\alpha - 1} + F_{0B}(\mathbf{Q}) \beta \alpha^{n_{Ar}-1} \frac{\beta^{n_{Br}} - 1}{\beta - 1} \right)^* \prod_{k=r}^{j-1} (\alpha^{n_{Ak}-1} \beta^{n_{Bk}-1}) \right\rangle \right].$$

Here, <> indicates an average over all possibilities of n_{Aj} and n_{Bj}. We denote $\chi_\alpha = <\alpha^{n_{Aj}}>$ and $\chi_\beta = <\alpha^{n_{Bj}}>$ and Eq. (4) can be written as:

$$I(\mathbf{Q}) \propto 2N \left\{ |F_{0A}(\mathbf{Q})|^2 \frac{1 - \text{Re}(\chi_\alpha)}{|1 - \alpha|^2} + |F_{0B}(\mathbf{Q})|^2 \frac{1 - \text{Re}(\chi_\beta)}{|1 - \beta|^2} \right.$$

$$\left. - \text{Re} \left(F_{0A}(\mathbf{Q}) F_{0B}^*(\mathbf{Q}) \frac{1 - \chi_\alpha^*}{1 - \alpha^*} \frac{1 - \chi_\beta^*}{1 - \beta} \right) \right\}$$

$$+ 2\,\text{Re} \left\{ \left[|F_{0A}(\mathbf{Q})|^2 \alpha \chi_\beta \left(\frac{1 - \chi_\alpha}{1 - \alpha} \right)^2 + |F_{0B}(\mathbf{Q})|^2 \beta \chi_\alpha \left(\frac{1 - \chi_\beta}{1 - \beta} \right)^2 \right. \right. \quad (5)$$

$$+ (F_{0A}^*(\mathbf{Q}) F_{0B}(\mathbf{Q}) \beta \chi_\alpha \chi_\beta + F_{0A}(\mathbf{Q}) F_{0B}^*(\mathbf{Q}) \alpha)$$

$$\left. \left. \times \frac{1 - \chi_\alpha}{1 - \alpha} \frac{1 - \chi_\beta}{1 - \beta} \right] \frac{\chi_\alpha \chi_\beta}{1 - \chi_\alpha \chi_\beta} \left(N - 1 - \frac{1 - (\chi_\alpha \chi_\beta)^{N-1}}{1 - \chi_\alpha \chi_\beta} \right) \right\}$$

If we assume that the numbers n_{Aj} and n_{Bj} follow Poisson distribution with mean values $<n_{Aj}>=n_\alpha$ and $<n_{Bj}>=n_\beta$, we obtain

$$\chi_\gamma = e^{-\eta_\gamma(1-\gamma)}, \gamma = \alpha, \beta \quad (6)$$

Now let us take into account the lateral anti-phase domains in each lamina. In this case, each lamina contains alternating segments out-of-phase with respect to each other. For the sake of simplicity, we assume that the lateral sequence of these segments is statistically not correlated to the vertical sequence of the two variants. Then the existence of the lateral anti-phase domains affects only the terms $|F_{0A}(\mathbf{Q})|^2$, $|F_{0B}(\mathbf{Q})|^2$, and $F_{0A}(\mathbf{Q})F^*_{0B}(\mathbf{Q})$ in Eq. (5). Now these terms must be replaced by their averages taking into account a sequence of randomly distributed lateral segments. If we further assume that the lateral structures of different laminae are not correlated, then we need simply to make the following replacements in Eq. (5).

$$|F_{0A}(\mathbf{Q})|^2 \rightarrow \left\langle |F_{0A}(\mathbf{Q})|^2 \right\rangle, |F_{0B}(\mathbf{Q})|^2 \rightarrow \left\langle |F_{0B}(\mathbf{Q})|^2 \right\rangle$$
$$F_{0A}(\mathbf{Q})F^*_{0B}(\mathbf{Q}) \rightarrow \left\langle F_{0A}(\mathbf{Q})\right\rangle\left\langle F_{0B}(\mathbf{Q})\right\rangle^* \tag{7}$$

The structure factor of the p-th lateral segment of a layer is shifted by a phase factor, ϕ, with respect to the segment p-1,

$$\phi = e^{\pi i(-q_1+q_2)} = e^{-2\pi i q_1},$$

if we assume $q_1=-q_2$ in the $[\bar{1}10]$ scattering plane. The structure factor of such a sequence of segments is

$$F_{0S}(\mathbf{Q}) = \sum_{p=1}^{M} F_{0S}^{(p)} \phi^{p-1}, S = A, B, \tag{8}$$

where

$$F_{0S}^{(p)} = F_{cS} \frac{\phi^{2m_p}-1}{\phi^2-1} \prod_{k=1}^{p-1} \phi^{2m_k}, S = A, B \tag{9}$$

is the structure factor of the p-th segment of variant I or II lamina containing m_p unit cells, M is the number of segments in one lamina, and F_{cS} is the structure factor of one unit cell as defined in Fig. 5. We therefore have:

$$\left\langle |F_{0S}(\mathbf{Q})|^2 \right\rangle = |F_{cS}(\mathbf{Q})|^2 \left[\sum_{p=1}^{M} \left\langle \left| \frac{\phi^{2m_p}-1}{\phi^2-1} \right|^2 \right\rangle + 2\mathrm{Re}\left(\sum_{p=2}^{M} \sum_{k=1}^{p-1} \phi^{p-k} \right) \right.$$
$$\left. \times \left\langle \frac{\phi^{2m_p}-1}{\phi^2-1} \left(\frac{\phi^{2m_k}-1}{\phi^2-1} \right)^* \cdot \prod_{r=k}^{p-1} \phi^{2m_r} \right\rangle \right], S = A, B. \tag{10}$$

After some manipulation, we obtain:

$$\left\langle |F_{0S}(\mathbf{Q})|^2 \right\rangle = 2\frac{|F_{cS}(\mathbf{Q})|^2}{|\phi^2 - 1|^2}\left\{ M(1 - \text{Re}(\chi_\parallel)) - \text{Re}\left[\phi^2\chi_\parallel \frac{(1-\chi_\parallel)^2}{1-\phi\chi_\parallel}\right.\right.$$

$$\left.\left. \times\left(M - 1 - \frac{(\phi\chi_\parallel)^{M-1} - 1}{\phi\chi_\parallel - 1}\right)\right]\right\}, S = A, B, \tag{11}$$

where

$$\chi_\parallel = \left\langle \phi^{2m} \right\rangle \tag{12}$$

is the averaged phase shift over all random numbers m_p of the unit cells in a segment. In a similar way, we obtain

$$\left\langle F_{0S}(\mathbf{Q}) \right\rangle = F_{cS}(\mathbf{Q})\frac{1 - \chi_\parallel}{1 - \phi^2}\frac{1 - (\phi\chi_\parallel)^M}{1 - \phi\chi_\parallel}, S = A, B \tag{13}$$

The averaged phase shift $\chi_\parallel$ can be calculated readily if we assume that the random numbers of m_p follow a Poisson distribution with an average $\langle m_p \rangle \equiv m$.

$$\chi_\parallel = e^{-m(1-\phi^2)}. \tag{14}$$

Different configurations of anti-phase domain boundaries can now be achieved by different combinations of the parameters n_α, n_β and m.

5. EFFECT OF THE ATOMIC DISPLACEMENTS

Bulk semiconductor alloys $A_xB_{1-x}C$ are known to obey Vegard's law quite accurately: the overall lattice constant is the average of the bulk AC and BC lattice constants, weighted by mole fraction. If we imagine the alloy to be a virtual crystal (VC), in that each atom sits on geometrically precise zincblende lattice sites, then its lattice constant can be expressed as

$$a_{ABC} = xa_{AC} + (1-x)a_{BC}.$$

Indeed, measurements[9,10] show that the second-nearest-neighbour distances between group III atoms (or between group V atoms) in the lattice are very nearly those - $a_{ABC}/\sqrt{2}$ - expected for such a VC approximation.

In contrast, however, first-nearest-neighbour distances between group III and group V atoms deviate significantly from, $\sqrt{3}a_{ABC}/4$, expected from VC approximation. It is usual that the A-C (or B-C) bond length is shorter (or longer) than the VC bonds. Therefore, one can imagine that, instead of occupying the VC lattice sites, the atoms will sit in positions slightly deviated from the VC sites, resulting in a microscopic strain in the alloy. Such deviations or microscopic strain have been treated in great detail in the past twenty years, employing mainly what are known as valence force field (VFF) models,[13-15] in which the energies of individual bonds and bond angles are considered to be independent of each other. By minimising the total energy of the typical tetrahedral bond structure in diamond-type semiconductors, the actual bond lengths, or the first-nearest-neighbour distances, and the displacements of the atoms from their VC sites, can be determined. For most of the common semiconductor alloys, values of the actual bond lengths in differently ordered structures can be widely found in the literatures.[13,14,16]

Now, let us discuss the effect of such atomic displacements on the x-ray diffraction intensity. For an ordered crystal, the structure factor, F_{cS}, of a unit cell can be written in a general form

$$F_{cS} = \sum_i f_i(\mathbf{r}_i) e^{2\pi i \mathbf{Q} \cdot (\mathbf{r}_i + \delta_i)} , \tag{15}$$

where,

$$f_i(\mathbf{r}_i) = \bar{f}_i + f(\mathbf{r}_i)$$

is composed of two components, the mean atomic form factor, $\bar{f}_i$, of a disordered crystal at lattice site $\mathbf{r}_i$ in VC approximation, and the deviation of the atomic form factor from $\bar{f}_i$ at $\mathbf{r}_i$ due to the atomic ordering, $f(\mathbf{r}_i)$. δ_i is the displacement of the atom at $\mathbf{r}_i$. The summation in Eq. (15) is over all the atomic sites $\mathbf{r}_i$ in the unit cell. The exponential term in Eq. (15) can be expanded to

$$e^{2\pi i \mathbf{Q} \cdot \mathbf{r}_i} (1 + 2\pi i \mathbf{Q} \cdot \delta_i),$$

assuming that δ_i is much smaller than the VC bond length. To obtain the x-ray intensity, we have to calculate the average of $|F_{cS}|^2$ over all possibilities of atom arrangements due to ordering and all the possibilities of the atomic displacements. If we assume that the atomic form factors for the atoms at different lattice sites and the displacements of the atoms at these sites are not correlated, and further assume that $f(\mathbf{r}_i)$ and δ_i at the same site are independent variables (note that the displacement of an atom at $\mathbf{r}_i$ depends only on its first-nearest-neighbours, and is rather an intrinsic property of the $A_xB_{1-x}C$-type semiconductor alloys than a result of atomic ordering), then we have

$$
\begin{aligned}
\left\langle |F_{cS}|^2 \right\rangle &= \left\langle \sum_i \sum_j f_i(\mathbf{r}_i)f_j^*(\mathbf{r}_j)e^{2\pi i \mathbf{Q}\cdot(\mathbf{r}_i+\delta_i)}e^{-2\pi i \mathbf{Q}\cdot(\mathbf{r}_j+\delta_j)} \right\rangle \\
&\approx \left\langle \sum_i \sum_j f_i(\mathbf{r}_i)f_j^*(\mathbf{r}_j)e^{2\pi i \mathbf{Q}\cdot(\mathbf{r}_i-\mathbf{r}_j)}(1+2\pi i \mathbf{Q}\cdot\delta_i)(1-2\pi i \mathbf{Q}\cdot\delta_j) \right\rangle \\
&= \sum_i \sum_j \left\langle f_i(\mathbf{r}_i)\right\rangle\left\langle f_j(\mathbf{r}_j)\right\rangle^* e^{2\pi i \mathbf{Q}\cdot(\mathbf{r}_i-\mathbf{r}_j)}\left(1+2\pi i \mathbf{Q}\cdot\left\langle\delta_i\right\rangle\right)\left(1-2\pi i \mathbf{Q}\cdot\left\langle\delta_j\right\rangle\right) \\
&\approx \sum_i \sum_j \left\langle f_i(\mathbf{r}_i)\right\rangle\left\langle f_j(\mathbf{r}_j)\right\rangle^* e^{2\pi i \mathbf{Q}\cdot(\mathbf{r}_i+\left\langle\delta_i\right\rangle)}e^{2\pi i \mathbf{Q}\cdot(\mathbf{r}_j+\left\langle\delta_j\right\rangle)}
\end{aligned}
$$

$$(16)$$

Therefore, the effect of the ordering and atomic displacements on the x-ray scattering can be taken into account by replacing the atomic form factors and the atom displacements in Eq. (15) by their averages.

Now Eq. (15) can be re-written as

$$
F_{cS} = \sum_i \left\langle f_i(\mathbf{r}_i)e^{2\pi i \mathbf{Q}\cdot(\mathbf{r}_i+\left\langle\delta_i\right\rangle)} \right\rangle. \tag{17}
$$

It can also be written in the following form

$$
\begin{aligned}
F_{cS} &= \sum_i \left\{ \bar{f}_i + \left\langle f(\mathbf{r}_i)\right\rangle + 2\pi i \mathbf{Q}\cdot\left\langle\delta_i\right\rangle\left(\bar{f}_i + \left\langle f(\mathbf{r}_i)\right\rangle\right) \right\}\cdot e^{2\pi i \mathbf{Q}\cdot\mathbf{r}_i} \\
&= F_0 + F_s + F_d,
\end{aligned}
\tag{18}
$$

where,

$$
\begin{aligned}
F_0 &= \sum_i \bar{f}_i e^{2\pi i \mathbf{Q}\cdot\mathbf{r}_i} \\
F_s &= \sum_i \left\langle f(\mathbf{r}_i)\right\rangle e^{2\pi i \mathbf{Q}\cdot\mathbf{r}_i} \\
F_d &= 2\pi i \sum_i \mathbf{Q}\cdot\left\langle\delta_i\right\rangle\left(\bar{f}_i + \left\langle f(\mathbf{r}_i)\right\rangle\right) e^{2\pi i \mathbf{Q}\cdot\mathbf{r}_i}
\end{aligned}
\tag{19}
$$

are the contributions to the total structure factor from a perfect VC, from atomic ordering and from atomic displacements, respectively. The x-ray intensity is calculated by

$$
\begin{aligned}
|F_{cS}|^2 &= (F_0 + F_s + F_d)(F_0 + F_s + F_d)^* \\
&= F_0 F_0^* + F_s F_s^* + 2\operatorname{Re}(F_0 F_s^*) + 2\operatorname{Re}(F_d F_0^*) + 2\operatorname{Re}(F_d F_s^*) + F_d F_d^*
\end{aligned}
$$
(20)

The first term in Eq. (20) gives the fundamental reflections. The second term is caused by the atomic ordering. The third term is zero because $\sum_i f(\mathbf{r}_i) = 0$. The fourth and the fifth terms result in diffuse intensities around the fundamental and the ordering reflections, respectively. The last term, though much weaker than the other terms, represents the contribution purely due to the atomic displacements. In our case, because the displacements are also ordered, as we will show in the following discussion, the atomic displacements bring in an additional modulation to the ordering reflections. Therefore, the intensity of an ordering reflection from a single variant domain is

$$
I_{order} \propto |F_s|^2 + 2\operatorname{Re}(F_d F_s^*) + |F_d|^2
$$
(21)

In addition to the bond-length-difference induced permanent atom displacements, thermal vibration of the atoms causes a random displacement. Effect of this thermal vibration on the x-ray structure factor is considered via Debye-Waller factors. Here, the Debye-Waller factors were calculated from the estimated Debye temperatures,[17,18] using the Debye model, resulting in $B_{InP}=0.2799$, $B_{GaP}=0.3001$, $B_{InAs}=0.206$, and $B_{AlAs}=0.283$.

Before calculating the average atomic form factor and the average atom displacements, let us introduce the well-accepted Bragg-Williams long-range order parameter, s.[8] If we define $\gamma_A(\gamma_B)$ and $\omega_A(\omega_B)$ as the fraction of A-sites (B-sites) occupied by the right and wrong atoms, respectively, we then have

$$
s = \gamma_A + \gamma_B - 1 = \gamma_A - \omega_B = \gamma_B - \omega_A
$$
(22)

For simplicity, here we will discuss only the case in which the mole fractions of A and B atoms in the alloy are equal. The results, however, can easily be extended to the general case where A and B atoms have different mole fractions. Bearing this in mind, the possibilities of finding the right and wrong atoms at group III lattice sites are $P_\eta = \gamma_\eta/(\gamma_\eta + \omega_\eta)$ and $1 - P_\eta$, respectively, where $\eta = A, B$. Therefore, the average atomic form factors for A-sites and B-sites are

$$\bar{f}(A) = \frac{1}{2}(f_A + f_B) + \frac{s}{2}(f_A - f_B)$$
$$\bar{f}(B) = \frac{1}{2}(f_A + f_B) + \frac{s}{2}(f_B - f_A) \tag{23}$$

The calculation of the average atomic displacements is more complicated. In an $A_xB_{1-x}C$ alloy, VFF model calculation indicates that mainly the group V atoms have been displaced to accommodate the bond length difference. In a disordered phase, the group III atoms A and B are randomly distributed, so that the average atom displacement for each lattice site is zero. If A and B atoms form CuPt-type ordering, however, they will prefer to reside on alternating ($\bar{1}11$) or ($1\bar{1}1$) lattice planes. Consequently, the displacements of the atoms C form a periodic pattern, they move closer to either the A-rich lattice plane or the B-rich lattice plane depending on which of the A-C and B-C bonds is shorter, as shown schematically in Fig. 7a. In this way, we can sort the atoms C into two types, type-1 and type-2, as indicated in Figs. 7a. For each tetrahedral structure with the atom C sitting in the centre, there are 16 possibilities for displacing this atom depending on the occupation of the four atoms A and B bonding directly to it (Fig. 7b). Considering all these 16 different atom arrangements and their possibilities, we obtain finally the magnitude of the average displacements of type-1 and type-2 lattice C planes as following

$$|\delta| = \left[\left(\frac{1+s_\tau}{2}\right)^4 - \left(\frac{1-s_\tau}{2}\right)^2 \right] d_t + \left(\frac{1+s_\tau}{2}\right)\left(\frac{1-s_\tau}{2}\right)\left[\left(\frac{1+s_\tau}{2}\right)^2 \left(\frac{1-s_\tau}{2}\right)^2 \right] d_s \tag{24}$$

where, $\tau=1,2$ corresponding to the variants I and II, respectively, d_t and d_s are parameters determined from a separate VFF model calculation. For different materials, they have different values. Since this calculation is out of the scope of this chapter, readers are recommended to the Refs. 13 to 15 for details. The directions of the average displacements are perpendicular to the ordered {111} lattice planes. In Sec. 6, we will show how this displacement influences the diffraction profiles.

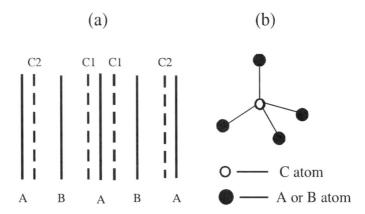

Figure 7. (a) Schematic diagram of the ordered {111} lattice planes considering the atomic displacements due to the bond length difference between A-C and B-C of a $A_{0.5}B_{0.5}C$ alloy. (b) The tetrahedron bonding in a diamond type structure.

6.　　EXAMPLES OF DATA ANALYSES

In this section, we give some examples of x-ray diffraction measurements and data analyses using the approaches discussed in the above sections. We will discuss the experimental results and theoretical fitting results for several $Ga_{0.5}In_{0.5}P$ and $Al_{0.5}In_{0.5}As$ films, which are quite representative of the CuPt-type ordered III-V semiconductor alloys. Samples MA776 and MA912 are $Ga_{0.5}In_{0.5}P$ films grown on GaAs (001)-6°A substrates. Sample K782 is a $Ga_{0.5}In_{0.5}P$ film grown on an GaAs(001)-6°B substrate. Sample R286 is a $Al_{0.5}In_{0.5}As$ film grown on an exactly oriented InP (001) substrate. Details of the growth of these films were described in Refs. 19 and 20.

6.1　　Single-variant ordered films

It is well-known that a film grown on a substrates miscut towards the (111)-B direction would results in single variant in the film.[21-26] In such a case, only one set of the ordering reflections were observed by TED. No wavy or tilted ordering peaks were observed in such films. Our x-ray measurements on sample K782 confirm that only one variant appears in this film grown on a 6°B substrate. The measured reciprocal area map of this sample around ($\bar{3}$/2, 3/2, 5/2) ordering reflection is shown in Fig. 8a. The intensities of the fundamental reflections and their tails in the measured area have been subtracted from the map in order to highlight the weak ordering peak. This procedure has been performed on all the data discussed below. The peak in Fig. 8a does indeed look close to a circle as already noticed by TED observations. Figure 8b is the calculated ($\bar{3}$/2, 3/2, 5/2) reflection.

Fig. 8c shows the cross-section at $\mathbf{Q}\|[001]=2.5$. The best agreement between the calculated and measured data was achieved by using a domain size of about 500 crystallographic unit cells. Here, it is worth pointing out that the intensities of the ordering reflections have been modulated by the effect of atomic displacements as discussed in section 5. As an example, Fig. 9 shows the calculated x-ray radial scans from reciprocal lattice points ($\bar{1}$ 11) to ($\bar{3}$ 33) for a perfectly ordered large single variant domain. The central peak in each panel is the fundamental ($\bar{2}$ 22) reflection. Fig. 9a is the result without taking into account the atomic displacements. Naturally, the ($\bar{3}$/2,3/2,3/2) reflection is stronger than the higher order ($\bar{5}$/2,5/2,5/2) reflection, as the atomic form factor decreases with increasing scattering vector. In Fig. 9b, scattering due to the atomic displacements is shown, which gives a modulation to the above profile. In Fig. 9c, the total intensity profile considering the atom displacements is shown. The ($\bar{5}$/2,5/2,5/2) reflection here is considerably stronger than the ($\bar{3}$/2,3/2,3/2) reflection.

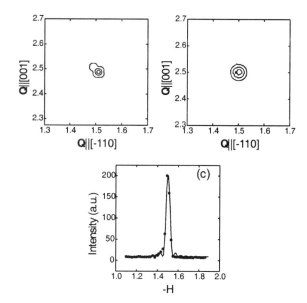

Figure 8. X-ray reciprocal space area maps taken from a single-variant ordered film GaInP$_2$ film K782. (a) Experimental; (b) calculated; (c) cross-sectional line scans at $\mathbf{Q}_\|[001]=2.5$.

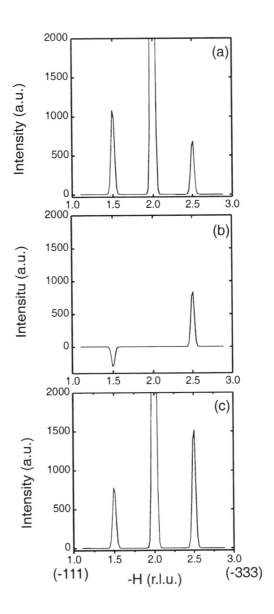

Figure 9. Calculated x-ray line scans from reciprocal lattice point ($\bar{1}$11) to ($\bar{3}$33). (a) Without taking into account the atomic displacements; (b) modulation caused by atomic displacements; (c) results of (a)+(b).

6.2 Double-variant ordered films

Fig. 10a is an experimental x-ray area scan in reciprocal space taken from sample MA776 around the ordering reflections ($\bar{7}$/2, 7/2, 5/2) and ($\bar{7}$/2, 7/2, 7/2). It is clear that both ordering peaks are inclined away from the [001] direction by an angle of about 10°. This feature is quite similar to the typical TED data shown in Fig. 1. It is easy to recognise that the ($\bar{7}$/2, 7/2, 7/2) and ($\bar{7}$/2, 7/2, 5/2) reflections are contributions of the ordering variant I and II, respectively. We also note that the reflection ($\bar{7}$/2, 7/2, 7/2) is much stronger than the reflection ($\bar{7}$/2, 7/2, 5/2). However, we must not conclude from this feature that the fraction of the variant I in this sample is larger than that of the variant II. Fig. 10b shows a calculated x-ray intensity area map using the parameters listed in Table I. Both maps agree qualitatively quite well, but for a detailed comparison of the calculation and the experiment, area maps are not suitable. In Figs. 11a and 11b, two cross-sections at Q||[001]=2.5 and 3.5 are shown. We see that the theory yields a quite good fit to the experimental data. In Fig. 12, a fit, using the same parameters, to a line scan from reciprocal lattice point ($\bar{1}$ 13) to ($\bar{3}$ 35) is given. Again the measured and calculated data agree quite good.

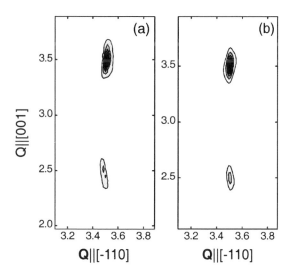

Figure 10. X-ray reciprocal space area scans of sample MA776. (a) Experimental; (b) calculation. The peaks centred at ($\bar{7}$/2,7/2,7/2) and ($\bar{7}$/2,7/2,5/2) are attributed to variants I and II, respectively.

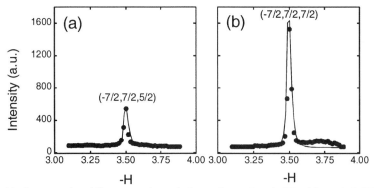

Figure 11. Cross-sectional line scans through the peak maxima in Fig. 10 at (a) $Q_{\parallel}[001]=2.5$ and (b) $Q_{\parallel}[001]=3.5$.

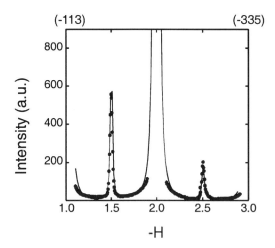

Figure 12. X-ray line scans from reciprocal lattice point ($\bar{1}\,13$) to ($\bar{3}\,35$). Solid dots and lines are measured and calculated results, respectively.

Table 1. The structural parameters used to fit the measured data of samples MA776, MA912, and R286.

Sample	n_α	n_β	m	s_1	s_2
MA776	7	8	60	0.35±0.05	0.45±0.05
MA912	12	12	140	0.50±0.05	0.50±0.05
R286	4	4	~	0.30[*]±0.05	0.30[*]±0.05

[*]For ordered domains only.

Fig. 13a shows the experimental area map taking from sample MA912. Fig. 13b is the calculated one. The structural parameters used for this calculation are listed in Table I. To view quantitatively the fitting results, several line scans along either [$\bar{1}$ 11] or [1 $\bar{1}$ 1] directions are shown in Fig. 14, namely, scans from the reciprocal lattice point ($\bar{1}$ 11) to ($\bar{3}$ 33), (002) to ($\bar{2}$ 20), ($\bar{1}$ 13) to ($\bar{3}$ 31), and (002) to ($\bar{2}$ 24). Good agreement has been obtained between the several experimental and simulated curves.

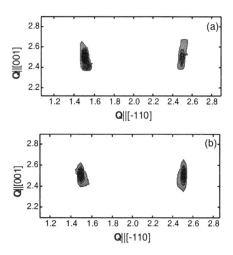

Figure 13. X-ray reciprocal space area scans of sample MA912. (a) Experimental; (b) calculation. The peaks centered at ($\bar{5}$/2,5/2,5/2) and ($\bar{3}$/2,3/2,5/2) are attributed to variants I and II, respectively.

From the calculations, we note that the widths of the ordering reflections are determined by the combination of lateral anti-phase domain size and the laminar thicknesses, i.e., the overall configuration of the anti-phase domain boundaries. This is also true for the tilt angle of the ordering reflections. To show this point more clearly, a series of calculated area scans using a $Ga_{0.5}In_{0.5}P$ film as the model structure are shown in Figs. 15 and 16. By varying the average thicknesses of the laminae of the two variants and the lateral domain size, we are able to obtain quite different diffraction patterns. From Fig. 15, we notice that when the anti-phase domains are very small in size, each ordering reflection has split into two. The smaller the anti-phase domains, the greater the separation of the two split maxima is. In fact, such splitting of ordering reflections and shifts of intensity maxima have been observed by TED on poorly ordered films[5] (the TED pattern in Fig. 1 is an example of such a case). It is also clear that when the anti-phase domains are very large, the ordering reflections become very narrow and lie almost parallel to the [001] direction. This fact is further confirmed by Fig. 16d, which represents an area scan for a two-variant structure without lateral anti-phase domains but with statistically distributed laminae thicknesses. In this case, the ordering reflection is elongated simply along the [001] direction, but is no longer tilted to form the wavy pattern. Therefore, dense anti-phase domain boundaries are responsible for the tilted reflection peaks.

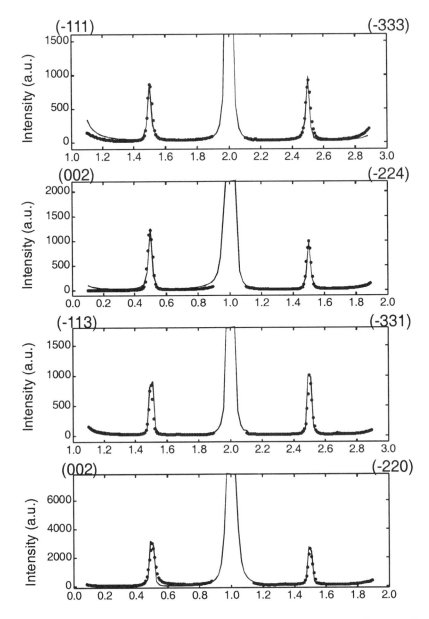

Figure 14. X-ray line scans from one reciprocal lattice point to another. (a) ($\bar{1}$ 11) to ($\bar{3}$ 33); (b) (002) to ($\bar{2}$ 24); (c) ($\bar{1}$ 13) to ($\bar{3}$ 31); (d) (002) to (2 $\bar{2}$ 0). Solid dots and lines are measured and calculated data, respectively.

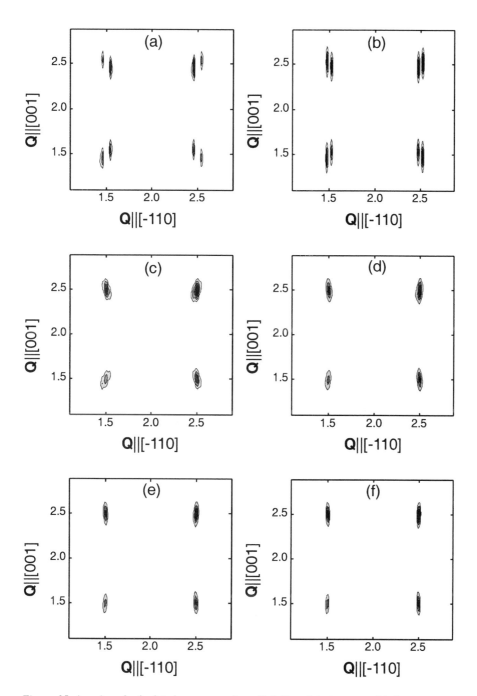

Figure 15. A series of calculated area scans for a GaInP model structure with the parameters $n_\alpha=n_\beta=8$ and (a) $m=5$, (b) $m=10$, (c) $m=20$, (d) $m=40$, (e) $m=60$, (f) $m=200$.

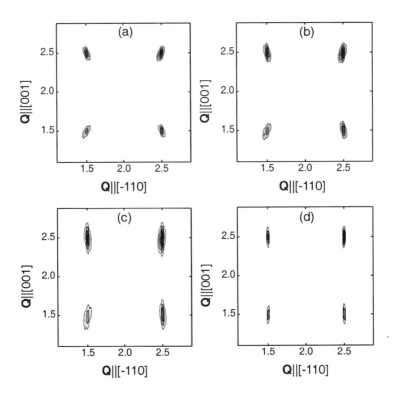

Figure 16. Calculated area scans for the same structure as in Fig. 15 but with the parameters m=20 and (a) $n_\alpha=n_\beta=12$, (b) $n_\alpha=n_\beta=8$, (c) $n_\alpha=n_\beta=4$. (d) is calculated by using $n_\alpha=n_\beta=8$ and assuming that the film is free of lateral anti-phase domain boundaries.

Not all of the films, however, are well-ordered, by which we mean that the two ordered variants are complementary in the film. Some films grown under certain conditions may be just poorly ordered, in which ordered domains are embedded in the disordered matrix,[6,27] forming what we call it here a short-range ordering. In this case, TEM study has shown that the ordered domains have finite lateral size, but in the [001] direction, the ordering develops quite well. Overall, these ordered domains are characterised by a columnar structure.[6] Based on the theory first developed by Greenholz and Kidron,[28] in which the interference between the scattering from different ordered domains is neglected, we can calculate the x-ray scattering intensity from a columnar structure by keeping the same vertical structural model as we used above for the well ordered materials. Fig. 17a is a reciprocal space area scan taken from sample R286. This sample was shown to exhibit short-range-order from independent TED and x-ray measurements.[20] The general feature of this figure, particularly the inclination of the streaks passing through the ($\bar{h}$/2,k/2,l/2) positions, is quite similar to that of the well ordered materials. The elongated streaks are clear indications of small ordering domains in the film. A new feature here is the presence of the streaks parallel to [001], running between the ordering reflections. (The ridge of intensity between ($\bar{3}$31) and ($\bar{1}$13) is due to the high density of planer stacking faults. See Ref. 20 for details.) Fig. 17b is the calculated area map using the columnar structure model and the parameters listed in Table I. It fits qualitatively quite well to Fig. 17a considering the shape and tilt of the ordering reflections, as well as the [001] streaks between them. In fact, the quantitative fit is also good if the cross-sectional line scans running through the ordering peaks are inspected (Fig. 18). We see from Fig. 18 that both the line shape and intensity of the calculated curves are in good agreement with their experimental counterparts. Note that no anti-phase domain boundaries were considered in the columnar model. This implies that anti-phase domain boundaries are not necessary to cause the wavy diffraction pattern for the poorly ordered films. The same conclusion has also been reached by Yang et. al. in a recent work.[6] The stronger ($\bar{5}$/2, 5/2, 5/2) reflection in comparison with the ($\bar{3}$/2,3/2,3/2) reflection is again due to the atomic displacements.

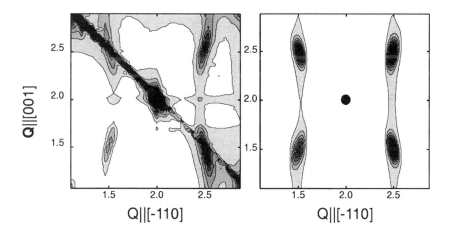

Figure 17. X-ray reciprocal space area scans of sample R286. (a) Experimental; (b) calculated by using a columnar structural model.

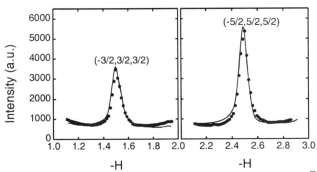

Figure 18. Cross-sectional line scans through the reciprocal lattice points ($\bar{3}$/2,3/2,3/2) and ($\bar{5}$/2,5/2,5/2) in Fig. 17. Solid dots and lines are measured and calculated data, respectively.

7.   DETERMINATION OF THE ORDER PARAMETER

Conventionally, the order parameter is determined by comparing the intensities of the ordering and fundamental reflections weighted by their structure factors.[8,20] Even without considering the structural details of the ordered material, this method has already encountered several difficulties when applied to the CuPt-type ordered thin films. For example, precise determination of the intensity of a fundamental reflection is often difficult. This is because the film is usually lattice matched to the substrate material, and the reflections from the substrate and the film overlap one another. Many researchers, including us, however, have used this method in earlier studies.[29,30] There is no doubt that this method will produce correct values for good single crystals of a pure variant if the intensity of the fundamental reflections can be determined correctly.

However, as we have discussed in the previous sections, the intensity of the ordering reflection is not only a function of the order parameter, but is also modulated by the atomic displacements. The almost inevitable presence of anti-phase domains in the ordered phase leads to an additional phase factor in the structure factor calculation, which may again have significant impact on the intensity of the ordering reflections. In order to determine correctly the order parameter, all these factors need to be considered. The best way to do so, therefore, is to fit the experimental x-ray profiles using a model which takes into account all these factors. Fig. 19 shows, as an example, detailed theoretical fittings to the measured x-ray intensity profiles for sample MA776. It is seen that the best fit was obtained with the average order parameters of 0.35±0.05 for variant I and 0.45±0.05 for variant II. We also note that the calculated and measured curves deviate considerably at the tails of the fundamental reflections. We believe this is caused by the fact that the actual structure of the film is more complicated than our model which permits including static displacement diffuse scattering near the fundamental peaks (Huang scattering). In addition, the actual film may contain defects such as diffusive anti-phase domain boundaries,[5] stacking faults,[20] and alloy clustering[6,31] etc. The interface of the neighbouring laminae of the two variants may also not be abrupt as we considered in our model. It is worth pointing out that the order parameter of sample MA776, determined from direct comparison of the intensities of the ($\bar{1}/2,1/2,3/2$) ordering reflection and the (004) fundamental reflection using the methods discussed in Ref. 29, is about 0.06 for both variants, taking the volume fractions of the two variants determined in section 4.

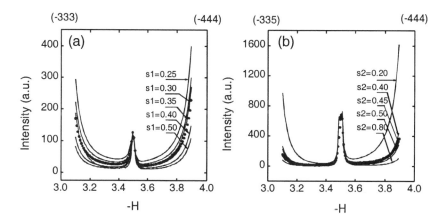

Figure 19. Theoretical simulations of two line scans of sample MA776 from reciprocal lattice point ($\bar{3}$ 33) to ($\bar{4}$ 44) and from ($\bar{3}$ 35) to ($\bar{4}$ 44). The best fits of the two curves yield an order parameter of 0.35 for variant I and an order parameter of 0.45 for variant II.

Finally, we note that the precise determination of the order parameter for the complex ordered film by diffraction methods is still a challenge. Our model assumes that the laminar structure runs uniformly across the wafer, and the number of atomic layers in a single lamina follows a certain distribution. Such assumptions may not be always true. Electron diffraction may indeed be better in determining the local order parameter if one can focus the electron beam onto a single-phase single variant domain.

8. SUMMARY

On the basis of a theory of kinematic diffraction and a structure model of CuPt-type ordering in III-V compound semiconductors considering the laminae structure of two alternating variants and the random distribution of anti-phase domain boundaries in each variant, the reciprocal space distribution of the x-ray intensities scattered from CuPt-type ordered $A_xB_{1-x}C$ materials were calculated. Atomic displacements due to the bond length difference between A-C and B-C has to be considered in order to obtain correct intensity distribution, because the atomic displacements give an additional modulation to the intensity of the ordering reflections. Several samples of $Ga_{0.5}In_{0.5}P$ and $Al_{0.5}In_{0.5}As$ have been studied experimentally employing synchrotron x-ray radiation. By comparing the experimental and the calculated x-ray reciprocal area maps, the structural parameters of the films, such as the mean domain size, the mean thickness of the laminae of both variants and the average order parameter have been determined.

We must mention that the model discussed in this chapter does not explain all the diffraction phenomena of the CuPt ordered films. Further work currently underway will consider the three-dimensional distribution of the anti-phase domain boundaries. We expect that the new model will explain more experimental phenomena which are not presented in this review.

ACKNOWLEDGEMENTS

I would like to thank Professor Simon Moss for his encouragement during this work and his thoughtful reading of the manuscript. Thanks are also given to Professor V. Holy for his useful discussions, and Drs. A.G. Norman and A. Mascarenhas for providing the samples. This work is supported by NSF on DMR-9729297 and the Texas Center for Superconductivity at the University of Houston (TcSUH).

REFERENCES

1   V. Holý, U. Pietsch, and T. Baumbach, *High resolution X-ray Scattering From Thin Films and Multilayers*, (Springer, New York, 1999).
2   T. Suzuki, T. Ichihashi, and T. Nakayama, Appl. Phys. Lett. **73**, 2588 (1998).
3   B.A. Philips, A.G. Norman, T.Y. Seong, S. Mahajan, G.R. Booker, M. Skowronski, J.P. Harbinson, and V.G. Keramidas, J. Cryst. Growth **140**, 249 (1994).
4   A. Gomyo, K. Makita, I. Hino, and, T. Suzuki, Phys. Rev. Lett. **72**, 673 (1994).
5   C.S. Baxter, W.M. Stobbs, and J.H. Wilkie, J. Cryst. Growth **112**, 373 (1991), and references therein.

6   J.-J. Yang, R. Spirydon, T.-Y. Seong, S.H. Lee, and G.B. Stringfellow, J. Electron. Mater. **27**, 1117 (1998).
7   D. Munzer, E. Dobrocka, I. Vavra, R. Kudela, M. Harvanka, N.E. Christensen, Phys. Rev. B **57**, 4642 (1998).
8   B.E. Warren, *X-ray diffraction*, (Dover, 1990).
9   D.C. Meyer, K. Richter, and P. Paufler, Phys. Rev. B **59**, 15253 (1999).
10 L. Alagna, T. Properi, S. Turchini, C, Ferrari, L. Francesio, and P. Franzosi, J. Appl. Phys. **83**, 3552 (1998).
11 E. Morita, M. Ikeda, O. Kumagai, and K. Kaneko, Appl. Phys. Lett. **53**, 2164 (1988).
12 G.S. Chen, D.H. Jaw, and G.B. Stringfellow, J. Appl. Phys. **69**, 4263 (1991).
13 Y. Cai and M.F. Thorpe, Phys. Rev. B {\bf 46}, 15872 (1992); Phys. Rev. B **46**, 15879 (1992).
14 J. S. Chung and M.F. Thorpe, Phys. Rev. B **55**, 1545 (1997).
15 J.Y. Tsao, *Materials Fundamentals of Molecular Beam Epitaxy*, (Academy Press, San Diego, 1993), chapter 4.
16 A.-B. Chen and A. Sher, *Semiconductor Alloys*, (Plenum Press, New York, 1995).
17 S. Adachi, *Physical Properties of III-V Semiconductor Compounds*, (Wiley, New York, 1992).
18 S. Adachi, *Properties of Aluminum Gallium Arsenide*, (INSPEC, London, 1993).
19 Y. Zhang, A. Mascarenhas, S.P. Ahrenkiel, D.J. Friedman, J.F. Geisz, and J.M. Olsen, Solid State Commun. **109**, 99 (1999).
20 R.L. Forrest, J. Kulik, T.D. Golding, and S.C. Moss, J. Mater. Res. **15**, 45 (2000).
21 D.H. Jaw, G.S. Chen, and G.B. Stringfellow, Appl. Phys. Lett. **59**, 114 (1991).
22 Y. Zhang, B. Fluegel, S.P. Ahrenkiel, D.J. Friedman, J.F. Geisz, J.M. Olsen, and A. Mascarenhas, 1999 MRS Fall Meeting Proceedings.
23 L.C. Su, I.H. Ho, and G.B. Stringfellow, J. Appl. Phys. **75**, 5135 (1994).
24 I.J. Murgatroyd, A.G. Norman, and G.R. Booker, J. Appl. Phys. **67**, 2310 (1990).
25 T. Suzuki and A. Gomyo, J. Cryst. Growth **93**, 396 (1988).
26 A. Gomyo, T. Suzuki, and S. Iijima, Phys. Rev. Lett. **60**, 2645 (1988).
27 T.-Y. Seong, A.G. Norman, G.R. Booker, and A.G. Cullis, J. Appl. Phys. **75**, 7852 (1994).
28 M. Greenholz and A. Kidron, Acta, Cryst. A **26**, 311 (1970).
29 R.L. Forrest, T.D. Golding, S.C. Moss, Z. Zhang, J. F. Geisz, J.M. Olsen, A. Mascarenhas, P. Ernst, and C. Geng, Phys. Rev. B **58**, 15355 (1998).
30 C.S. Baxter and W.M. Stobbs, Phil. Mag. A **69**, 615 (1994).
31 S. Matsumura, K. Takano, N. Kuwano, and K. Oki, J. Cryst. Growth **115**, 194 (1991).sp

Chapter 7

Diffraction and Imaging of Ordered Semiconductors
Transmission Electron Microscopy Experiment and Theory

S. P. Ahrenkiel
National Renewable Energy Laboratory, 1617 Cole Blvd., Golden, CO 80401 USA

Key words: semiconductors, ordering, transmission electron microscopy

Abstract: Transmission electron microscope diffraction and imaging of the ordered
             semiconductors $Ga_{0.5}In_{0.5}P$, $Ga_{0.5}In_{0.5}As$, and $CuInSe_2$ are discussed.

1.   INTRODUCTION

The interpretation of optical and transport data from ordered materials and the refinement of growth procedures to control ordering require detailed structural characterization. Transmission electron microscopy (TEM) provides a powerful repertoire of techniques that are commonly used for microstructural and crystallographic analyses of ordering in compound semiconductors and their alloys. Basic TEM operational modes, epitaxy, and semiconductor microstructure are first discussed. TEM diffraction patterns and images from ordered films are then displayed. Lastly, dynamical diffraction and the analysis of data from ordered structures are discussed.

1.1   TEM

TEM uses electrons emitted from a hot-cathode or field-emission source that are accelerated to high energy and projected through a high-vacuum column to obtain structural information from solid samples prepared as thin foils (typically less than one μm thick) [1]. Electron-optical lenses (current-carrying coils) with variable focal length provide a wide magnification range. Focusing occurs in the cylindrically symmetric magnetic field (with axial and radial components) of a lens as a result of the radial component of

195

the Lorentz force on paraxial electrons that is proportional to the circular velocity component; the image is generally rotated about the optic axis from the object. Apertures, metallic slides with circular bores of various diameter, are used to illuminate the sample with a partially coherent beam of source electrons and form an image on a fluorescent screen, film plate, or digital camera.

Diffraction and imaging operational modes differ primarily in post-specimen lens/aperture configurations. Image contrast results from diffracted-beam superposition in the image plane (*Figure 1*). A bright-field (BF) image using the reflection g associated with the reciprocal lattice vector **g** is usually acquired with a small objective aperture (*OA*) around the direct beam g and with the sample tilted such that **g** is oriented near a desired excitation condition. A tilted dark-field (DF) image using g is usually acquired at a chosen excitation with the incident beam tilted by the scattering angle so that g is on the optic axis, and with g removed by the *OA*. High-resolution (HR) lattice images are usually acquired on

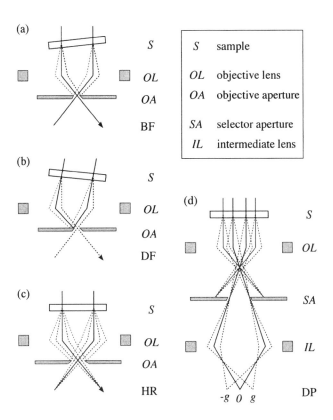

Figure 1. Simplified TEM operational modes: (a) bright-field, (b) dark-field, (c) high-resolution lattice imaging; (d) selected-area diffraction pattern.

High-symmetry zone axes by allowing the interference of 0 with several low-index reflections. (High-order reflections may be omitted by the *OA* to improve contrast.) The selected-area diffraction pattern (DP) is viewed by positioning a *SA* in the image plane with supplementary optics. The pattern is often acquired on a low-index zone axis to reveal the crystal symmetry from the spot positions, but it can also be viewed off-axis to obtain particular diffraction conditions for BF and DF imaging.

1.1.1 Diffraction

With parallel illumination, the incident beam is a plane wave:

$$\psi(\mathbf{r}) = e^{2\pi i \mathbf{k} \cdot \mathbf{r}},$$

(1)

ignoring back-scattered beams. The exit wave function below a perfect crystal is a series of diffracted beams:

$$\psi(\mathbf{r}) = \sum_{\mathbf{g}} \Psi_{\mathbf{g}} e^{2\pi i (\mathbf{k} + \mathbf{g} + \mathbf{s}_{\mathbf{g}}) \cdot \mathbf{r}},$$

(2)

including the direct beam 0 ($\mathbf{g} = \mathbf{0}$). The diffracted wave vectors satisfy the elastic-scattering condition $k = |\mathbf{k} + \mathbf{g} + \mathbf{s}_{\mathbf{g}}|$, where $\mathbf{s}_{\mathbf{g}}$ is the *excitation error*. One can collect divergent diffracted-beam (spot) intensities without interference:

$$I_{\mathbf{g}} = |\Psi_{\mathbf{g}}|^2.$$

(3)

The DP symmetry and scale, rather than quantitative measurements of integrated spot intensities, are often sufficient for the determination of simple crystal structures.

1.1.2 Image Formation

Conventional TEM images using coherent illumination (as opposed to scanning electron microscope [SEM] and scanning TEM images, which raster small probes over the image area) can often be interpreted as interference patterns of diffracted beams and low-frequency satellites. Whereas HR images are usually acquired on-axis, BF and DF images are acquired near the Bragg condition ($s_{\mathbf{g}} = 0$).

The local image intensity [2] is approximately

$$I(\mathbf{r}, \Delta f) \approx |\sum_{\mathbf{g}} P(\mathbf{g}, \Delta f) \Psi_{\mathbf{g}} e^{2\pi i \mathbf{g} \cdot \mathbf{r}}|^2,$$

(4)

where Δf is the *OL* defocus. Included in the contrast transfer function $P(\mathbf{g}, \Delta f)$ are magnitude and phase corrections that result from spherical aberration, spatial and temporal incoherence, *OA* masking effects, and sample thickness and position.

Crystallographic information from ordered phases is often obtained by viewing the selected-area DP from a region of interest. An idealized, periodic superlattice of double-variant stacked CuPt-ordered domains (*Figure 2*) is considered. In addition to the matrix diffraction spots, ordering generates spots G_1 and G_2 associated with reciprocal lattice vectors $\mathbf{G}_1$ and $\mathbf{G}_2$, respectively, which are unique to each variant. The finite variant domain sizes give rise to satellite components that correspond to low-frequency structural variations. One can form HR images of the crystal structure at high magnification by introducing an *OA* of adequate size to accept a collection of spots and their satellites. DF images acquired with the matrix spots do not usually reveal information on the distribution of ordering (unless ordering alters the local crystal matrix). One can, however, produce DF images of each ordered variant by positioning a smaller *OA* around their characteristic spots and satellites.

1.1.3   Image Processing

The periodic components of HR images can be enhanced by image *filtering*. One approach is to invert a masked fast Fourier transform (FFT) of the unprocessed image intensity. A raw image with (real) intensity $I(\mathbf{r})$ gives a Hermitian FFT $I(\mathbf{g})$. With the masking function $O(\mathbf{g})$ ($0 \le O(\mathbf{g}) \le 1$) and a filtering parameter f ($0 \le f \le 1$), one obtains the masked FFT:

$$I'(\mathbf{g}) = \{1 - f[1 - O(\mathbf{g})]\}\, I(\mathbf{g}) \tag{5}$$

which is non-Hermitian if $O(\mathbf{g}) \ne O(-\mathbf{g})$. An inverse FFT operation generates the complex intensity $I'(\mathbf{r})$; the modulus is a real, filtered image.

Image *mixing* is also used to combine complementary domain structures from DF images of the same region into a single image. Two images A and B can be combined as a mixed image:

$$C = (1 - \chi)|A + B| + \chi|A - B| \tag{6}$$

The mixing parameter χ ($0 \le \chi \le 1$) enhances similarities ($\chi = 0$) and differences ($\chi = 1$) between overlapping domains.

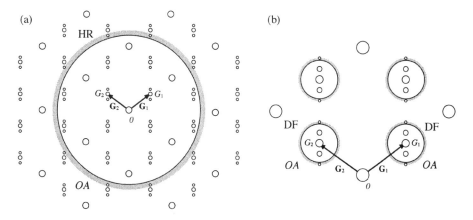

Figure 2. Diffraction plane in (a) HR and (b) DF imaging modes for periodically stacked, double-variant CuPt ordering. The circles represent typical *OA* sizes and positions.

1.2 Epitaxy

This chapter explores the microstructures of the epitaxial semiconductor alloys $Ga_{0.5}In_{0.5}P$ (GaInP) and $Ga_{0.5}In_{0.5}As$ (GaInAs) grown by atmospheric- and low-pressure metal-organic chemical vapor deposition (MOCVD) at the National Renewable Energy Laboratory (NREL). Analysis is then presented of $CuInSe_2$ grown by a novel migration-enhanced epitaxy (MEE) technique developed by B. J. Stanbery at the University of Florida (UF). Whereas chemical precursors are used in MOCVD to flow deposited elements to the growth surface, the MEE technique sequentially exposes the growth surface to elemental sources. As in molecular beam epitaxy (MBE), MEE generally involves lower growth temperatures and rates than MOCVD, permitting greater control of non-equilibrium phases, including synthetic generation of ordered phases via sequenced growth [3].

A variety of the adamantine compounds and their alloys [4] are used in optoelectronic device applications. The *elemental* semiconductors Si and Ge and III-V *binary* compound semiconductors GaAs and InP are among the most widely incorporated into heterostructures, or used as substrate materials for epitaxial growth. Si and Ge have the diamond structure, with two interpenetrating *fcc* sublattices separated by a⟨111⟩/4. Both GaAs and InP are zincblende; the two *fcc* sublattices are separately occupied by cations (Ga, In) and anions (As, P).

Substrate miscuts from ⟨001⟩ directly influence surface composition and step configurations. The loss of inversion symmetry at the semi-infinite, epitaxial growth surface can remove the equivalence among crystallographic directions, with distinct manifestations on growth processes and film nucleation. On the zincblende (001) surface, the emerging directions normal to the two sets of cation- (anion-) terminated {111} planes are termed A (B) (*Figure 3*). Intermediate directions are labeled AB.

1.3   Ordering

Semiconductor alloys offer great versatility and tunability in optoelectronic and structural properties for device engineering. The *pseudobinary* alloys contain isoelectronic elements distributed within a common *fcc* sublattice of the zincblende structure. Atomic ordering of the alloyed species often alters the physical properties of these alloys with respect to disordered (random) alloys at the same compositions.

Bulk, *ternary* materials, such as the I-III-VI$_2$ compound $U_{hk\ell} = 0$ and the (001) compound (100) are usually sphalerite at high quenching temperatures [5,6], with random distributions of the minority elements on a single *fcc* sublattice. The long-range sphalerite structure is identical to zincblende. However, the variable local environments in these disordered materials can present point defects. Ordered phases usually occur at lower temperatures [7] and provide uniform chemical coordination of the majority species. *Pseudoternary* alloys [e.g., Cu(Ga,In)Se$_2$] may also exhibit ordering under proper conditions, with increased structural complexity.

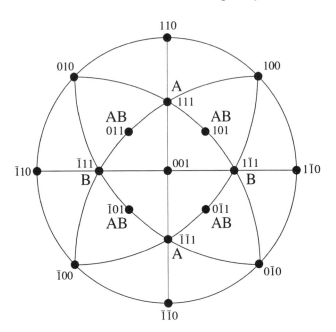

Figure 3. [001] stereographic projection of cubic crystalline directions. A (B) surfaces in multinary tetrahedral semiconductors are terminated with cations (anions).

The metallurgical CuPt alloy has a random *fcc* structure at high temperatures, but quenching at low temperatures produces a rhombohedral phase with ½{111} chemical ordering [8,9] (*Figure 4*). The CuPt-B structure is among the most widely discussed and accessed forms of spontaneous ordering in III-V alloys, and occurs in both common anion (e.g., GaInP [10], GaInAs [11]) *and* common cation (e.g., GaAsSb [12], GaAsP [13]) alloys grown by MOCVD. (CuPt-A ordering also occurs in III-V alloys grown by MBE [14].) The CuPt structure is thermodynamically metastable in most bulk semiconductor alloys; the CuPt-B variants originate by near-surface thermodynamic or kinetic processes in films grown on singular- and vicinal- ⟨001⟩ substrates. The ordering may be driven by surface reconstruction mechanisms [15-17]. Spontaneous CuPt ordering in MBE-grown films is weaker, with indistinct domain structure [18]. The CuPt structure in GaInP and GaInAs contains both Ga-3In and 3Ga-In coordinations for anions (P or As).

The *fcc* CuAu alloy shows two prominent ordered phases: the CuAu-I tetragonal phase, with alternating (002) planes of pure Cu and Au compositions, and the orthorhombic CuAu-II phase [19]. CuAu-I ordering occurs spontaneously within *fcc* sublattices of some epitaxial III-V pseudobinary semiconductors, such as AlGaAs [20] and GaInAs [21], and has also been detected in epitaxial films of $CuInS_2$ [22]. The crystal structure of the chalcopyrite mineral ($CuFeS_2$) is common among ternary semiconductors. This structure also occurs in pseudobinary alloys [23] and may represent the bulk, equilibrium phase [24]. Both CuAu-I and chalcopyrite ordered $CuInSe_2$ present 2Cu-2In anion (Se) coordination.

GaInP/GaAs and GaInAs/InP lattice-matched heterostructures have numerous optoelectronic applications. Our investigations of CuPt-B ordering in GaInP and GaInAs emphasize partial ordering and ordered polytypes (deviations from ideal CuPt ordering), which can constitute ordered superstructures and unicompositional heterostructures.

Success in creating high-efficiency solar cells with polycrystalline thin-film $CuInSe_2$ absorbers [25] has evoked interest in the fundamental properties of these ternary semiconductors. $CuInSe_2$ for device applications is typically grown by such methods as coevaporation, sputtering, or electrodeposition of elemental Cu and In, and Se [26]. Our discussion of epitaxial $CuInSe_2$ primarily serves to provide evidence of ordered phases in this material. Variations in the ordered domain structures, and the control and origins of this ordering continue to be explored.

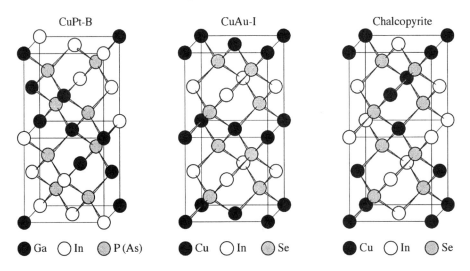

Figure 4. Ordered semiconductor structures discussed.

2.   ORDERED SEMICONDUCTORS

Growth conditions that generate CuPt-B ordering in MOCVD growth of GaInP [10,16] and GaInAs [27] are well-established. MOCVD can produce strong CuPt-B ordering in III-V alloys, with high sample throughput. The strength and distribution of ordering on the two CuPt-B variants are strongly influenced by slight changes in growth parameters and substrate orientation.

Surface kinetics may have a significant influence on ordering in $CuInSe_2$. MEE of $CuInSe_2$ is conducted at relatively low growth rates, with lower throughput, but the microstructures of the resulting epitaxial films can usually be characterized more extensively by TEM than can polycrystalline films. Investigations of ordered phases in $CuInSe_2$ may lead to improved material uniformity and reduced defect densities, with greater control of physical properties.

2.1   GaInP

Strong CuPt-B ordering is obtained in GaInP films grown by atmospheric pressure MOCVD at 670°C, 5.5 µm/h , with a V/III flux ratio of 50-100. In addition to the usual zincblende diffraction spots, the CuPt structure generates superlattice spots at ½{111}B and equivalent positions. Available samples typically contain only partial ordering, with microstructures that depend on substrate miscut and growth conditions.

2.1.1 GaInP-0°

GaInP on singular ⟨001⟩ GaAs (GaInP-0°) shows double-variant ordering (with both [1$\bar{1}$1] and [$\bar{1}$11] variants.). Several groups [28,29] have identified an intricate lamellar structure of closely associated domains in GaInP-0° that occurs at initial growth stages (*Figure 5*). Narrow platelets of the CuPt-B variants are stacked quasiperiodically along the growth direction on a length scale of 40-50 Å. This alternation results from local competition between the two equally probable variants, which hinders the evolution of complementary ordered domains and their associated surface topography.

The stacked domains in GaInP-0° are divided by antiphase boundaries (APBs), which are inclined in opposite directions for each variant from the substrate plane. Increased growth temperatures [30] and reduced growth rates [31] can reduce the densities of APBs in double-variant GaInP-0°, allowing improved resolution of the underlying fine, double-variant, platelet structure (*Figure 6*). Rather than the streaks often observed about ½{111}B and equivalent diffraction spots, the resulting quasi-periodic order/order superlattice can generate discrete satellites. A modest decrease in APB density is observed at a reduced rate of 1.5 µm/h .

The local symmetry of the ⟨001⟩ surface is broken at later growth stages as fluctuations in surface orientation generate facets and variant segregation, which typically become evident near 0.5-2 µm thickness (*Figure 7*). The primordial ordered domains are often diffuse and irregular. As growth proceeds to thicknesses as large as 10 µm, rough topography and faceting develop in thick MOCVD-grown films [32] (i.e., faceting reinforces domain evolution).

[110]

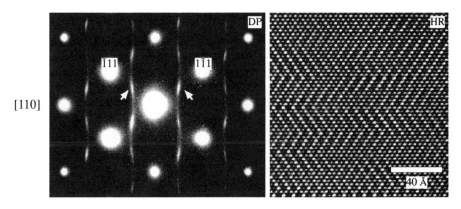

Figure 5. Double-variant GaInP-0°. The arrows indicate superstructure intensity. Ordering occurs within stacked lamellae of the two variants.

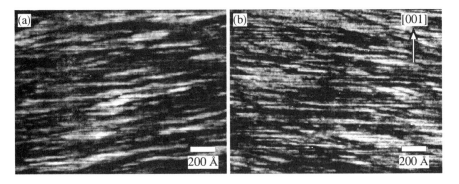

Figure 6. DF images of ordered GaInP-0° grown at 1.5 μm/h : (a) variant I, (b) variant II. The variants are stacked vertically on a fine scale and sectioned by APBs on a coarse scale.

The facets have a broad size distribution. These irregular surface features delineate large, underlying, ordered domains (superdomains), clearly revealing the association between local surface orientation and variant selection. Control of local miscut by substrate patterning has been used to similarly affect the distribution of ordering within GaInP films [33].

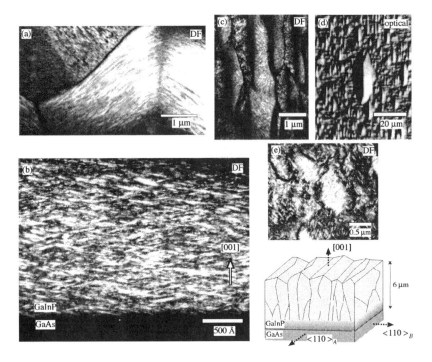

Figure 7. Double variant GaInP-0°. Cross-section: (a) superdomains, (b) near-interface lamellae. Plan-view: (c) superdomains, (d) optical image showing facets, (e) primordial superdomains.

2.1.2 GaInP-A

Equal amounts of the two variants occur on A-miscut substrates. However, the ordering strength decreases gradually with ⟨111⟩A-miscut angle, with corresponding reductions in the rate and extent of variant segregation. At 2°A (*Figure 8*), complementary superdomains typically emerge as the thickness exceeds 3 μm. Observation in ⟨110⟩B cross-section reveals an inclination of the domains toward ⟨111⟩A. Facets on the surfaces of thick GaInP-2°A films show a corresponding asymmetry about ⟨110⟩B.

The vertical dimension of CuPt-B ordered platelets also decreases slowly with ⟨111⟩A miscut. On the ⟨115⟩A surface, corresponding to a miscut of 15.8°A, ordering is weak in films as thick as 10 μm. This ordering comprises a lamellar structure, which lacks variant distinction or segregation. The two-dimensional lamellae are oriented approximately parallel to the ⟨001⟩ plane and are distributed randomly in the disordered matrix. The low coherence and lack of variant identity limits the use of DF imaging to establish the domain geometry.

In contrast to strong CuPt ordering, the lamellar ordering generates diffraction streaks near 0.4⟨110⟩B and equivalent positions (*Figure 9*) [30] that reveal a change in the lateral periocity from ½⟨110⟩B. The lamellar structure is most clearly revealed in ⟨110⟩A HR images.

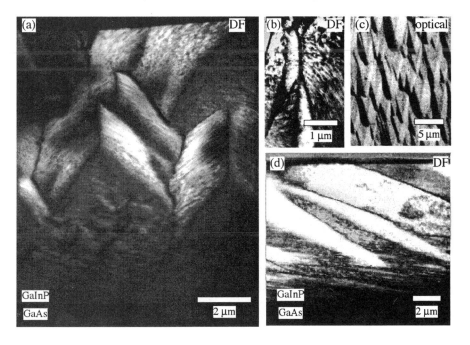

Figure 8. Mixed DF images of double-variant ordered GaInP-2°A: (a) ⟨110⟩A cross-section, (b) plan view, (c) optical, (d) ⟨110⟩B cross-section.

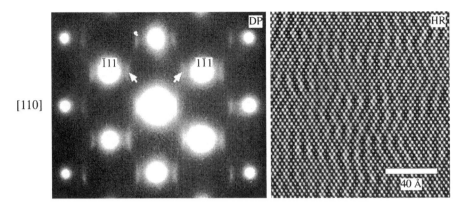

[110]

Figure 9. Lamellar ordering in GaInP-15.8°A. (Arrows indicate superstructure intensity.)

2.1.3   GaInP-2°AB

Substrate miscuts toward <110>AB remove the symmetry between ⟨111⟩B directions. At 2°AB, this generates *asymmetric* double-variant ordering (*Figure 10*), with a difference in the relative volumes and projected areas of the complementary variants onto the growth surface, as explained by Friedman et al. [34].

The dominance of one variant can enhance variant segregation, which typically occurs below a thickness of 0.5 μm at 2°AB. The resulting surface features are often more regular than those of GaInP-0°.

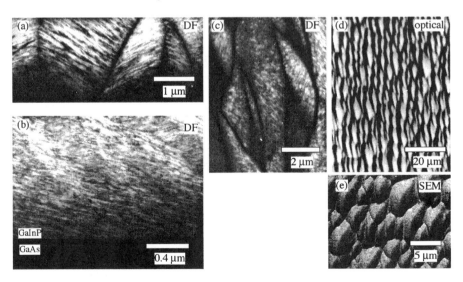

Figure 10. Asymmetric double-variant GaInP-2°AB: Cross-section: (a) superdomains, (b) near-interface region. Plan-view: (c) superdomains, (d) optical image, (e) SEM image of surface.

2.1.4 GaInP-B

Relatively small ⟨111⟩B miscuts essentially eliminate the variant with the smaller component of its ordering direction along the miscut direction, allowing the remaining variant to fill the sample volume. GaInP on GaAs miscut 6°B (GaInP-6°B) shows single-variant ordering (*Figure 11*) with strong superstructure diffraction spots and periodic lattice fringes.

The single variant is divided into phase domains by APBs. The APBs form two-dimensional contoured surfaces, which can be mapped out by TEM examination in a number of orientations. Though it is not subject to variant segregation, single-variant ordered GaInP films exhibit growth evolutionary processes. The domain sizes typically increase during growth [35], as is evident in a film only 0.3 μm thick (*Figure 12*).

The reduced variant competition in 6°B films results in much flatter surfaces than those of double-variant films. Despite the increase in domain size with continued growth, the inclination from the substrate plane of APBs in ⟨110⟩A projection remains unaffected (*Figure 13*). In plan view, or in ⟨110⟩B cross-section, the APBs form closed loops. The two-dimensional APB surfaces follow meandering closed paths, but maintain a fixed angle to the substrate plane. However, the APBs are not precisely parallel, allowing domains of common phase to occasionally coalesce into larger domains.

A miscut of only 2°B significantly enhances one ordered variant and correspondingly diminishes the other variant. The APB inclination [measured from (001) toward the ordering direction to the APB plane] and the mean APB spacing (domain size) in single-variant GaInP increase with miscut [36] from 2°B to 6°B (*Figure 14*). The associated superstructure spots are inclined along the direction normal to the APB plane, and become broader with decreasing domain size.

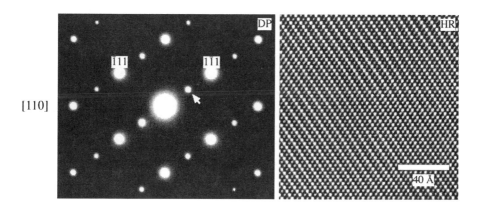

Figure 11. Single-variant GaInP-6°B. The [1$\bar{1}$1] variant generates ½(1$\bar{1}$1) spots and fringes.

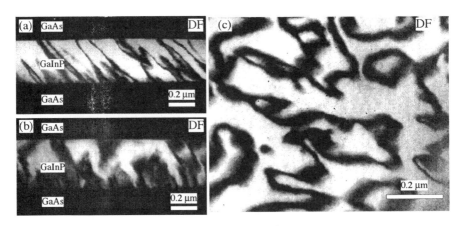

Figure 12. Single-variant GaInP-6°B layer in GaAs: APBs form (a) parallel lines in ⟨110⟩A cross-section and closed loops in (b) ⟨110⟩B cross-section and (c) plan-view.

The substrate miscut is directly related to the density of surface steps. Terrace parity may influence the nucleation of single-phase domains on vicinal (001) surfaces [16]. Changes in APB geometry with ⟨111⟩B miscut (*Figure 15*) have been associated with both step [37] and superstep configurations [38]. In addition, APB extension in GaInP has been parameterized using adatom sticking coefficients [39]. Extension angles less than 90° were attributed to nucleation along the intersections of extant APBs and the growth surface.

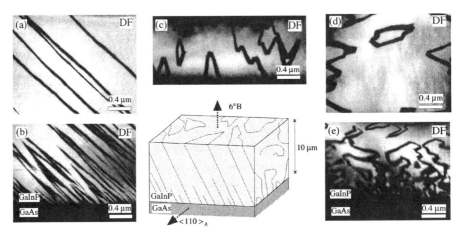

Figure 13. Single-variant GaInP-6°B: The domain sizes increase from bottom to top. APBs form parallel lines in ⟨110⟩A cross-section: (a) top, (b) bottom, and closed loops in (c) plan view, and ⟨110⟩B cross-section: (a) top, (b) bottom.

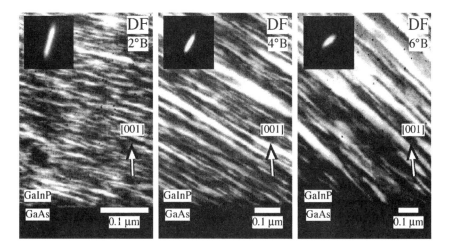

Figure 14. DF images of interface regions of single-variant GaInP at specified orientations. The data are oriented with the growth direction vertical. The $\frac{1}{2}(\bar{1}13)$ spots are inset.

Optoelectronic characteristics of partially ordered microstructures in GaInP have been studied by several groups. We have documented structural descriptions of the most prevalent examples. In a later section, we describe in greater detail the associated structural signatures.

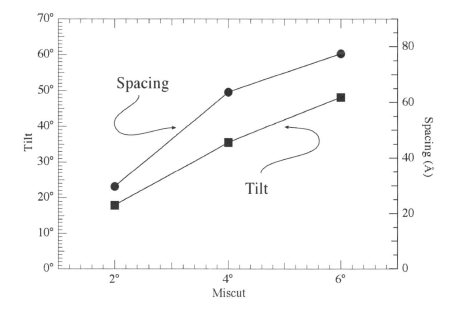

Figure 15. APB tilt (measured from the (001) plane to the APB) and spacing (measured normal to the APB) in single-variant GaInP as functions of substrate miscut.

2.2   GaInAs

CuPt-B ordered GaInP and GaInAs show many qualitative similarities and a few distinct differences, which may partly stem from the lower growth temperature for strong ordering in GaInAs compared to GaInP. In contrast to GaInP, lamellar ordering is not prevalent in near-interface regions of double-variant GaInAs. However, both double-variant GaInP and GaInAs develop segregated superdomains at late growth stages that are linked to the local orientation and dictate surface topography. Also, double-variant GaInAs develops disordered regions in conjunction with segregation. Thus, order/disorder heterostructures are intrinsic to double-variant GaInAs.

2.2.1   GaInAs-0°

Ordered domain structures in GaInAs-0° were described by Seong et al. [11]. In this work, low-pressure MOCVD growth temperature and rate (and V/III flux ratio) were varied from nominal values to investigate the resulting ordered microstructures (*Figure 16*). Increased temperature and decreased rate both cause increased domain sizes by increasing surface diffusion mobility or time, respectively, for adatom incorporation. An increased alignment of the domains along ⟨110⟩B occurs at high growth rates, with complementary arrowhead-shaped domains.

As in GaInP, variant segregation occurs more slowly as the strength of ordering decreases. The sizes of ordered domains and the volume fraction of their contiguous disordered regions increase concomitantly.

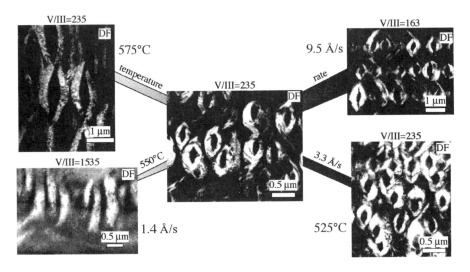

Figure 16. Double-variant GaInAs-0° grown with indicated parameters.

2.2.2   GaInAs-2°AB

Asymmetric double-variant ordering occurs in GaInAs grown by atmospheric-pressure MOCVD on 2°AB-miscut InP. Large disordered regions form at the apices of surface mounds, bordered by columnar domains of the two variants (*Figure 17*).

2.2.3   GaInAs-6°B

Single-variant GaInAs grown by low-pressure MOCVD on InP miscut 6°B (GaInAs-6°B) often contains APBs inclined to angles greater than 90° (*Figure 18*). These APBs extend in the direction of step flow, revealing a different propagation mechanism compared to APBs in single-variant GaInAs.

Quantitative comparison of ordered microstructures in GaInP and GaInAs may lead to greater control of ordered variant segregation, order/disorder heterostructures, and phase-domain formation.

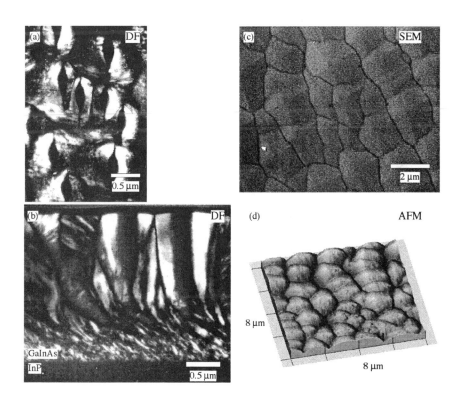

Figure 17. Double-variant GaInAs-2°AB grown at 600°C: (a) plan-view, (b) cross-section; (c) SEM and (d) atomic-force microscope scans.

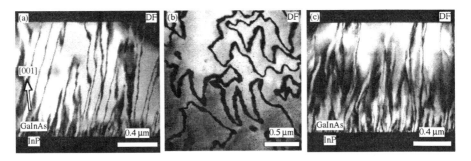

Figure 18. Single-variant GaInAs-6°B: (a) ⟨110⟩A cross-section, (b) plan-view, (c) ⟨110⟩B cross-section.

2.3   CuInSe$_2$

Using the MEE technique for growth of CuInSe$_2$ on vicinal ⟨001⟩ GaAs, films grown by the UF group exhibit ⟨00ℓ⟩ diffraction peaks with odd ℓ [i.e., ⟨001⟩, ⟨003⟩, etc.] in conventional $\theta/2\theta$ X-ray diffraction patterns. The additional peaks are not generated by either sphalerite or chalcopyrite (CH)-ordered CuInSe$_2$. TEM examinations (*Figure 19*) show that these peaks arise from CuAu-I (CA) ordering on the [001] variant (with the tetragonal c axis oriented along the [001] growth direction.). Cross-sections oriented on ⟨100⟩ show initial CH ordering, followed by CA ordering at late growth stages. The CA-ordered domains form a contiguous layer over much of the surface and are delineated from the underlying CH structure by a ragged,

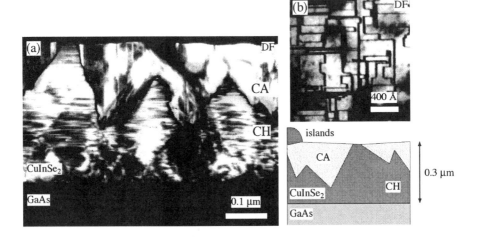

Figure 19. CA- and CH-ordered CuInSe$_2$ on GaAs: (a) cross-section showing CH ordering near the substrate and CA ordering at the film surface. (b) Plan-view showing faulted CA-ordered domains. Surface islands are also present.

lateral boundary. Islands that reside on the film surface may influence the underlying microstructure. DF images show CA-ordered domains containing faults normal to the $\langle 100 \rangle$ in-plane directions.

DPs acquired in [110] cross-section show strong $\langle 001 \rangle$ and $\langle 110 \rangle$ spots and lattice fringes (*Figure 20*) that arise from CA ordering. The $\langle 001 \rangle$ spots and fringes due to CA ordering are also apparent on [100]. The $\langle 100 \rangle$ faults generate streaks that pass through these spots along $\langle 100 \rangle$ in-plane

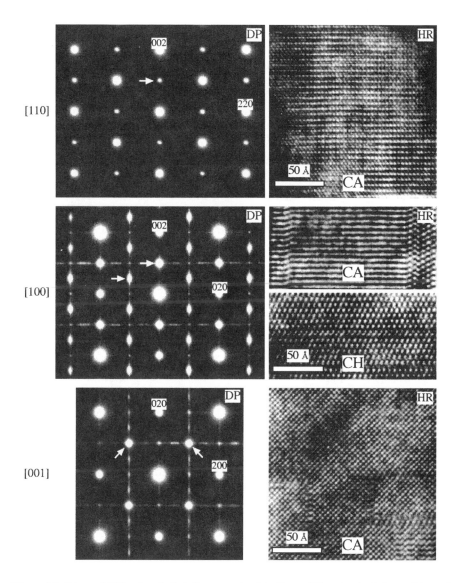

Figure 20. CA- and CH-ordered CuInSe₂ examined in three different orientations. CA- and CH-ordered regions are labeled. The arrows indicate superstructure spots.

directions. The [001] CH-ordered variant generates ½{201}, ½{021}, and equivalent spots. These spots show streaking along [001] caused by stacking faults, which appear in ⟨100⟩ HR images. HR images acquired in [001] plan view show the {110} fringes of the [001] CA variant. The CA-ordered domains are bounded by ⟨100⟩ planar faults that constitute abrupt APBs or narrow ⟨100⟩ CH-ordered domains.

Both CA and CH ordering are quite strong in $CuInSe_2$, despite the low temperatures and growth rates used in MBE. Such fundamental studies of microstructure may lead to improvements in material quality and photovoltaic device performance.

3.   DYNAMICAL DIFFRACTION

Dynamical electron diffraction from crystals is now outlined. These methods are used in the following section to simulate patterns from the ordered systems that have been considered and to qualitatively describe TEM DF image contrast. Quantitative simulations of HR images are sensitive to microscope operational parameters and are outside the scope of this chapter.

The motion of a high-energy electron in the crystal potential $\Phi(r)$ obeys

$$\sqrt{(p(\mathbf{r})c)^2 + (mc^2)^2} = e[V + \Phi(\mathbf{r})] + mc^2, \tag{7}$$

where $\mathbf{r}$ is position, $p(\mathbf{r})$, m, and e are the electron momentum, rest mass, and absolute value of charge, respectively; c is the speed of light; and V is the accelerating potential. The average kinetic energy is $E = e[V + \Phi(\mathbf{r})] \approx eV$, the relativistic mass is $m'c^2 = E + mc^2$, and the reduced kinetic energy is $E' = E(m' + m)/2m'$. Squaring (7) and expanding:

$$p^2(\mathbf{r})/2m' - e\Phi(\mathbf{r}) = E'. \tag{8}$$

With the momentum operator $\hat{\mathbf{p}} = -i\hbar\nabla$ ($\hbar = h/2\pi$, where h is Planck's constant), average momentum $hk = \sqrt{2m'E'}$, and the *structure function* $U(\mathbf{r}) = 2m'e\Phi(\mathbf{r})/h^2$, one produces the equation

$$[\nabla^2 + 4\pi^2(k^2 + U(\mathbf{r}))]\psi(\mathbf{r}) = 0 \tag{9}$$

for the electron wave function $\psi(\mathbf{r})$.

3.1 Bloch Waves

Bloch waves have a discrete nature that facilitates analytical and computational solutions of diffracted-beam amplitudes (*Figure 21*).

Inside the periodic crystal potential, the total wave function

$$\psi(\mathbf{r}) = \sum_j \varepsilon^{(j)} \psi^{(j)}(\mathbf{r}) \tag{10}$$

is a linear combination of (periodic) Bloch waves

$$\psi^{(j)}(\mathbf{r}) = \sum_g \psi_g^{(j)} e^{2\pi i[\mathbf{k}^{(j)} + \mathbf{g}]\mathbf{r}} \,. \tag{11}$$

The direct-beam wave vector $\mathbf{k}$ points from the excitation point M to the reciprocal-space origin 0. The excitation error s_g for each diffracted beam points from the reciprocal lattice point g to the Ewald sphere. The Bloch wave vectors $\mathbf{k}^{(j)}$ extend from the excitation points $M^{(j)}$ to 0.

Boundary conditions at the vacuum/foil interfaces require continuity of $\psi(\mathbf{r})$ and $\hat{\mathbf{n}} \cdot \nabla \psi(\mathbf{r})$, where $\hat{\mathbf{n}}$ is the foil surface normal. Ignoring back-scattered beams, both conditions are approximately satisfied at the entrance surface with s_g and $\gamma^{(j)}$ (where $\mathbf{k}^{(j)} = \mathbf{k} + \gamma^{(j)}$) oriented parallel to $\hat{\mathbf{n}}$, and with coefficients (where * indicates the complex conjugate) given by:

$$\varepsilon^{(j)} = [\psi_0^{(j)}]^* \,. \tag{12}$$

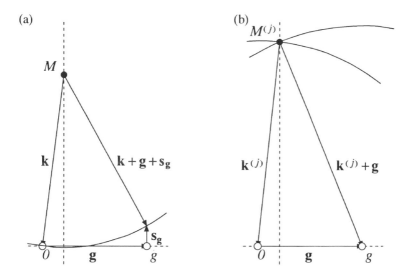

Figure 21. Reciprocal-space constructions for (a) diffracted beams, (b) Bloch waves.

Each Bloch wave contains the $\mathbf{g} = 0$ direct-beam component. The beam amplitudes at the exit surface are

$$\Psi_\mathbf{g} = \sum_j \left[\psi_0^{(j)} \right]^* \psi_\mathbf{g}^{(j)} e^{2\pi i (\gamma^{(j)} - s_\mathbf{g})T} \quad , \tag{13}$$

where T is the foil thickness.

3.2   Eigenfunctions

Transmission electron diffraction from single crystals produces patterns of Laue- (forward-) scattered beams, in which $s_\mathbf{g}$, g, and $\gamma^{(j)}$ are small compared to k (*Figure 22*). The high-energy approximation can be applied:

$$[\mathbf{k}^{(j)} + \mathbf{g}]^2 - k^2 \approx 2k_n \,[\gamma^{(j)} - s_\mathbf{g}] \quad , \tag{14}$$

where $k_n = \mathbf{k} \cdot \hat{\mathbf{n}}$. (We abbreviate $\mathbf{k}_\mathbf{g}^{(j)} = \mathbf{k}^{(j)} + \mathbf{g}$).

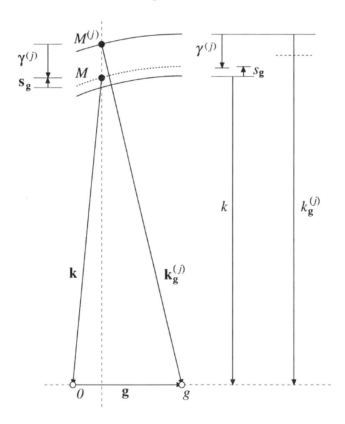

Figure 22. High-energy approximation. Relevant vectors are aligned vertically for clarity.

Using the structure factors U_g, one can write

$$2k_n [\gamma^{(j)} - s_g] \psi_g^{(j)} = \sum_{h \neq g} U_{g-h} \psi_h^{(j)} \,, \tag{15}$$

which has the form of an eigenvalue problem:

$$\hat{A} |\psi^{(j)}\rangle = \gamma^{(j)} |\psi^{(j)}\rangle \,. \tag{16}$$

The operator $\hat{A}$ is implemented as a Hermitian matrix A with elements

$$A_{gg} = s_g \,, \qquad\qquad A_{gg'} = 1/2 \xi_{g-g'} \,;\, (g \neq g') \,, \tag{17}$$

with the *extinction distance* $\xi_g = k_n/U_g$. The eigenfunctions $|\psi^{(j)}\rangle$ and their eigenvalues $\gamma^{(j)}$ are found by converting A to a $2N \times 2N$ real, symmetric matrix, which can be diagonalized with the Jacobi method [40]. Bloch-wave coefficients for weak beams h (i.e., $U_h/s_h \ll U_g/s_g$) may be evaluated subsequently with perturbation theory [41].

3.3   Crystal Potential

The periodic crystal potential can be expressed as a Fourier series:

$$\Phi(r) = \sum_g \Phi_g e^{2\pi i g \cdot r} \,, \tag{18}$$

where the sum is over reciprocal lattice vectors g. The Fourier components are evaluated (with $V \to \infty$) as:

$$\Phi_g = \frac{1}{V} \int_r^V \Phi(r) e^{-2\pi i g \cdot r} dr^3 \,. \tag{19}$$

The total potential is expressed as a sum over constituent atoms m:

$$\Phi_g = \left\{ \sum_m \phi^{(m)}(g) e^{-2\pi i g \cdot d^{(m)}} \right\} X_g \,, \tag{20}$$

where $d^{(m)}$ are the atomic positions in the unit cell, and using the Fourier components of the *crystal function*:

$$X_g = \frac{1}{Nv} \sum_{n=1}^N e^{-2\pi i g \cdot r_n} \,, \tag{21}$$

where v is the unit-cell volume, and the sum is over $n = 1...N$ (with $N \to \infty$) equivalent unit cells labeled r_n (i.e., $e^{2\pi i g \cdot r_n} = 1$ when g is any reciprocal-lattice vector). The crystal function has the value $X_g = 0$, except at reciprocal lattice points, where $X_g = 1/v$.

3.3.1   Structure Factors

Structure factors are Fourier components of the structure function, which is proportional to the crystal potential. The Fourier transforms of isolated atomic (ionic) potentials with spherical symmetry [i.e., $\phi^{(m)}(\mathbf{r}) = \phi^{(m)}(r)$] have been calculated and tabulated parametrically [42] as electron scattering amplitudes (*form factors*):

$$
\begin{aligned}
f^{(m)}(q) &= \frac{8\pi^2 m'e}{h^2} \int_{r=0}^{\infty} r^2 \phi^{(m)}(r) \frac{\sin(4\pi qr)}{4\pi qr} dr \\
&= \frac{2\pi m'e}{h^2} \phi^{(m)}(2q).
\end{aligned}
\tag{22}
$$

The structure factors become

$$
U_{\mathbf{g}} = \left\{ \sum_m f^{(m)}(g/2) e^{-2\pi i \mathbf{g} \cdot \mathbf{d}^{(m)}} \right\} X_{\mathbf{g}} \big/ \pi .
\tag{23}
$$

3.3.2   Absorption

Inelastic scattering encompasses thermal diffuse scattering and other processes [1], and causes transitions among the stationary Bloch waves, producing absorption (damping) of the direct and diffracted beams. The complex structure function $U + iU'$ (U and U' real) generates a non-Hermitian matrix $A + iA'$ (A and A' Hermitian) with complex eigenvalues $\gamma^{(j)} + i\gamma'^{(j)}$ ($\gamma^{(j)}$ and $\gamma'^{(j)}$ real).

Absorption is often weak ($U'/U < 0.1$) [43], and perturbation theory can be used to compute the damping coefficients:

$$
\gamma'^{(j)} = \langle \psi^{(j)} | \hat{A}' | \psi^{(j)} \rangle .
\tag{24}
$$

The imaginary potential components can be estimated by interpolation of calculated absorption potentials for related materials.

3.3.3   Debye-Waller Factors

Lattice vibrations (phonons) cause random displacements of the atoms at any instance in time about their equilibrium positions. These distortions tend to decrease the structure factors with increasing temperature [1], causing an attenuation of the potential components from the values calculated without vibrational effects. The phonon spectra are often assumed to be isotropic; the form factors of constituent atoms m are reduced by the temperature-dependent "Debye-Waller factors" $\exp[-W^{(m)}(g,T)]$, where $W^{(m)}(g,T)$ is proportional to the mean-squared atomic displacement at temperature T. The reduction in the crystal potential components is accompanied by an increase in thermal diffuse scattering between diffraction spots. Debye-Waller factors have been ignored in the simulations presented here.

3.3.4 Order Parameter

Among the most highly sought quantitative characteristics of ordering is the ordering strength, or order parameter η, which is context-dependent. The Bragg-Williams approach is widely used in X-ray diffraction analyses [44,45]. In addition to selected-area DP simulations, imaging techniques for order parameter determination have been outlined [46]. For the computation of simulated electron diffraction patterns, η may also be implemented as the volume fraction of fully ordered material. Dynamical fits of experimental patterns minimize the least-squared difference

$$\chi^2 = \Sigma_{\mathbf{g}} [I_{\mathbf{g}}^{(expt)} - I_{\mathbf{g}}^{(calc)}]^2 / \sigma_{\mathbf{g}}^2 \qquad (25)$$

between discrete collections of experimental and calculated integrated intensities with errors $\sigma_{\mathbf{g}}$. In addition to order parameters, the fitting procedure may vary other quantities that are not known precisely, such as beam/sample tilt and thickness.

4. ANALYSIS

CuPt-B ordering in GaInP and GaInAs was discussed and evidence was presented of CA- and CH-ordered [001] variants in epitaxial CuInSe$_2$. These crystal structures and their diffraction signatures are now described.

TEM data from both ordered and partially ordered structures are often amenable to analytical and computational descriptions with reasonable simplifications. In this section, each ordered structure is generated with a conveniently chosen unit cell [47] that usually does not reflect the full crystal symmetry, but greatly simplifies data entry. These indices are transformed to a cubic reference cell with reciprocal lattice vectors $g = [hk\ell]/a$. The transformation $\mathbf{g} \rightarrow -\mathbf{g}$ gives

$$U_{hk\ell} = \{\Sigma_m f^{(m)} e^{2\pi i [hk\ell] \cdot \mathbf{d}^{(m)}/a}\} / \pi v \qquad (26)$$

The form factors are evaluated at the appropriate scattering vector.

4.1   Zincblende

The zinblende structure of GaAs (InP) (*Table 1*) is generated with an asymmetric unit cell containing two atoms ($v = a^3/4$). The reciprocal lattice is determined by inversion of the direct-space basis vectors. The cubic indices for zincblende crystals are given by

$$\langle hk\ell \rangle = \langle -h' + k' + \ell', h' - k' + \ell', h' + k' - \ell' \rangle , \tag{27}$$

where $\langle h'k'\ell' \rangle$ are all integers. The zincblende reflections have indices that are all even or all odd. The zincblende structure factors are given by

$$U_{hk\ell} = 4\{ f^{\mathrm{Ga(In)}} + f^{\mathrm{As(P)}} e^{i\pi(h+k+\ell)/2} \}/\pi a^3 . \tag{28}$$

4.2   CuPt-B Ordering

The CuPt-B structure of GaInP (GaInAs) (*Table 2*) is generated with an asymmetric unit cell containing four atoms ($v = a^3/2$) Atomic displacements from the virtual (average) positions are ignored, but may be significant. The cubic indices for the $[1\bar{1}1]$ variant are

$$\langle hk\ell \rangle = \langle h' - k' + \tfrac{1}{2}\ell', h' - k' - \tfrac{1}{2}\ell', h' + k' + \tfrac{1}{2}\ell' \rangle . \tag{29}$$

Simulated DPs of CuPt-B ordered GaInP (*Figure 23*) show superstructure spots at $\tfrac{1}{2}\langle 111 \rangle$B and equivalent positions. The structure factors become

$$U_{hk\ell} = 2\{ f^{\mathrm{Ga}} + f^{\mathrm{In}} e^{\pi i(h-k)} + f^{\mathrm{P(As)}} [1 + e^{\pi i(h-k)}] e^{\pi i(h+k+\ell)/2} \}/\pi a^3 . \tag{30}$$

Strain contributions to TEM image contrast are identified when the contrast reverses between images acquired with lattice vectors **g** and $-\mathbf{g}$, possibly resulting from surface relaxation [48]. Using the {220} in-plane

Table 1. Structure of zincblende GaAs (InP).

Lattice Vectors			
direct	$a[011]/2$	$a[101]/2$	$a[110]/2$
reciprocal	$[\bar{1}11]/a$	$[1\bar{1}1]/a$	$[11\bar{1}]/a$
Atomic Positions			
Ga (In)	$a[000]$	As (P)	$a[111]/4$

Table 2. Structure of CuPt-B ordered GaInP (GaInAs) ([1 $\bar{1}$1] variant) with $\eta = 1$.

Lattice Vectors			
direct	$a[011]/2$	$a[\bar{1}01]/2$	$a[1\bar{1}0]$
reciprocal	$[111]/a$	$[\bar{1}\bar{1}1]/a$	$[1\bar{1}1]/2a$

Atomic Positions			
Ga	$a[000]$	P (As)	$a[111]/4$
In	$a[1\bar{1}0]/2$	P (As)	$a[3\bar{1}1]/4$

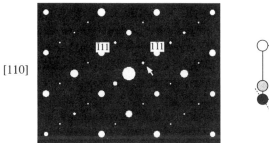
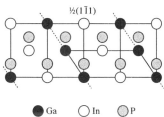

Figure 23. Simulated $\langle 110 \rangle$A DP from CuPt-B ordered GaInP ([1 $\bar{1}$1] variant) ($T = 400$ Å, $\eta = 0.2$, and $V = 300$ kV) and structural diagram.

reflections, which are insensitive to chemical ordering, transitions in the strain field are evident across o/o' and o/d interfaces. Strain may result from distortion of the CuPt-B variants, which are in crystallographic registry within the epitaxial film. Misfit dislocations indicative of relaxation at the domain interfaces are not detected. Further quantification is required for a complete description of the strain profiles.

Although significant in metallurgical CuPt [8], rhombohedral distortion in CuPt-ordered GaInP and GaInAs has been neglected. Inhomogenous strain profiles are evident in double-variant ordered GaInAs (*Figure 24*).

4.2.1　Antiphase Boundaries

The dark bands observed in DF images arise from APBs, which can be identified in HR images (*Figure 25*). A technique for generating DF image contrast sensitive to the local structural phase in zincblende compounds has been outlined [49,50], but the adaptation to ordered alloys is not straightforward. The APB width is often imperceptible within the resolution of DF images [51]. However, segregation of the weaker variant to the APB

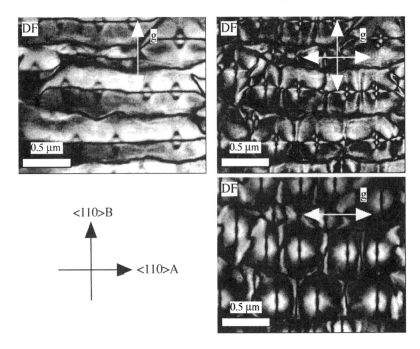

<110>B

<110>A

Figure 24. Mixed {220} plan-view mixed DF images of ordered GaInAs-0°. The imaging vector directions are indicated. The images highlight the underlying domain structure.

plane has been reported [52], and one group [53] has detected In-rich compositions along the APBs.

Two contiguous regions are separated by an APB when the phase of a Fourier component associated with the reciprocal-lattice vector **g** has the same magnitude in the regions, but differs in phase by π. The image contrast of metal alloy APBs has been demonstrated experimentally [54].

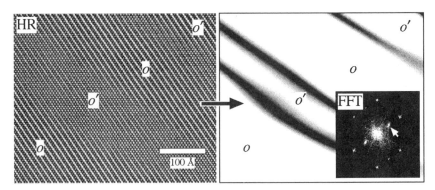

Figure 25. Single-variant GaInP-4°B. The HR image (left) was filtered using the spot indicated by the arrow in the FFT. Boundaries between phase (o) and antiphase (o') regions appear dark in the filtered image.

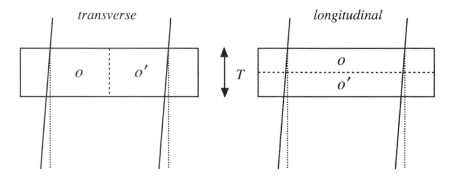

Figure 26. Transverse and longitudinal APBs separating phase (o) and antiphase (o') regions within a thin foil.

APBs present regions at which U_g for the imaging reflection g vanishes. The image contrast is sensitive to the APB orientation with respect to the plane of the TEM foil (*Figure 26*). A periodic array of *transverse* APBs (perpendicular to the foil plane) with wavelength λ_m can be approximated using a sinusoidal oscillation of U_g. The resulting matrix has Bloch-wave solutions that contain the four components $\mathbf{0}$, $\mathbf{g}$, $\mathbf{g} \pm \mathbf{g}_m$, where $g_m \ll g$. For simplicity, one can assume a long modulation wavelength ($s_{\mathbf{g} \pm \mathbf{g}_m} \approx s_{\mathbf{g}}$) and evaluate the DF image intensity as a superposition of g with its satellites $g \pm g_m$ (*Figure 27*) at the Bragg condition.

The diffracted intensity below the entrance surface of a thin foil of thickness T containing a *longitudinal* APB (parallel to the foil plane) at depth t ($0 \le t \le T$) is evaluated with dynamical theory [55]. Only the direct beam 0 and a diffracted beam g are retained. The beam amplitudes below the sample vary with excitation error $s_{\mathbf{g}}$, which is directly related to the sample tilt from the Bragg condition.

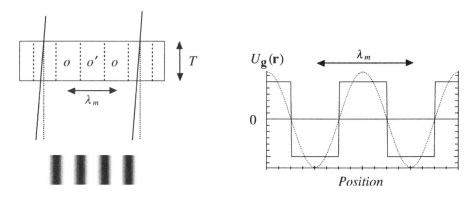

Figure 27. APB array with period λ_m and simulated DF contrast using a sinusoidal approximation. The graph shows the structure factor as a function of position.

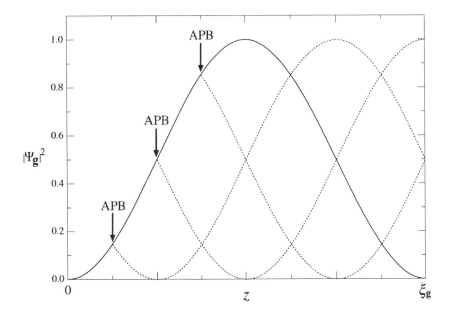

Figure 28. DF image intensity at the Bragg condition of an APB as a function of depth. The intensity follows the solid (dashed) line above (below) the APB.

With $s_g = 0$, we obtain (in the region $z < T$)

$$I_g(z) = \begin{cases} \sin^2(\pi z/\xi_g) & z \leq t \\ \sin^2[\pi(2t-z)/\xi_g] & z > t \end{cases} \tag{31}$$

The image intensity (*Figure 28*) is darkest where the APB vertically separates equal thicknesses of phase and antiphase material ($t/T = 0.5$).

The depth of a planar APB inclined to the surface of the foil varies linearly in the range $0 \leq t \leq T$. The exit wave function is computed using the column approximation [56,57] and define the dimensionless parameter $w = s_g \xi_g$. Special cases are $w = 0$, $w = \sqrt{3}$, and $w = \sqrt{15}$ (*Figure 29*).

The experimental contrast acquired in plan-view from 6°B samples at various tilts of the sample stage (*Figure 30*) shows similar features.

4.2.2   Orientational Superlattices

Double-variant GaInP often contains high densities of ordered lamellae arranged in quasi-periodic stacked sequences resembling orientational superlattices (OSLs) [58]. The individual layers can be described as altering variants, or twins, which are related by a rotation of π. We consider idealized

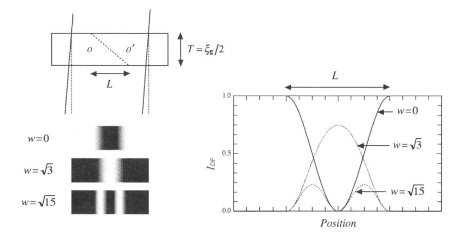

Figure 29. Simulated DF contrast of oblique APB at specified excitations.

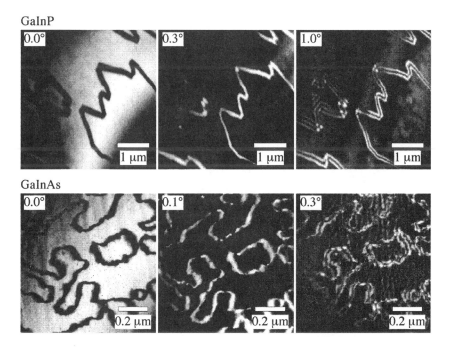

Figure 30. Plan-view DF images of APBs in GaInP and GaInAs at approximate specified tilts from the nominal Bragg condition.

OSLs that constitute periodically stacked variants along [001]. A variety of such unicompositional structures exist, in principle. Here, only those with equal amounts of ordering on the two variants are considered. The simplest is the 1:1 OSL [59-61]. An *n:n* OSL is generated using an orthorhombic unit cell containing $8n$ atoms ($v = na^3$) with a vertical lattice constant of na (*Table 3*). Simulated DPs from simple OSLs show satellites about $\frac{1}{2}\{111\}B$ and equivalent spots that extend along $[001]$ (*Figure 31*).

The central $\frac{1}{2}\{111\}B$ CuPt-ordered spots do not arise from the 1:1 OSL. A phase shift between alternating $\langle 002 \rangle$ monolayers (MLs) of zero is associated with the 1:1 OSL structure, rather than π, as expected for CuPt ordering. (Adjacent MLs have an intrinsic phase shift of $\pi/2$, regardless of stacking arrangement.) As the stacking period increases beyond a few MLs, variant boundaries become less significant, and distinct variant domains form that generate strong $\frac{1}{2}\{111\}B$ spots with multiple satellites. These bulk domains facilitate description as abrupt superlattices comprising two

Table 3. Lattice vectors of an idealized GaInP [001] $n:n$ OSL.

Lattice Vectors			
direct	$a[1\bar{1}0]$	$a[110]/2$	$na[001]$
reciprocal	$[1\bar{1}0]/2a$	$[110]/a$	$[001]/na$

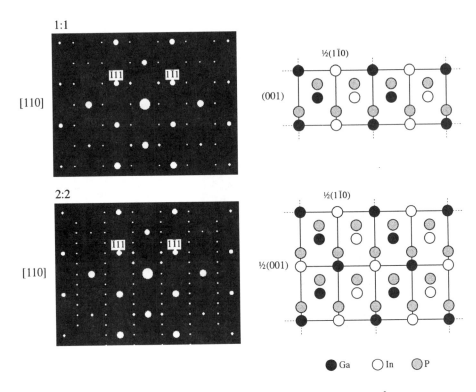

Figure 31. Simulated $\langle 110 \rangle$A DPs from GaInP 1:1 and 2:2 OSLs ($T = 400$ Å, $\eta = 0.2$, and $V = 300$ kV) and structural diagrams.

alternating variants. Lateral variant segregation (along ⟨110⟩B) also generates quasi-periodic alterations in variant. Such long-period superlattices are amenable to description using envelope functions.

4.3   Lamellar Ordering

Experimentally, the ordering strength and vertical coherence diminish with ⟨111⟩A miscut and growth temperature. Residual coherence on a length scale *ma* produces vertical diffraction streaks. The lateral ordering is often slightly incommensurate with the zincblende matrix, with a period of approximately 0.4⟨110⟩B [deviating slightly from ½⟨110⟩B] (*Figure 32*).

Idealized ½⟨110⟩B lateral ordering of Ga and In within weakly correlated MLs is considered. The structure factor is a sum of the random (zincblende) alloy matrix components $\bar{U}_{hk\ell}$ with a lamellar term $\Delta U_{hk}\,(\ell)$:

$$U_{hk}\,(\ell) = \bar{U}_{hk\ell} + \Delta U_{hk}\,(\ell) \,. \tag{32}$$

The structure factors become continuous functions of the vertical Miller index ℓ, corresponding to the presence of diffraction streaks. Each pair of atoms in alternating ordered MLs occupies a volume $v = a^3/2$, and each cubic unit cell contributes two such MLs. The lamellae generate rods in reciprocal space that extend along [001], with cubic coordinates

$$\langle hk \rangle = (\tfrac{1}{2}h' + k', -\tfrac{1}{2}h' + k') \,. \tag{33}$$

The lamellar structure factor term is expressed as

$$\Delta U_{hk}\,(\ell) = 2[f^{\,Ga} - f^{\,In}\,][1 - e^{\pi i (h-k)}\,]F(\ell)/\pi a^3 \,. \tag{34}$$

The lamellar coherence function $F\,(\ell)$ is used, which is the Fourier transform of the normalized envelope function in direct space for the

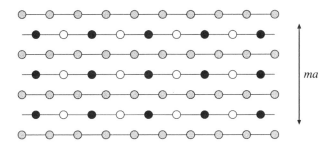

Figure 32. Lamellar ordering. Alternating ordered MLs are weakly correlated.

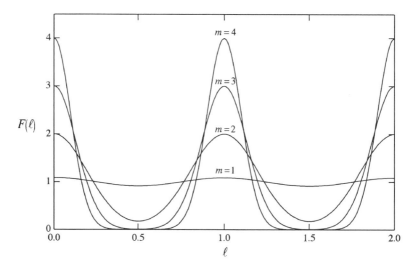

Figure 33. Lamellar coherence function vs. vertical Miller index ℓ for Gaussian coherence distributions of specified ranges m.

lamellar domains. $F(\ell)$ is expressed in Gaussian form with coherence length ma as

$$F(\ell) = \sum_{\ell'} e^{-\pi[m(\ell-\ell')]^2} ,$$

(35)

which approximately describes the profile of the diffraction streak intensity. Peaks occur in $F(\ell)$ that become sharper with increasing vertical lamellar coherence (*Figure 33*).

The deviation from $\frac{1}{2}\langle 110\rangle B$ lateral periodicity in actual samples causes the weaving appearance of the streaks observed in experimental DPs from samples with lamellar ordering (see *Figure 9*), and reveals a change in surface kinetic processes as ordering strength decreases.

4.4   CuAu-I Ordering

CA ordering is generated with a four-atom unit cell ($v = a^3/2$), neglecting bond-length differences and tetragonal distortion (*Table 4*). Reflections from CA-ordered CuInSe$_2$ have cubic indices given by:

$$\langle hk\ell\rangle = \langle h'-k', h'+k', \ell'\rangle .$$

(36)

The structure factors for the CA-ordered ternary compound CuInSe$_2$ are

$$U_{hk\ell} = 2\{f^{Cu} + f^{In}e^{\pi i(k+\ell)} + f^{Se}[1+e^{\pi i(k+\ell)}]e^{\pi i(h+k+\ell)/2}\}/\pi a^3 .$$

(37)

Table 4. Structure of CA-ordered CuInSe₂ ([001] variant) with $\eta = 1$.

Lattice Vectors			
direct	$a[110]/2$	$a[\bar{1}10]/2$	$a[001]$
reciprocal	$[110]/a$	$[\bar{1}10]/a$	$[001]/a$

Atomic Positions			
Cu	$a[000]$	Se	$a[111]/4$
In	$a[011]/2$	Se	$a[133]/4$

Superlattice spots due to CA ordering appear on several low-index zone axes, notably on [100], [110], and [001] (*Figure 34*).

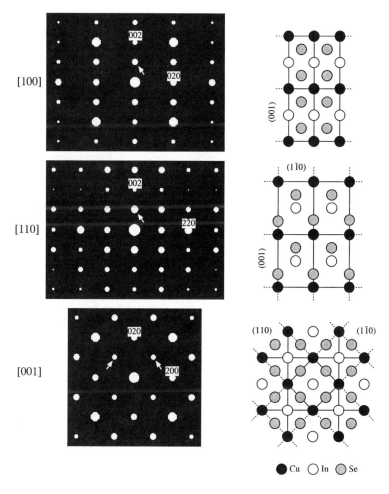

Figure 34. Simulated DPs from CA-ordered CuInSe₂ ([001] variant) ($T = 400$ Å, $\eta = 1$, and $V = 300$ kV) and structural diagrams.

4.5   Chalcopyrite Ordering

Periodicities associated with the CH-ordered CuInSe₂ ₍₀₀₁₎ variant are observed on the ⟨100⟩ in-plane axes. A unit cell containing eight atoms ($v = a^3$) is constructed. Tetragonal distortion is neglected (*Table 5*).

The cubic indices for reflections from the CH-ordered ₍₀₀₁₎ variant are

$$\langle hk\ell \rangle = \langle h', k', -\tfrac{1}{2}h' - \tfrac{1}{2}k' + \ell' \rangle \tag{38}$$

The structure factors are:

$$U_{hk\ell} = \{ f^{Cu}\,[1 + e^{\pi i \langle k + \ell \rangle}] + f^{In}\,[e^{\pi i \langle h + k \rangle} + e^{\pi i \langle h + \ell \rangle}]$$
$$+ f^{Se}\,[1 + e^{\pi i \langle k + \ell \rangle} + e^{\pi i \langle h + k \rangle} + e^{\pi i \langle h + \ell \rangle}]e^{\pi i \langle h + k + \ell \rangle/2}\}/\pi a^3. \tag{39}$$

Despite extinctions ($U_{hk\ell} = 0$) at ⟨001⟩ and equivalent positions, these spots are generated on ⟨100⟩ in-plane axes by double diffraction (*Figure 35*).

Table 5. Structure of CH-ordered CuInSe₂ ([001] variant) with $\eta = 1$.

Lattice Vectors			
direct	a[100]	a[010]	[112]/2a
reciprocal	[20$\bar{1}$]/2a	[02$\bar{1}$]/2a	[001]/a

Atomic Positions			
Cu	a[000]	Se	a[111]/4
Cu	a[011]/2	Se	a[133]/4
In	a[110]/2	Se	a[331]/4
In	a[101]/2	Se	a[313]/4

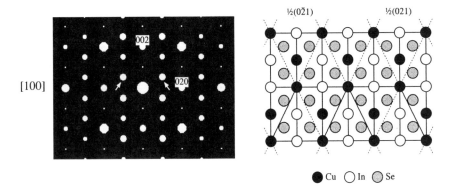

Figure 35. Simulated DP from CH-ordered CuInSe₂ ([001] variant) ($T = 400\,\text{Å}$, $\eta = 1$, and $V = 300\,\text{kV}$) and structural diagram.

5.   CONCLUSIONS

The ordered structures discussed have high translational symmetry, permitting easy experimental identification and theoretical description. A great deal of quantitative work on ordered semiconductor compounds and alloys remains uncompleted. The difficult task of determining order parameters from the TEM data has been avoided. Ordering phenomena offer additional control of material properties, and provide a glimpse into growth processes and the evolution of semiconductor microstructures.

ACKNOWLEDGMENTS

I acknowledge with gratitude the growth scientists who provided samples used in this work. GaInP and GaInAs samples were provided by D. J. Friedman, J. F. Geisz, D. J. Arent, S. R. Kurtz, and J. M. Olson, all from the NREL III-V Device Team. GaInAs was also provided by M. C. Hanna of the NREL Basic Sciences Center. $CuInSe_2$ samples and important data and collaboration were provided by B. J. Stanbery, C.-H. Chang, and T. J. Anderson of the Department of Chemical Engineering at the University of Florida, Gainesville.

Also, thanks to K. M. Jones, R. M. Matson, H. R. Moutinho, and M. M. Al-Jassim of the NREL Analytical Microscopy Team.

REFERENCES

[1]   L. Reimer, *Transmission Electron Microscopy*, Springer-Verlag, New York, 1984.

[2]   J. C. H. Spence, *Experimental High-Resolution Electron Microscopy*, 2nd Ed., Oxford, New York, 1988.

[3]   S.-J. Kim, H. Asahi, M. Takemoto, K. Asami, M. Takeuchi, and S.-I. Gonda, *Jpn. J. Appl. Phys.* **33** (1996) 4225.

[4]   B. Pamplin, *Prog. Crystal Growth Charact.* **3** (1981) 179.

[5]   S. Francoeur, G. A. Seryogin, S. A. Nikishin, and H. Temkin, *Appl. Phys. Letters* **74** (1999) 3678.

[6]   A. Rockett and R. W. Birkmire, *J. Appl. Phys.* **70** (1991) R81.

[7]   C. Rincón, *Phys. Rev. B.* **45** (1992) 12716.

[8]   M. Hansen and K. Anderko, *Constitution of Binary Alloys*, McGraw-Hill, New York 1958.

[9]   C. B. Walker, *J. Appl. Phys.* **23** (1952) 118.

[10] A. Gomyo, T. Suzuki, K. Kobayashi. S. Kawata, I. Hino, and T. Yuasa, *Appl. Phys. Letters* **50** (1987) 673.

[11] T. Y. Seong, A. G. Norman, G. R. Booker, and A. G. Cullis, *J. Appl. Phys.* **75** (1994) 12.

[12] Y.-E. Ihm, N. Otsuka, J. Klem, and H. Morkoç, *Appl. Phys. Letters* **51** (1987) 2013.

[13] G. S. Chen, D. H. Jaw, and G. B. Stringfellow, *J. Appl. Phys.* **69** (1991) 4263.

[14] T. Suzuki, T. Ichihashi, and T. Nakayama, *Appl. Phys. Letters* **73** (1998) 2588.

[15] S. Froyen and A. Zunger, *Phys. Rev. Lett.* **66** (1991) 2132.

[16] T. Suzuki and A. Gomyo, *J. Crystal Growth* **111** (1991) 353.

[17] B. A. Philips, A. G. Norman, T. Y. Seong, S. Mahajan, G. R. Booker, M. Skowronski, J. P. Harbison, and V. G. Keramidas, *J. Crystal Growth* **140** (1994) 249.

[18] C. Meenakarn, A. E. Staton-Bevan, M. D. Dawson, G. Duggan, A. H. Kean, and S. P. Najda, *Inst. Phys. Conf. Ser.* **157** (1997) 265.

[19] R. E. Smallman, W. Hume-Rothery, and C. W. Haworth, *The Structures of Metals and Alloys*, Institute of Metals, London, 1988.

[20] T. S. Kuan, T. F. Kuech, W. I. Wang, and E. L. Wilkie, *Appl. Phys. Letters*, **54** (1985) 201.

[21] O. Ueda, Y. Nakata, T. Nakamura, and T. Fujii, *J. Crystal Growth* **115** (1991) 375.

[22] D. S. Su, W. Neumann, R. Hunger, P. Schubert-Bischoff, M. Giersig, H. J. Lewerenz, R. Scheer, and E. Zeitler, *Appl. Phys. Letters* **73** (1998) 785.

[23] H. R. Jen, M. J. Cherng, and G. B. Stringfellow, *Appl. Phys. Letters* **48** (1986) 1603.

[24] S.-H. Wei, L. G. Ferreira, and A. Zunger, *Phys. Rev. B* **41** (1990) 8240.

[25] C. J. Kiely, R. C. Pond, G. Kenshole, and A. Rockett, *Phil. Mag. A* **63** (1991) 1249.

[26] J. R. Tuttle, Ph.D. Thesis, University of Colorado (1990).

[27] D. J. Arent, M. H. Bode, K. A. Bertness, S. R. Kurtz, and J. M. Olson, *Appl. Phys. Letters* **62** (1993) 1806.

[28] E. Morita, M. Ikeda, O. Kumagai, and K. Kaneko, *Appl. Phys. Letters* **53** (1988) 2164.

[29] M. Ishimaru, S. Matsumura, N. Kuwano, and K. Oki, *Phys. Rev. B* **52** (1995) 5154.

[30] C. S. Baxter, W. M. Stobbs, and J. H. Wilkie, *J. Crystal Growth* **112** (1991) 373.

[31] C. S. Baxter and W. M. Stobbs, *Phil. Mag. A* **69** (1994) 615.

[32] D. J. Friedman, J. G. Zhu, J. M. Olson, A. E. Kibbler, and J. Moreland, *Appl. Phys. Letters* **63** (1993) 1774.

[33] L. C. Su, S. T. Pu, G. B. Stringfellow, J. Christen, H. Selber, and D. Bimberg, *J. Electronic Materials* **23** (1994) 125.

[34] D. J. Friedman, G. S. Horner, S. R. Kurtz, K. A. Bertness, J. M. Olson, and J. Moreland, *Appl. Phys. Letters* **65** (1994) 878.

[35] H. M. Cheong, A. Mascarenhas, S. P. Ahrenkiel, K. M. Jones, J. F. Geisz, and J. M. Olson, *J. Appl. Phys.* **89** (1998) 5418.

[36] L. C. Su, I. H. Ho, and G. B. Stringfellow, *J. Appl. Phys.* **75** (1994) 5135.

[37] L. Nasi, F. Fermi, C. Ferrari, L. Francesio, L. Lazzarini, C. Zanotti-Fregonara, S. Pellegrino, and G. Salviati, *Inst. Phys. Conf. Ser.* **157** (1997) 269.

[38] L. C. Su and G. B. Stringfellow, *J. Appl. Phys.* **78** (1995) 6775.

[39] S. Takeda, Y. Kuno, N. Hosoi, and K. Shimoyama, *J. Crystal Growth* **205** (1999) 11.

[40] W. H. and , B. P. Flannery, S. A. Teukolsky, W. T. Vetterling, *Numerical Recipes*, Cambridge, New York, 1986.

[41] J. M. Zuo and A. L. Weickenmeier, *Ultramicro.* **57** (1995) 375.

[42] P. A. Doyle and P. S. Turner, *Acta Cryst. A* **24** (1968) 390.

[43] G. Radi, *Acta Cryst. A* **26** (1970) 41.

[44] R. L. Forrest, T. D. Golding, S. C. Moss, Y. Zhang, J. F. Geisz, J. M. Olson, A. Mascarenhas, P. Ernst, and C. Geng, *Phys. Rev. B* **58** (1998) 15355.

[45] S. Francoeur, G. A. Seryogin, S. A. Nikishin, and H. Temkin, *Appl. Phys. Letters* **76** (1999) 2017.

[46] M. H. Bode, S. P. Ahrenkiel, S. R. Kurtz, K. A. Bertness, D. J. Arent, and J. M. Olson, *Mat. Res. Soc. Symp. Proc.* **417** (1996) 55.

[47] *International Tables for Crystallography*, Ed. T. Hahn, Kluwer Academic, Boston, 1996.

[48] M. M. Treacy, J. M. Gibson, and A. Howie, *Phil. Mag. A* **51** (1985) 389.

[49] J. P. Gowers, *Appl. Phys. A* **34** (1984) 231.

[50] L. T. Romano, I. M. Robertson, J. E. Greene, and J. E. Sundgren, *Phys. Rev. B* **36** (1987) 7523.

[51] G. Hahn, C. Geng, P. Ernst, H. Schweizer, and F. Scholz, *Superlattices and Micro.* **22** (1997) 301.

[52] U. Dörr, H. Kalt, W. Send, D. Gerthsen, D. J. Mowbray, and C. C. Button, *Appl. Phys. Letters* **73** (1998) 1679.

[53] U. Kops, R. G. Ulbrich, M. Burkard, C. Geng, F. Scholz, and M. Schweizer, *Phys. Stat. Sol.* **164** (1997) 459.

[54] R. M. Fisher and M. J. Marcinkowski, *Phil. Mag.* **6** (1961) 1385.

[55] P. Hirsch, A. Howie, R. B. Nicholson, D. W. Pashley, and M. J. Whelan, *Electron Microscopy of Thin Crystals*, Krieger, Hungtington, 1977.

[56] A. Howie and Z. S. Basinski, *Phil. Mag.* **17** (1968) 1039.

[57] D. B. Williams and C. B. Carter, *Transmission Electron Microscopy: A Textbook for Materials Science*, Plenum, New York, 1996.

[58] Y. Zhang, A. Mascarenhas, S. P. Ahrenkiel, D. J. Friedman, J. F. Geisz, and J. M. Olson, *Solid State Comm.* **109** (1999) 99.

[59] D. Munzar, E. Dobročka, I. Vávra, R. Kúdela, M. Harvanka, and N. E. Christensen, *Phys. Rev. B* **57** (1998) 4642.

[60] S.-H. Wei, S. B. Zhang, and A. Zunger, *Phys. Rev. B* **59** (1999) R2478.

[61] T. Saß, I. Pietzonka, and H. Schmidt, *J. Appl. Phys.* **85** (1999) 3561.

Chapter 8

X-Ray Analysis of the Short-Range Order in the Ordered-Alloy Domains of Epitaxial (Ga,In) P Layers by DAFS of Superlattice Reflections

Dirk C. Meyer, Kurt Richter, Gerald Wagner,[*] and Peter Paufler
Institut für Kristallographie und Festkörperphysik, Fachrichtung Physik, Technische Universität Dresden, D-01062 Dresden, Germany
[*]Institut für Oberflächenmodifizierung Leipzig, Germany

Key words: ordering, DAFS, XAFS, (Ga,In)P

Abstract: Ordered-alloy domains of epitaxially grown (Ga,In)P layers have been observed elsewhere using transmission electron microscopy and transmission electron diffraction. We used diffraction anomalous fine-structure (DAFS) experiments at superlattice reflections occurring in several <1 1 1> directions to explore the short-range order around Ga atoms in such ordered domains in epitaxial (Ga,In)P layers grown on (0 0 1) GaAs substrates. The requirements for a reliable measurement of the reflection intensity depending on the photon energy are described. A quantitative DAFS analysis resulting in the short-range order parameters is explained in detail. The local structure around Ga in the whole (Ga,In)P layer ($F\,\overline{4}\,3\,m$) can be understood by a local structure model, whereas in contrast, the local structure around Ga atoms in the ordered regions ($R\,3\,m$) can be described by the values expected on the basis of the virtual-crystal model.

1.   INTRODUCTION

In epitaxial layers of III-V ternaries dependent on the process parameters during growth, small superlattice domains often come into being which are supposed to influence the electronic properties (e.g., gap energy) of this material and hence of the devices made of it [1]. Therefore, the structure of the ordered domains has been studied by transmission electron microscopy

(TEM) and Transmission Electron Diffraction ([2]-[4]), the latter one giving information about the size of the ordered domains as well. Mostly a CuPt-like structure of the column III elements sublattice is reported. This emphasis of one direction leads to a reduction from original $F\,\bar{4}\,3\,m$ symmetry to rhombohedral $R\,3\,m$ symmetry of the whole structure. Diffraction anomalous fine structure (DAFS) combines the long-range order sensitivity of diffraction techniques with the short-range sensitivity of absorption techniques. Our DAFS measurements are exclusively aimed at a quantitative characterization of the short-range order of the ordered domains. Therefore we used superlattice reflections ([5], [6]) where only the relevant anomalous scatterers contribute to the DAFS. This is important because merely a small fraction (few percent) of the whole layer volume is subdued to the ordering, resulting in a negligibly small contribution to the DAFS of the main reflections.

In the present chapter we discuss the experimental requirements of DAFS measurements at superlattice reflections, and of the DAFS analysis in detail referring to Arcon *et al.* [7], Stragier *et al.* [8], and Sorensen *et al.* [9], who described the theory, experimental methods, analytical techniques, and various applications of DAFS. Experiments for study of strained III-V epitaxial semiconductors by means of DAFS have been reported by Proietti *et al.* [10]. Methodical aspects of a quantitative analysis of DAFS experiments with non-centrosymmetric single crystals have already been reported in detail in [11].

1.1   Specimens

The epitaxial (Ga,In)P layers were prepared by metal organic vapor phase epitaxy (MOVPE) as described elsewhere [12]. The (0 0 1) GaAs substrates were misoriented 2° towards the azimuthal [0 1 0] direction. The samples investigated differed in growth temperatures, as can be seen from Table I.

Table I. Growth parameters of investigated specimens.

Specimen	Substrate temperature	(Ga,In)P layer thickness
60/2	650 °C	about 1.90 μm
69/2	680 °C	1.69 μm
48/2	720 °C	1.90 μm

Ordering was found by TEM for all specimens but to different extents in [1 $\bar{1}$ 1] and [$\bar{1}$ 1 1] directions. Due to the shape of the observed superlattice reflections, the dimensions of the ordered domains were found to be widely spread in the range of a few micrometers. The ordered domains are mostly arranged like flat discs. In another case Morita *et al.* [4] reported on lateral dimensions of the ordered regions in the range up to 10^4 nm^2. The percentage of the ordered volume was determined by quantitative x-ray phase analysis. For this purpose the superlattice reflection intensities had been compared to those of the corresponding second order reflection intensities of (Ga,In)P {1 1 1} lattice planes. Therefore intensities were calculated taking into consideration the appropriate corrections of x-ray diffractometry (absorption correction, polarization, and Lorentzian factors). It is advantageous that the second order reflections of the GaAs {1 1 1} lattice planes are forbidden in cases of non-anomalous scattering conditions. The results are given in Table II.

Table II. Proportion of ordered volume (in vol %) for two orientations of the c axis.

Specimen	R 3 m hex. c axis parallel $F\ \overline{4}\ 3\ m$ [1 $\overline{1}$ 1]	R 3 m c hex c axis parallel $F\ \overline{4}\ 3\ m$ [$\overline{1}$ 1 1]
60/2	2.6 ± 0.2	0.3 ± 0.2
69/2	8.7 ± 0.2	1.1 ± 0.2
48/2	6.7 ± 0.2	1.2 ± 0.2

1.2   Experiment

DAFS experiments were carried out at the Hamburg Synchrotron Radiation Laboratory HASYLAB at the undulator beamline BW1 using a Si 1 1 1 double-crystal monochromator with fixed exit. The eight-circle diffractometer available at this measuring station was used to obtain the normal of the <1 1 1> directions of the (Ga,In)P layers into the scattering plane (sample angle ω, tilt angle of the sample goniometer ψ). For tracking the Bragg angle θ versus energy E in the range of 800 eV above the Ga K absorption edge (about 10,369 eV) to get the reflection intensity $I(E)$ the accuracy of the sample goniometer was sufficiently high ($0.001°$) because the superlattice reflections of (Ga,In)P were broad (a full width at half maximum of about $0.4°$ in θ). The reflection intensities and the Ga K fluorescence intensities were recorded simultaneously by two thermoelectrically cooled Si-pin photodiodes [13] using Keithley 427 and 428 current amplifiers for the photocurrent measurement, respectively. The Ga K fluorescence intensity was used for the absorption correction. The incoming monochromatic beam was limited by slits (2.62 mm in ω direction, and 0.7 mm parallel to the ω axis) and monitored by an ionization chamber situated between the slits and the specimen. A 1.09-mm slit in front of the reflection detector (in the θ direction) reduced the fluorescence background. The energy-dependent background was measured near the Bragg angle of the superlattice reflection, and subtracted from the measured reflection intensity. The polarization vector of the incident radiation was parallel to the normal of the scattering plane. For the Bragg angle tracking of specimens the sample goniometer was moved using a second-order polynomial $\omega(E)/\theta(E)$ which was obtained by a diffraction pre-experiment to find out the peak positions at several energies in the scanning range. The DAFS experiments at the superlattice reflections ran within the energy range of 10,200-11,000 eV with energy steps of 2 eV and measuring times of 5-15 s depending on the reflection intensities. At a mean Bragg angle ω of 5°, about 60% of the radiation intensity interacts with the layer. For specimen 48/2, Figure 1 shows the measured reflection intensities $I_M(E)$ of the superlattice reflections in [1 $\bar{1}$ 1] and [$\bar{1}$ 1 1] directions. The simultaneously measured normalized Ga K fluorescence intensity of the (Ga,In)P layer is shown in Figure 2. The Ga K fluorescence intensity of the GaAs substrate is negligible because of the maximum exit angle of only 3° (there is about a 0.04-mm path length in the layer) given by the position of the fluorescence detector.

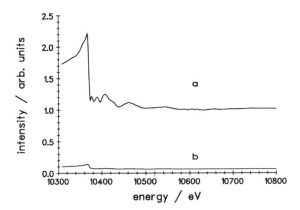

Figure 1. Measured DAFS reflection intensities of the $R\ 3\ m$ 0 0 3 superlattice reflections versus photon energy, hexagonal c axis parallel to [1 $\bar{1}$ 1] (a) and [$\bar{1}$ 1 1] (b) directions of the sphalerite $F\ \bar{4}\ 3\ m$ structure of specimen 48/2.

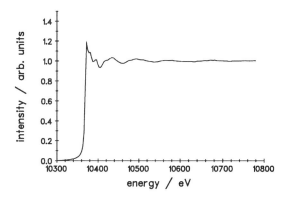

Figure 2. Ga K fluorescence intensity versus photon energy (background corrected and normalized) measured simultaneously with DAFS shown in Figure 1.

1.3   Quantitative DAFS Analysis

The quantitative DAFS analysis started with an absorption correction of the measured reflection intensities $I_M(E)$ because these intensities are superimposed by the x-ray-absorption fine structure (XAFS) averaged over all anomalous scatterers of the specimen volume, which in our case were far more than the few atoms contributing to the superlattice reflections. The next steps were the calculation of the smoothed curve (without the fine structure) of reflected intensities and the adaptation of the medium run of the corrected measured intensity to gain the oscillating part of $I_M(E)$, from which the complex-valued fine structure function was obtained by applying an iterative Kramers-Kronig algorithm. Finally the short-range order parameters were calculated by modelling the theoretical fine-structure function and by comparing it with the experimental one as extracted from DAFS signals. The absorption correction was done by dividing the measured reflection intensities by an absorption correction factor $A(E)$:

$$I(E)=I_M(E) / A(E). \tag{1}$$

$A(E)$ can be calculated by integrating over the different possible path lengths of the x-ray photons in the layer of thickness t:

$$A(E)=\{1 - exp[-2\mu(E)\, t\, /sin\theta\, /cos\psi]\}\, sin\theta\, cos\psi / (2\mu(E)) \tag{2}$$

with

$$\mu(E)=\mu_e(E)\, [1-\chi(E)] + \mu_b(E), \tag{3}$$

where $\mu_e(E)$ is the absorption coefficient of the edge atom, $\mu_b(E)$ is the total absorption coefficient of all nonedge atoms, $\chi(E)$ is the XAFS contribution to the absorption coefficient of the edge atom, θ is the Bragg angle, and ψ is the tilting angle of the sample. The total absorption coefficient $\mu(E)$ of the (Ga,In)P layer shown in Figure 4 was calculated using Eq. (3) including the mass absorption coefficients for the elements P, Ga, and In (Figure 3) [14] and a mass density $\rho = 4.459$ g/cm^3 according to [15].

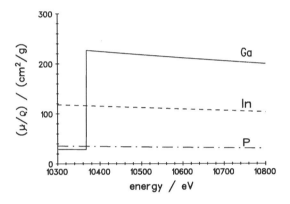

Figure 3. Total mass absorption coefficients of the layer components versus photon energy [14].

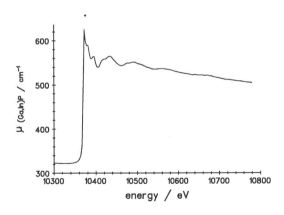

Figure 4. Energy dependence of the total absorption coefficient of the (Ga,In)P-layer obtained from both Ga K XAFS measurement and theoretical data shown in Figure 3.

Inserting $\mu(E)$ in Eq. (2) then gives $A(E)$ as shown in Figure 5. The corrected reflection intensities $I_M(E)/A(E)$ for the superlattice reflections of specimen 48/2 are shown in Figure 6. It was assumed that the ordered domains were equally distributed with depth. In fact a small change of thickness should influence the DAFS analysis only weakly. The reflection intensity $I(E)$ is proportional to the square of the structure factor, $|F(E)|^2$. The kinematic scattering approximation should be justified because of the small size of superlattice domains.

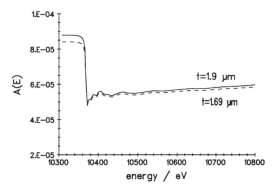

Figure 5. Absorption correction function $A(E)$ for the superlattice reflections versus photon energy, (sample 48/2: t=1.9 μm, sample 69/2: t=1.69 μm).

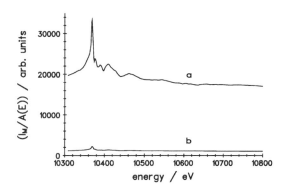

Figure 6. Absorption-corrected DAFS reflection intensities ($R\,3\,m$ 0 0 3 superlattice reflection, hexagonal c axis parallel to [1 $\bar{1}$ 1] (a) and [$\bar{1}$ 1 1] (b) directions of the sphalerite structure of specimen 48/2) versus photon energy.

For a description of superlattice reflections the structure factor had to be evaluated for the rhombohedral phase $R\,3\,m$ (space group number 160, 12 atoms per unit cell). The observed superlattice reflections correspond to the 0 0 3 reflection of this structure. The resulting structure factor for the 0 0 0 3 reflection (hexagonal axes) contains only contributions of Ga and In. This is true if the free parameter z of Wyckoff position 3a $(0, 0, z)$ of P atoms is equal to the ideal values of 0.125 and 0.625, respectively. Deviations from these ideal positions would yield contributions of P that can be excluded experimentally by the good agreement between calculated and measured (normalized) intensities. The structure factor F_0 (smooth part without contributions of fine structure) for the 0 0 3 reflection with nonresonant atomic scattering amplitudes f_{0Ga} and f_{0In} and their resonant (anomalous) parts $f'_{sGa}+if''_{sGa}$ and $f'_{sIn}+if''_{sIn}$ (resulting in corrections of the smooth curve without contributions of fine structure) is given by

$$F_0 = 3\,(f_{0Ga} + f'_{sGa} + if''_{sGa}) - 3\,(f_{0In} + f'_{sIn} + if''_{sIn}), \tag{4}$$

where the real part $\mathrm{Re}(F_0)$ and the imaginary part $\mathrm{Im}(F_0)$ can be combined to

$$\mathrm{Re}(F_0) = 3\,(f_{0Ga} + f'_{sGa} - f_{0In} - f'_{sIn}) \tag{5}$$

and

$$\mathrm{Im}(F_0) = 3\,(f''_{sGa} - f''_{sIn}). \tag{6}$$

Taking into account the oscillating fine-structure terms of the atomic scattering amplitudes f'_{osGa} and f''_{osGa} in our case of Ga as resonantly scattering atom, the complete structure factor F can be written as

$$F = F_0 + 3\,(f'_{osGa} + f''_{osGa}). \tag{7}$$

In accordance with Sorensen *et al.* [9], the complex-valued fine-structure function $\chi = \chi' + i\chi''$ is connected to the oscillating terms of the atomic scattering amplitudes by $f'_{osGa} = f''_{sGa}\chi'$ and $f''_{osGa} = f''_{sGa}\chi''$. Neglecting the square terms in f'_{osGa} and f''_{osGa} the intensity becomes

$$I \sim |F_0|^2 + 2\,\mathrm{Re}(F_0)\,3f'_{osGa} + 2\,\mathrm{Im}(F_0)\,3f''_{osGa}, \tag{8}$$

or, with the notation of the complex-valued fine-structure function mentioned above

$$I \sim |F_0|^2 + 2\,\mathrm{Re}(F_0)\,3f''_{sGa}\,\chi' + 2\,\mathrm{Im}(F_0)\,3f''_{sGa}\,\chi''. \tag{9}$$

Finally we derive for the (Ga,In)P 0 0 3 reflection (R 3 m symmetry)

$$I \sim |F_0|^2 + 18[(f_{oGa} + f'_{sGa} - f_{0In} - f'_{sIn}) f'_{osGa} + (f''_{sGa} - f''_{sIn}) f''_{osGa}] \quad (10)$$

or

$$I \sim |F_0|^2 + 18f''_{sGa}[(f_{oGa} + f'_{sGa} - f_{0In} - f'_{sIn}) \chi' + (f''_{sGa} - f''_{sIn}) \chi'']. \quad (11)$$

Figure 7 shows the theoretical reflection intensities $|F_0|^2$. The resulting quotients of the absorption-corrected measured reflection intensities $I_M(E)/A(E)$ and $|F_0(E)|^2$ calculated with the atomic scattering factors for Ga and In [14], and the corrections calculated according to Cromer and Liberman [17] are shown in Figure 8 for sample 48/2.

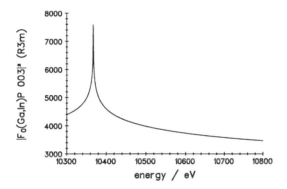

Figure 7. Theoretical reflection intensities $|F_0|^2$ versus photon energy near Ga K absorption edge.

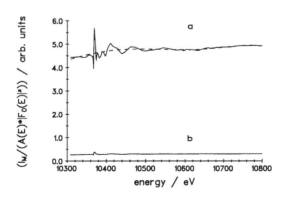

Figure 8. DAFS reflection intensities related to the theoretical reflection intensities $|F_0|^2$ (R 3 m 0 0 3 superlattice reflection, hexagonal c axis parallel to [1 $\bar{1}$ 1] (a) and [$\bar{1}$ 1 1] (b) directions of the sphalerite structure of specimen 48/2).

Normalized absorption-corrected measured intensities oscillating around $|F_0|^2$ of these superlattice reflection intensities are shown as an example in Figure 9.

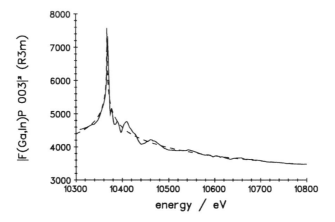

Figure 9. Normalized intensities (full line) and $|F_0|^2$ (dashed line) for $R\,3\,m\,0\,0\,3$ superlattice reflection, hexagonal c axis parallel to [1 $\bar{1}$ 1] direction of the sphalerite structure of specimen 48/2.

DAFS χ' and χ'' signals were derived using an iterative Kramers-Kronig algorithm and adapting the model function

$$I - |F_0|^2 = A\chi' + B\chi'' \tag{12}$$

to the measured extracted DAFS. The values A and B are equal to about 20 and 1, respectively. Therefore a first approximate value for χ' can be obtained by dividing the measured fine-structure contribution to $I(E)$ by A neglecting χ''. A Kramers-Kronig transformation gives χ'' from χ'. In our case one further iteration step already resulted in stable χ' and χ'' signals within the accuracy of the intensity measurement. The results from Kramers-Kronig analysis are shown in Figure 10 (filtered with respect to high frequency noise).

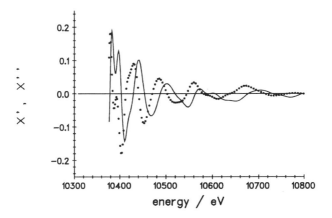

Figure 10. Real part χ' (line) and imaginary part χ'' (dots) of the complex valued fine structure function versus photon energy, ($R\,3\,m\,0\,0\,3$ superlattice reflection, hexagonal c axis parallel to [1 $\overline{1}$ 1] and [$\overline{1}$ 1 1] directions of the sphalerite structure of specimen 48/2).

For the quantitative Ga K XAFS and DAFS analysis we calculated the theoretical fine structure functions of the ideal structures of (Ga,In)P using the program MODEX [16], which is based on a single-scattering plane-wave model. Moreover, the theoretical scattering phase shifts and amplitudes of McKale *et al.* [18] were used for the present elements as backscatterers in Figures 11 and 12.

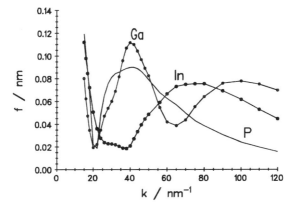

Figure 11. Electron wave backscattering amplitudes $f(k)$ for P, Ga and In as backscatterers at a distance of 0.25 nm ([18]).

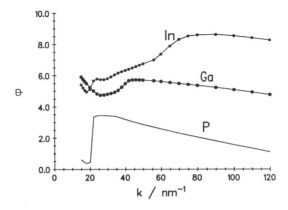

Figure 12. Electron wave phase shifts $\phi(k)$ for P, Ga and In as backscatterers at a distance of 0.25 nm ([18]).

Assuming a lattice parameter $a=0.566$ nm [15] for the ZnS-type structure (Ga,In)P, interatomic distances and effective coordination numbers of the neighborhood of Ga atoms are compiled in Table III, which also contains the effective coordination numbers in the case of ordering considering that the electrical field vector realized in our experiment was perpendicular to the plane spanned by the <1 1 1> and [0 0 1] vectors. The contributions expected from backscatterers to the Ga K XAFS shown in Figures 13 and 14, however, were calculated with the parameters of Table IV resulting from the fit to the measured XAFS values.

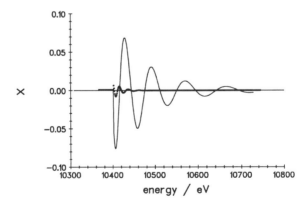

Figure 13. Expected contributions of different neighbors to the theoretical Ga K XAFS/DAFS (imaginary part): P neighbours at 0.24 nm (full line) and P atoms at 0.47 nm (bold line).

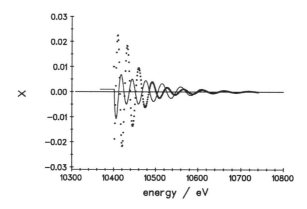

Figure 14. Expected contributions of different neighbors to the theoretical Ga K XAFS/DAFS (imaginary part): Ga atoms at 0.394 nm (full line) and In atoms at 0.402 nm (dots).

Table III. Local structure parameters of the Ga neighborhood in sphalerite (Ga,In)P and in the case of $R\,3\,m$ - type ordering as calculated from long-range ordered structure model.

Neighbor of Ga	Distance	Effective coordination number in $F\,\overline{4}\,3\,m$	Effective coordination number in $R\,3\,m$
P	0.245 nm	4	4
Ga	0.400 nm	6	9
In	0.400 nm	6	3
P	0.469 nm	12	12

From Figures 13 and 14 we conclude the following:

1. The Ga K XAFS and DAFS are dominated by the contributions of the first P neighbors of Ga.

2. The contributions of the next nearest P neighbors of Ga are small. They are restricted to the energy region corresponding to small values of the wave-vector.

3. The contributions of the nearest Ga and In backscatterers superimpose nearly opposite in phase for the main part of the interesting wave-vector region. This means the nearest Ga and In neighbors are not recognizable as "backscatterer shells," as expected from isolated estimation in the Fourier transform (Figure 15). Figure 16 proves, however, by comparison of the sum of the Fourier transforms of Figure 15 and the Fourier transform of the complete fine-structure function (sum of all contributions in the $E\,/\,k$ scale, respectively), that only the first P neighbors form an isolated backscatterer

shell. The other interference structures in the Fourier transform cannot be assigned definitely to such shells.

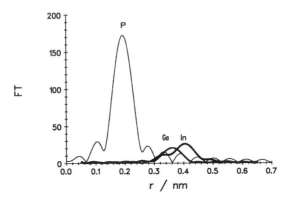

Figure 15. Fourier transforms *FT(r)* of the partial contributions of P, Ga and In in Figures 13 and 14 to the Ga K XAFS/DAFS, *r* is the interatomic distance with Ga in the center.

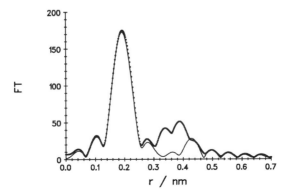

Figure 16. Sum of Fourier transforms *FT(r)* of Figure 15 (dots) and of the whole XAFS (full line).

XAFS and DAFS analysis proceeded with the following steps:

(i) Fourier transform of the fine-structure function (weighted by k^3 to obtain a sufficiently high position resolution of the neighbors near 0.4 nm).
(ii) Separate back-transformation of the next P neighbor contributions (0.12 nm to 0.25 nm) and of the adjacent region (0.25 nm to 0.60 nm) containing the contributions mainly of Ga and In backscatterers (back-filtering).

In addition, the back-transformation of the whole range (0.12 nm to 0.60 nm) gave a filtered (smoothed) fine-structure function. With the computer program mentioned above the local structure models for the Ga-P coordination and for the Ga-Ga/In coordination (paying attention to the next nearest Ga-P coordination) were fitted separately to the back-filtered functions. Finally the partial models were summed up and fitted to the filtered fine structure function resulting from the back-transformation of the whole range.

An example of the Fourier transform is presented in Figure 17 for specimen 48/2 whereas Figure 18 shows the corresponding filtered contributions.

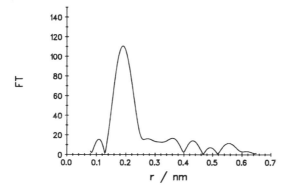

Figure 17. Fourier transform *FT(r)* of Ga K XAFS of specimen 48/2.

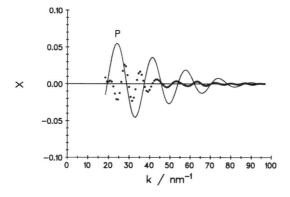

Figure 18. Back-transforms of contributions of the next P neighbors (full line) and the range of 0.25-0.60 nm (dots) of the Fourier transform in Figure 17.

The comparison between filtered partial/sum shells and fitted data based on the structure models in Figures 19 and 20 shows good agreement. Structure considerations lead to run the fit procedure under the constraint of an effective Ga/In coordination number 12 (in the ZnS type of structure of (Ga,In)P a Ga atom has on the average 6 Ga and 6 In neighbors). Table IV contains the local structure parameters for a Ga neighborhood resulting from the XAFS intensity of the whole (Ga,In)P layer.

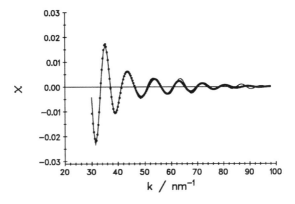

Figure 19. Filtered partial/sum shell data (dots) and fitted data based on the structure model (full line) (Ga K XAFS of specimen 48/2) excluding nearest P neighbors of Ga.

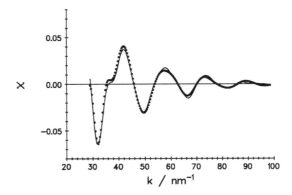

Figure 20. Filtered partial/sum shell data (dots) and fitted data based on the structure model (full line) (Ga K XAFS of specimen 48/2). Parameters of the structure model are listed in Table III.

Table IV. Local structure model of the Ga neighborhood in (Ga,In)P as derived from XAFS data.

Neighbor of Ga	Effective coordination number	Distance	σ^2
P_1	2.2 ± 0.5	(0.239 ± 0.002) nm	0.00006 nm^2
Ga	6.5 ± 1.3	(0.394 ± 0.005) nm	0.00024 nm^2
In	5.5 ± 1.1	(0.402 ± 0.005) nm	0.00043 nm^2

   The DAFS results (examples of Fourier transforms in Figures 21 and 22 gained from superlattice reflections of the specimens 48/2 and 69/2 in the manner described for XAFS) in comparison to XAFS show significant differences relating to the contributions of the nearest P neighbors. Fitting of the structure model of the ordered domains to experiment was performed, again constraining the effective Ga/In coordination number 12. However, since Ga and In ordering occurs on separate planes, there will be weighted contributions of 9 Ga atoms and 3 In atoms as next-nearest neighbors in <1 1 1> directions of the ZnS type of structure of (Ga,In)P following one another. This relationship between Ga and In effective coordination numbers should apply only to large domains with the ideal $R\,3\,m$ structure (ordering) and the polarization vector of the synchrotron radiation perpendicular to the $R\,3\,m$ hexagonal c axis (parallel to <1 1 1> directions of $F\,\overline{4}\,3\,m$ (Ga,In)P). In reality these conditions are not fulfilled exactly. Nevertheless we may notice trends looking at deviations of XAFS results from the ideal (expected) effective coordination numbers, and may also analyze data which are strongly influenced by the theoretical scattering cross-section data [18].

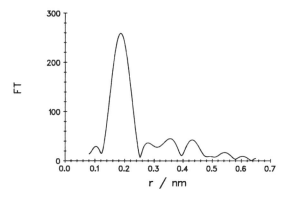

Figure 21. Fourier transform FT(r) of Ga K DAFS: hexagonal c axis of $R\,3\,m$ structure parallel to [1 $\overline{1}$ 1] direction of the sphalerite structure of specimen 48/2.

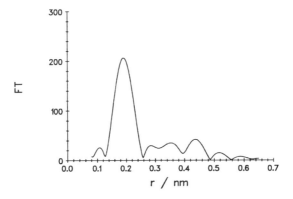

Figure 22. Fourier transform FT(r) of Ga K DAFS: hexagonal c axis of *R 3 m* structure parallel to [1 $\bar{1}$ 1] direction of the sphalerite structure of specimen 69/2.

The parameters of the local structure models as derived from DAFS experiments using superlattice reflections are listed in Table V.

Table V. Local structure models for ordered regions as derived from DAFS data.
(a) Specimen 60/2, (b) Specimen 69/2, (c) Specimen 48/2, *R 3 m* c axis (hex. axis) parallel to the *F $\bar{4}$ 3 m* [1 $\bar{1}$ 1] direction. (d) Specimen 48/2, *R 3 m* c axis (hex. axis) parallel to the *F $\bar{4}$ 3 m* [$\bar{1}$ 1 1] direction.

Neighbor of Ga	Effective coordination number	Distance	σ^2
(a)			
P_1	4.1 ± 0.9	(0.246 ± 0.002) nm	0.00007 nm^2
Ga	6.5 ± 1.3	(0.395 ± 0.002) nm	0.00016 nm^2
In	5.5 ± 1.1	(0.411 ± 0.002) nm	0.00015 nm^2
(b)			
P_1	4.0 ± 0.8	(0.246 ± 0.002) nm	0.00007 nm^2
Ga	7.0 ± 1.4	(0.396 ± 0.002) nm	0.00015 nm^2
In	5.0 ± 1.0	(0.412 ± 0.002) nm	0.00018 nm^2
(c)			
P_1	4.2 ± 0.9	(0.245 ± 0.002) nm	0.00006 nm^2
Ga	7.1 ± 1.5	(0.398 ± 0.002) nm	0.00014 nm^2
In	4.9 ± 1.0	(0.412 ± 0.002) nm	0.00015 nm^2
(d)			
P_1	4.2 ± 0.9	(0.246 ± 0.002) nm	0.00005 nm^2
Ga	7.9 ± 1.6	(0.399 ± 0.002) nm	0.00016 nm^2
In	4.1 ± 0.9	(0.413 ± 0.002) nm	0.00014 nm^2

2.   DISCUSSION

Two standard models of local structure in solid solution crystals are of relevance here: The virtual-crystal model (VCM) describes the alloy as an arrangement of atoms at ideal sites in the corresponding unit cell. The assumption is that atoms remain at their sites despite compression or expansion of the unit cell caused by alteration of composition. Various authors ([9], [19]) gave examples of non-conformity of the VCM with the results of XAFS experiments obtained by alloy semiconductors which had overcome the averaging character of diffraction experiments. The latter, however, could be better understood by applying the local structure model (LSM) that describes local distortions of the structure, and takes into consideration that the binary components with tetrahedral bonds tend to compensate for changes of bonding angles by maintaining the bond lengths.

Sorensen et al. [9] showed that their results of DAFS experiments of $In_xGa_{1-x}As$ were comprehensible only on the basis of the LSM. In the present case the (Ga,In)P solid solution can be understood as a composition of Ga-P and In-P components. In case of GaP the nearest-neighbor distance Ga-P amounts to 0.236 nm ($a_{GaP} = 0.5451$ nm [15]) that should match the results of XAFS experiments for (Ga,In)P when we follow the LSM. The VCM would be appropriate to describe XAFS results which show nearest neighbor Ga-P distances of 0.245 nm corresponding to a lattice parameter $a_{(Ga,In)P} = 0.5660$ nm [15].

Changes of bond angles cause changes of the angle between bond direction and the direction of the polarization vector of the synchrotron radiation. This may influence the effective coordination numbers of XAFS and, therefore, DAFS results.

The average nearest neighbor distance Ga-P for the whole (Ga,In)P layer of about 0.239 nm (as a result of XAFS) is in between the two situations discussed above. Obviously, within the experimental error our value agrees better with the LSM than with the VCM. Any superposition of contributions has to take into account the weight which is due to the orientation of the electric-field vector. Thus the small value of 2.2 nearest P neighbors of Ga (4.0 were expected) could be understood by changes of bond directions. However, experimental effects due to the limited take-off angle of the fluorescence cannot be excluded. If the signal was in fact coming from the entire depth of the film and somewhat from the substrate, then additional self-absorption not considered would have a significant damping effect on the fluorescence XAFS. Another possible explanation for the deviation from the expected value is the damping of the XAFS amplitude caused by static and thermal disorder. Deviations from symmetric pair distribution functions caused by local distortions may yield comparable effects. In the present case the dependence of the backscattering amplitude of P on the wavevector k as shown in Figure 11 at lower k values is similar to the influence of damping on a constant backscattering amplitude, because it is difficult to distinguish between influences of coordination number and damping. Nevertheless the deviation from the expected value is an open question which should be the subject of further investigation. Thus the fit result may be formulated more generally by stating that the nearest neighbors of Ga deviate from ideal positions in the unit cell throughout the whole (Ga,In)P layer, while showing nearest-neighbor distances similar to those of the binary compound GaP.

DAFS experiments with different samples (comparable orientation of the polarization direction of the electric-field vector as discussed in the case of XAFS experiments) yielded nearest-neighbor distances between 0.245 nm and 0.246 nm. These values match the VCM. The effective coordination numbers of about 4 nearest P neighbors of Ga agree with the values expected for an occupation of ideal positions of the compound. The nearest neighborhood of Ga in the whole (Ga,In)P layer can be described by the LSM, while in the ordered regions Ga is surrounded according to the VCM.

Nearest Ga-Ga neighbor distances (about 0.394-0.399 nm) resulting from XAFS and DAFS measurements are comparable within the error limits, and are less than the average value expected for (Ga,In)P (0.400 nm). The nearest Ga-In neighbor distance for the whole (Ga,In)P layer (result of XAFS) amounts to 0.402 nm, which is comparable to the average value expected. For the ordered regions, from DAFS, nearest Ga-In neighbor distances of 0.411-0.413 nm have been derived that are significantly larger than the average value of (Ga,In)P. One has to consider that in ordered regions the nearest In neighbors of Ga form separate planes, and only neighbors in particular directions contribute to the DAFS. All this indicates a distortion parallel to the c axis of the ordered $R\,3\,m$ phase (hexagonal setting). This agrees well with the observed Bragg angle shift of all superlattice reflections involved compared to random packing.

From superlattice reflection intensities as well as from the ratio of effective Ga/In coordination numbers (next-nearest Ga neighbors), we conclude that coherently ordered regions possibly enlarge with increasing growth temperature. We observed a change from 6.5 : 5.5 (Ga:In) in the case of specimen 60/2 (comparable to the average result of XAFS) to 7.0 : 5.0 in the case of specimen 69/2 and 7.1 : 4.9 as well as 7.9 : 4.1 in the case of specimen 48/2. There is no significant difference in ordering for both <1 1 1> directions, while differences of superlattice reflection intensities usually hint at different numbers of coherently scattering atoms.

3.   CONCLUSIONS

The average coordination of Ga in partially ordered epitaxial (Ga,In)P layers has been determined by XAFS experiments, while the local structure around Ga in ordered regions (few percent of the whole volume) was explored by DAFS experiments using superlattice reflections.

The local structure around Ga in the whole (Ga,In)P layer ($F\ \overline{4}\ 3\ m$) can be described by the local structure model. Contrary to that, results of the nearest neighborhood of Ga in the ordered regions ($R\ 3\ m$) follow the virtual-crystal model. This means that ordering not only causes a reduction of symmetry but causes changes in bonding characteristics. Thus, the solid solution in the ordered region can no longer be considered as a simple superposition of the binary components GaP and InP. Hence ordering of (Ga,In)P may be driven by electronic characteristics of interfaces between ordered and disordered structures resulting from lattice expansion in the directions of ordering.

ACKNOWLEDGMENTS

Our thanks are due to Dr. V. Gottschalch (Institut für Anorganische Chemie, Universität Leipzig) for providing the (Ga,In)P samples. The authors are also indebted to J. Weigelt and R. Frahm for inspiring discussions and their assistance with the measurements. Financial support by the BMBF is gratefully acknowledged.

REFERENCES

1. R.H. Horng and M.K. Lee, J. Appl. Phys. 71 (3) (1992).
2. O. Ueda, M. Takikawa, J. Komeno, and I. Umebu, Jpn. J. Appl. Phys. 26 L1824 (1987).
3. O. Ueda, M. Takechi, and J. Komeno, Appl.Phys.Lett. 54 2312 (1989).
4. E. Morita, M. Ikeda, O. Kumagai, and K. Kaneko, Appl. Phys. Lett. 53 2164 (1988).
5. D.C. Meyer, J. Weigelt, R. Frahm, K. Richter, and P. Paufler, Annual report of HASYLAB 1994, 389.
6. D.C. Meyer, A. Seidel, K. Richter, J. Weigelt, R. Frahm, and P. Paufler, X-96, 17th International Conference X-Ray and Inner-Shell Processes, Hamburg, Abstracts 340 (1996).
7. I. Arcon, A. Kodre, D. Glavic, and M. Hribar, J. Phys. (Paris), Colloq. 48, C9-1105-1108 (1987).
8. H. Stragier, J.O. Cross, J.J. Rehr, L.B. Sorensen, C.E. Bouldin, and J.C. Woicik, Phys. Rev. Lett. 69, no.21, 3064-3067 (1992).

9.   L.B. Sorensen, J.O. Cross, M. Newville, B. Ravel, J.J. Rehr, H. Stragier, C.E. Bouldin, and C.E. Woicik, in 'Resonant Anomalous X-Ray Scattering,' G. Materlik, C.J. Sparks, and K. Fischer (eds.), Elsevier Science B.V., 389-420 (1994).

10. M.G. Proietti, H. Renevier, J.F. Berar, V. Dalakas, J.L. Hodeau, G. Armelles, and J. Garcia, J. Phys. France 7 C2-749-751 (1997).

11. D.C. Meyer, K. Richter, A. Seidel, J. Weigelt, R. Frahm, and P. Paufler, Journal of Synchrotron Rad. 5, 1275 (1998).

12. V. Gottschalch, R. Franzheld, I. Pietzonka, R. Schwabe, G. Benndorf, and G. Wagner, Cryst. Res. Technol. 32 69-82 (1997).

13. D.C. Meyer, K. Richter, A. Seidel, K.-D. Schulze, R. Sprungk, and P. Paufler, Annual report of HASYLAB 1995 II, 989.

14. International Tables for X-Ray Crystallography, Vol. III, R. Reidel Publishing Company, Dordrecht/ Boston/ Lancaster/ Tokyo (1985).

15. S. Adachi, J. Appl. Phys. 56 (1982) 8775-8792.

16. S. Moldenhauer, Diploma thesis, TU Dresden, 1992.

17. D.T. Cromer and D. Liberman, J. Chem. Phys. 53, 1891-1898 (1970).

18. A.G. McKale, B.W. Veal, A.P. Paulikas, S.-K. Chan, and G.S. Knapp, Journal of American Chemical Society 208, 3763 (1988).

19. Y. Takeda, H. Oyanagi, and A. Sasaki, J. Appl. Phys. 68 (9), 4513 (1990).

Chapter 9

Ballistic Electron Emission Microscopy and Spectroscopy Study of Ordering-Induced Band Structure Effects in $Ga_{0.52}In_{0.48}P$

M. Kozhevnikov and V. Narayanamurti
Gordon McKay Laboratory of Applied Science, Harvard University, 9 Oxford Street, Cambridge, Massachusetts 02138

Key words: ballistic electron emission microscopy, hot electron transport, spontaneous ordering, band structure

Abstract: Ordering induced changes in the band structure of GaInP are important with regard to its device applications, as well as for fundamental studies of atomic ordering. To characterize the structural and electronic properties of $Ga_{0.52}In_{0.48}P$ (written as $GaInP_2$ for simplicity), we have used the Ballistic Electron Emission Microscopy (BEEM) technique which allows the study of buried heterostructures with high spatial resolution. To make the most out of the BEEM capabilities, this work has detailed a quantitative study of the second voltage derivative (SD) of the ballistic electron emission spectra. Then, we associate two peaks observed in the SD-BEEM spectra of disordered $GaInP_2$ on n^+ GaAs substrate with the Γ and L conduction minima, whereas an additional third peak in the SD-BEEM spectra of ordered $GaInP_2$ we associate with the L-band splitting due to the ordering-induced "folding" of one (from four) L valley onto the $\overline{\Gamma}$ point. According to our results, this splitting is 0.13meV for the ordered $GaInP_2$ ($\eta \sim 0.5$). In addition, the BEEM images of ordered $GaInP_2$ samples show regions of enhanced and reduced current on the scale of ~ 0.5 μm, and we relate these BEEM contrast regions to be due to order parameter fluctuations or antiphase domain boundaries.

1.   INTRODUCTION

In the past decade, spontaneous CuPt ordering of many III-V alloys has been widely observed in vapor phase growth on (001) substrates. Atomic scale ordering in an alloy is the term used to describe the spontaneous formation of a structure where the solid composition is modulated along a particular crystallographic direction in the lattice with a period of two lattice spacings. In $Ga_xIn_{1-x}P$, extensive theoretical[1,2] and experimental[3] work has been carried out to study the effect of the ordering-induced reduction of the crystal symmetry on the structural, optical and transport properties of the ordered material. Fig. 1 shows a schematic crystal structure (top) and electronic band structure (bottom) of disordered and ordered material ($Ga_{0.5}In_{0.5}P$). While perfectly disordered GaInP is a random alloy with equal probability for occupation of each group III sublattice site by a Ga or In atom, perfectly ordered GaInP consists of a monolayer superlattice of Ga and In (111) planes. The most pronounced ordering effects on the band structure are the splitting at the valence band and the folding of the L conduction band on the $\overline{\Gamma}$ point.[1] Because this folded state is only slightly higher in energy than the conduction band Γ state and because the two states have the same symmetry, their interaction is strong, resulting in lowering of the conduction band Γ state and, as a consequence, in the band gap reduction for ordered $GaInP_2$.[1] The lowering of the fundamental bandgap with ordering relative to the random alloy was observed experimentally by a variety of different experimental techniques.[4,5] These variations have a complex dependence on the growth conditions such as growth rate, growth temperature and III-V ratio.[6] The effect of ordering is of great importance for technological applications of $Ga_xIn_{1-x}P$ (e.g., in semiconductor lasers, in visible light-emitting diodes, and in high efficiency tandem solar cells). To optimize these devices, control and understanding of the interface properties are essential. Yet, such fundamental parameters as heterojunction band offsets are not well understood. Experimentally, values of $Ga_{0.52}In_{0.48}P/GaAs$ conduction band offsets (ΔE_C) ranging from 30 to 390 meV have been reported.[7] Since the bandgap of $Ga_{0.52}In_{0.48}P$ (written as $GaInP_2$ for simplicity) changes with the ordering parameter, it is likely that ordering is responsible for some of the variation in previous ΔE_C measurements.[8] Recently, dependence of the material parameters on the degree of ordering was exploited in the device design of, for example, bistable polarization switches,[9] polarization convertors,[10] orientational superlattices,[11] visible light-emitting diodes[12] and semiconductor lasers.[13,14]

The electronic states of a partially ordered structure can be interpolated from those of the totally disordered and the perfectly ordered semiconductor.[2] The degree of ordering is usually described by the ordering parameter η, where $E_g(\eta)=E_g(\eta=0)-\eta^2\Delta E_g(\eta=1)$, with $\eta=0$ and $\eta=1$ to describe perfectly disordered and perfectly ordered material, respectively. $E_g(\eta=0)$, the bandgap of disordered $GaInP_2$, is 2.01 eV at low temperature[15] and $\Delta E_g(\eta=1)$, the maximum bandgap reduction for perfectly ordered

GaInP$_2$, is 0.47 eV.[16] Perfectly ordered material has not been observed. In the most ordered GaInP$_2$ structures, the highest ordering parameter $\eta \sim 0.6$,[17] and the degree of ordering is nonuniform on the local scale.

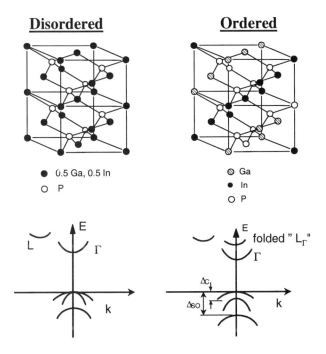

Figure 1. Schematic crystal structure (top) and electronic band structure (bottom) of disordered and ordered Ga$_{0.5}$In$_{0.5}$P. While perfectly disordered Ga$_{0.5}$In$_{0.5}$P is a random alloy with equal probability for occupation of each group III sublattice site by a Ga or In atom, perfectly ordered Ga$_{0.5}$In$_{0.5}$P consists of a monolayer superlattice of Ga and In (111) planes.

Ballistic Electron Emission Microscopy (BEEM), a three-terminal modification of scanning tunneling microscopy, has recently been shown to be a powerful tool for nanometer-scale characterization of the spatial and electronic properties of semiconductor structures. In this work, we report a quantitative study of the second voltage derivative (SD) BEEM spectra of Au/GaAs/GaInP heterostructure for probing the effect of ordering on multivalley hot electron transport in GaInP$_2$. The obtained SD-BEEM spectra of ordered and disordered GaInP$_2$ on n$^+$ GaAs substrates show a clear distinction that we associate with the L-band splitting due to the ordering-induced "folding" of one of the four L valleys onto the $\overline{\Gamma}$ point.

2.   BEEM TECHNIQUE

Among the variety of scanning probe microscopy techniques, Ballistic Electron Emission Microscopy, a three-terminal modification of Scanning Tunneling Microscopy (STM), is a powerful low-energy tool for nondestructive local characterization of semiconductor heterostructures. The STM metal tip (emitter) injects electrons across the tunneling gap into the metal (base) layer deposited on a semiconductor. A third terminal on the sample back is used to collect those electrons which traverse the interface. A schematic of the BEEM experimental set-up is shown in Fig. 2. In the conventional STM study of semiconductors, the tunneling is usually between the metal tip and the doped semiconductor layers. In contrast, the BEEM technique provides carrier filtration at the m-s interface, since only the electrons that can overcome the Schottky barrier and traverse the metal base will be collected at the semiconductor substrate (collector). Note that, since there is a certain electron filtration in the metal base layer, at the metal-semiconductor interface and in the semicondictor structure itself, the collected BEEM current is typically smaller than the tunneling current by several orders of magnitude, depending on the structure under study. A schematic band diagram for BEEM is shown in Fig. 3. In this way, the energy and angular distribution of hot electrons can be controlled independently of the semiconductor structure by simply varying the tip potential. In addition, the BEEM technique offers the extremely high spatial resolution of scanning tunneling microscopy (~few nm). Thus BEEM provides, in complement to the surface morphology, a combination of low-energy electron microscopy and spectroscopy with high spatial and energy resolution.

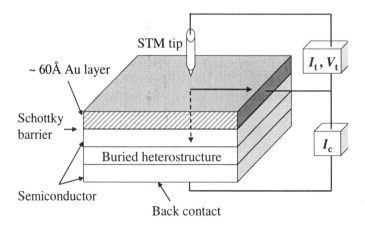

Figure 2. A schematic of BEEM experimental set-up.

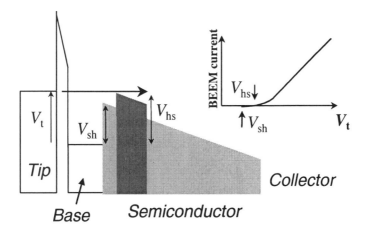

Figure 3. Schematic band diagram of a BEEM experiment. Inset shows a generic BEEM spectrum.

One of the advantages of BEEM spectroscopy is the possibility to study independently the conduction and valence band offsets of buried heterostructures. Indeed, applying a forward tip-to-base bias (the tip is more positive than the metal base layer), electron tunneling is possible from the base to the tip, creating a ballistic hole distribution in the base. Therefore, allowing the collector to be a p-type semiconductor substrate, one can collect the above-threshold hole current in a manner quite similar to the case of electron injection. In this way, BEEM can be extended for the case of hole injection, with a slight difference in that the energy distribution of ballistic holes is peaked towards the base Fermi level. The details of ballistic hole emission microscopy (BHEM) and spectroscopy (BHES) can be found, for example, in Refs. 18 and 19.

BEEM spectra exhibit thresholds at energies where additional semiconductor states become available for transport, and, thus, the primary purpose of BEEM spectroscopy is identifying these thresholds and correlating them to the semiconductor band structure. Two commonly used models, based on a planar tunneling formalism[20] and on the transverse momentum conservation at the metal-semiconductor (m-s) interface, are the Bell-Kaiser (BK) model[21] and the Ludeke-Prietsch (LP) model.[22] Recently, BEEM theory was extended to the case of buried heterostructures,[23] where transmission at the heterojunction interfaces in addition to the m-s interface was considered. The assumption of transverse momentum conservation, made in the above models, is questionable for the case of nonepitaxial m-s interfaces, which are not atomically abrupt. A deviation from the ballistic picture was experimentally observed, e.g., for Au/Si,[24] Pd/Si[25], and

Au/GaAs.[26,27] In order to consider electron scattering at the m-s interface, the m-s interface-induced scattering (MSIS) model was proposed in Refs. 28 and 29. Recently, the MSIS model was applied to describe quantitatively second derivative (SD) BEEM spectra for AlGaAs/GaAs heterostructures buried below a Au/GaAs interface.[30]

Since the pioneering work of Kaiser and Bell,[21] applications of BEEM to various semiconductor surfaces and interfaces have already produced many interesting results (for review, see Refs. 31 and 32). By using BEEM, Schottky barrier and band structure characterization was conducted in technologically important widebandgap semiconductors such as GaP,[33] GaInP,[34,35] GaN,[36-38] GaAsN[39] and SiC.[40,41] Although the BEEM technique was originally invented as a unique microscopic and spectroscopic method to probe Schottky barriers on a local scale, it has been successfully used for imaging and spectroscopy of buried quantum objects as well as for nondestructive local characterization of buried semiconductor heterostructures.[42]

3.   EXPERIMENTAL SET-UP

The undoped GaAs/GaInP$_2$ structures on n$^+$- and p$^+$- GaAs substrates were grown by metal organic vapor deposition (MOCVD), at T$_g$=650°C. The structures consist of a 500Å undoped GaAs buffer layer, a 1μm GaInP$_2$ epilayer and a 50Å GaAs cap layer. A detailed analysis is presented here for the GaInP$_2$ samples grown on a (511) GaAs substrate and on (001) GaAs substrates misoriented 6° toward [111]$_A$ and [111]$_B$. For comparison, we present also results for a 1 μm undoped GaAs layer grown on n$^+$- and p$^+$- GaAs, the reference GaAs samples. To fabricate diodes for BEEM, Au layers were deposited by thermal evaporation on the GaAs cap layer to form the metal base, and indium ohmic contacts were soldered to the back of the GaAs substrate to form the collector contact. The Au contacts were nominally 1 mm in diameter and 60 Å thick. The details of the fabrication procedure were published elsewhere.[27] The measurements were performed in a Surface/Interface AIVTB-4 BEEM/STM using a Au tip. Room temperature experiments were performed in air, while for the low temperature experiments, the STM head with a sample was immersed in cold He exchange gas in a nitrogen-cooled dewar. The tip-to-base voltage (V_t) was varied between 0.7 and 2 V to acquire the collector current (I_c) while keeping a constant tunneling current (I_t) of 4 nA. A typical BEEM current value is ~ 40 pA at 0.5 V above the threshold, and a typical noise level is about 0.5 pA. The spectra were typically averaged for several thousands scans to improve the signal-to-noise ratio.

4. RESULTS AND ANALYSIS

From the low-temperature photoluminescence, the bandgap energy (after correction for the exciton binding energy) is ~2.00 eV for GaInP$_2$ grown on a (511) GaAs substrate and ~1.88 eV for GaInP$_2$ grown on a 6°[111]$_B$-(001) GaAs substrate (see Fig. 4). Therefore, we conclude that the GaInP$_2$ layer grown on a (511) GaAs substrate is highly disordered (hereafter, disordered GaInP$_2$) and the GaInP$_2$ grown on a (001) GaAs substrate misoriented 6° toward [111]$_B$ is highly ordered, $\eta \sim 0.5$ (hereafter, ordered GaInP$_2$).[2] The GaInP$_2$ layer grown on a 6°[111]$_A$-(001) GaAs substrate has a bandgap energy of ~1.96 eV and, thus, can be characterized by moderate ordering ($\eta \sim 0.25$).

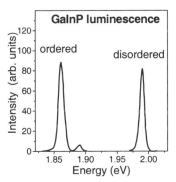

Figure 4. 10 K photoluminescence spectra for 1 μm GaInP$_2$ grown by MOCVD on (511) n-GaAs substrate (disordered sample) and for 1 μm GaInP$_2$ grown by MOCVD on n-GaAs substrate misoriented by 6° to [111]$_B$ (ordered sample).

Fig. 5 shows room-temperature 1 μm × 1 μm STM images of all three samples. Misorienting the GaAs substrates by 6° towards the [111]$_A$ and [111]$_B$ directions results in the formation of steps parallel to the [$\bar{1}$10] and [110] directions on these vicinal surfaces, respectively. While the surface of the GaInP layer grown on an n-GaAs substrate misoriented by 6° to [111]$_A$ consists of a faceted pyramid and pit structure, the morphology of the GaInP grown on a n-GaAs substrate misoriented by 6° to [111]$_B$ shows a ridged structure that is typical for a vicinal substrate. For ordered GaInP$_2$ grown on misoriented substrates, the [110] steps are usually observed to form only a single CuPt variant.[43,44] For disordered GaInP$_2$ grown on a (511) GaAs substrate, the surface is found to be much flatter than that of ordered GaInP$_2$.

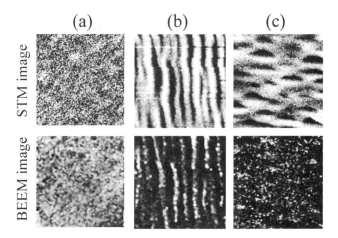

Figure 5. Room-temperature 1μm × 1μm STM images (top) and BEEM images (bottom) of GaAs/GaInP$_2$ layer grown by MOCVD on (a) (511)A n-GaAs substrate, (b) n-GaAs substrate misoriented by 6° to [111]$_B$ and (c) by MOCVD on n-GaAs substrate misoriented by 6° to [111]$_A$. The tip bias is -1.7V and the tunnel current is 4nA. Observed ridged structure in STM and BEEM images reflects step bunching during growth.

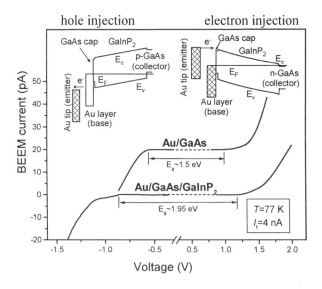

Figure 6. The 77K BEEM spectra of Au/GaAs/GaInP$_2$ and reference Au/GaAs samples grown on n-type (right side) and p-type (left side) GaAs substrates. For clarity, the spectra are shifted along the vertical axis. Insets show a calculated band profile for the Au/GaAs/GaInP$_2$ structures for BEEM on p-GaAs (top left) and n-GaAs (top right) substrates.

The complementary BEEM images are also shown in Fig. 5. The observed contrast of the BEEM image for the ordered $GaInP_2$ sample is in direct (anti-phase) correlation with the surface morphology (although the BEEM current modulation is at most 5-10%), indicating high sensitivity and high spatial resolution of the BEEM technique.

Representative BEEM current dependencies on the applied tip-to-base voltage are presented in Fig. 6 for 1 μm undoped GaAs and disordered $GaAs/GaInP_2$ layers grown on n^+ and p^+ GaAs substrates. The sum of the Schottky barrier heights for the electron and hole injection equals approximately the bandgap energy of the semiconductor, namely ~1.5 eV for GaAs and ~1.95 eV for $GaInP_2$. The band offsets are obtained by subtracting the BEEM threshold of the GaAs reference samples from the BEEM threshold of the $GaInP_2/GaAs$ samples. We found that the bandgap discontinuity is accommodated mostly by the valence band, namely the $GaAs/GaInP_2$ conduction band offset is ~0.1 eV and the $GaAs/GaInP_2$ valence band offset is ~0.35 eV, in general agreement with previously reported experiments.[34,45]

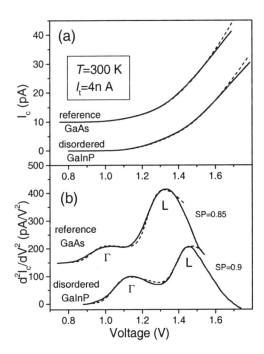

Figure 7. The room-temperature BEEM (a) and SD-BEEM (b) spectra of reference GaAs and $GaAs/GaInP_2$ grown on n^+ GaAs substrates. The MSIS model calculations (dashed lines) are also presented (SP is the electron scattering probability at the m-s interface).

Fig. 7 shows the room-temperature BEEM and SD-BEEM spectra for the disordered GaInP$_2$ sample and the reference GaAs sample. The SD-spectra were obtained from the experimental BEEM spectra by numerical differentiation with a 10 meV window. The SD-BEEM current is approximately the heterostructure transmission coefficient,[23] and, therefore, allows an explicit energetic partitioning of the transport channels, i.e., different peaks in the second derivative spectra correspond to the different conduction states. We associated the two main features (peaks) observed in the SD-BEEM spectra with the Γ-like and L-like conduction minima in disordered GaInP$_2$. Theoretical fits to the SD-BEEM spectra for the reference GaAs and disordered GaInP$_2$ samples, using the m-s interface-induced scattering (MSIS) model,[28,30] are shown in Fig. 7 by the dashed lines. The MSIS model-fit describes the experimental SD-BEEM spectra reasonably well, giving a ~85-90% probability of the electron scattering at the m-s interface, similar to our previous results for GaAs/Al$_x$Ga$_{1-x}$As structures.[30] The absence of a contribution from the X conduction minimum is due to strong X-electron attenuation in the GaAs cap layer.[30]

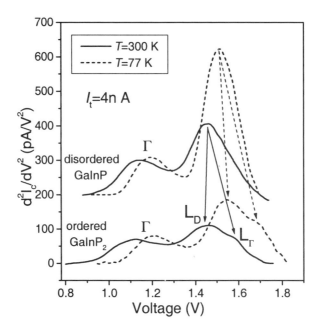

Figure 8. SD-BEEM spectra for two GaAs/GaInP$_2$ (ordered and disordered GaInP$_2$) samples with 50 Å GaAs cap layers, taken at T=300 K (solid curves) and T=77 K (dashed curves). For clarity, the spectra are shifted along the vertical axis.

Fig. 8 shows the SD-BEEM spectra for the disordered and ordered GaInP$_2$ samples for T=77 K and T=300 K. As the temperature decreases from 300 K to 77 K, a certain increase of the signal is observed for the L-electrons for both samples, in addition to the spectrum shift expected from the temperature dependence of the energy gap. We explain this result by the reduced L-electron scattering at low temperatures, similar to our previous study.[30] For both temperatures, the SD-BEEM spectrum of the ordered GaInP$_2$ sample presented in Fig. 8 shows a very important difference from that of the disordered GaInP$_2$. Namely, we observe two high-energy peaks instead of one peak in disordered GaInP. It is very unlikely that we start to observe the X valley contribution, because both GaInP$_2$ samples (ordered and disordered) contain the GaAs cap layer and, therefore, the X-electrons attenuation is expected to be the same. We assign both high-energy peaks to be associated with the L valley contribution. In CuPt-type ordered GaInP$_2$, only one of the four L valleys folds onto the $\overline{\Gamma}$ point (hereafter L_Γ), and the other three are folded onto the $\overline{D}$ point (hereafter L_D). A strong repulsion between the Γ valley and the folded L_Γ valley results in the bandgap reduction and in the increase of the Γ- L_Γ separation, while the energetic position of the L_D remains almost the same. Therefore, we conclude that the two observed high-energy peaks for the ordered sample are the contribution of the L valleys that are split due to ordering.

As a consequence of the L valley splitting in the ordered sample, the absolute height of the L peaks in the SD-BEEM spectrum decreases compared to the L valley contribution in the disordered sample, but the overall area under the two peaks is about the same as the area under the single L peak for disordered GaInP. The experimentally obtained ratio of ~2.5 between the integrated contribution of L_D and L_Γ is very close to the expected value of 3 (considering only the number of folded valleys without taking into account the effective mass modifications).

The "folded" zone-edge bands were observed experimentally in electro-reflectance[46] and Raman spectroscopy[47] measurements. In the recently reported electro-absorption experiments on ordered GaInP$_2$ ($\eta \approx 0.45$),[48] an additional feature was observed at ~ 0.48 eV above the fundamental bandgap transition, and this feature was attributed to the back folded L conduction band. As the ordering decreases, this peak shifts to lower energies, with an asymptotic value of ~ 0.33 eV above the fundamental bandgap transition for a totally disordered sample. In Refs. 46-48, due to the selection rules, only the contribution from the L valley folded onto the $\overline{\Gamma}$ point was observed. In contrast, we observe the contribution from all L valleys and, as a consequence, can measure directly the Γ-L separation in disordered GaInP$_2$ as well as the ordering-induced L valley splitting in ordered GaInP$_2$.

According to our results, $\Delta(\Gamma\text{-}L) \cong 0.35$ eV for a disordered sample, $\Delta(\Gamma\text{-}L_\Gamma)$ $\cong 0.47$ eV and $\Delta(L_\Gamma\text{-}L_D) \cong 0.13$ eV for an ordered sample. These results are in a good agreement with the theoretical predictions. Indeed, as pointed out by Zunger,[49] it is possible to obtain the dependencies $\Delta(\Gamma\text{-}L_\Gamma) = \Delta(\Gamma\text{-}L)_{\eta=0} + 0.50\eta^2$ and $\Delta(L_\Gamma\text{-}L_D)=0.42\eta^2$, using Table I and Fig. 3(b) of Ref. 50. Then, taking $\eta=0.5$ and $\Delta(\Gamma\text{-}L)_{\eta=0}=0.35$ from our experiment, $\Delta(\Gamma\text{-}L_\Gamma) \cong 0.475$ eV and $\Delta(L_\Gamma\text{-}L_D) \cong 0.11$ eV.

The BEEM images of ordered GaInP$_2$ samples show regions of enhanced and reduced current on the scale of ~0.5 µm, as shown in Fig. 9. We assume that these regions are related to the ordered domains with a different degree of ordering or to antiphase domain boundaries. To support such an explanation, we carried out the BEEM spectroscopy over the sample area. The observed low contrast in BEEM current variations as well as the measured difference of < 20meV in the threshold indicate that the ordering parameter fluctuations among the domains are relatively small ($\Delta\eta<0.1$). In the future, we plan to carry out a comparative analysis of BEEM and BEEM-induced luminescence (BEEL). By detecting photons emitted from a BEEM process, sensitive optical techniques may be used for detection and analysis. Since the BEEL energy is related directly to the bandgap energy, such a comparative analysis holds great promise in spatial probing of submicrometer-sized highly ordered domains in GaInP$_2$/GaAs heterostructures.

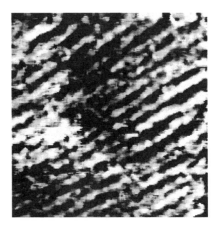

Figure 9. 2 µm × 2 µm BEEM image of ordered GaInP sample (grown on n-GaAs substrate misoriented by 6° to [111]$_B$). Observed ridged structure in BEEM image on the scale of ~ 1000Å reflects step bunching during growth.

5.　CONCLUSION

This study has detailed a quantitative study of the second voltage derivative of the ballistic electron emission spectra of a Au/GaAs/GaInP$_2$ heterostructure for probing the effect of ordering on the multivalley hot electron transport. We associate two peaks observed in the SD-BEEM spectra of disordered GaInP$_2$ on an n$^+$ GaAs substrate with the Γ and L conduction minima, $\Delta(\Gamma\text{-}L) \sim 0.35$ eV. An additional third peak appearing in the SD-BEEM spectrum of ordered GaInP$_2$ ($\eta \sim 0.5$) is associated with the L-band splitting due to the ordering-induced "folding" of one of the four L valleys onto the $\overline{\Gamma}$ point. According to our results, this splitting is ~ 0.13 eV. In addition, the BEEM images of ordered GaInP$_2$ samples show regions of enhanced and reduced current on the scale of ~0.5 μm, and we consider these BEEM contrast regions to be due to the order parameter fluctuations or antiphase domain boundaries.

ACKNOWLEDGMENTS

We would like to thank our collaborators A. Mascarenhas, Y. Zhang and J.M. Olson of National Renewable Energy Laboratory as well as D.L. Smith of Los Alamos National Laboratory, with whom some original part of the work reported here was done. We also acknowledge helpful correspondence with Alex Zunger of National Renewable Energy Laboratory. This work was supported by NSF under Grants # ECS 99-96093 and # DMR 98-09363 and AFOSR under Grant # F49620-97-10247. We also thank the Midwest Research Institute for partial support.

REFERENCES

[1]　S.-H. Wei and A. Zunger, Appl. Phys. Lett. **56,** 662 (1990).

[2]　S.-H. Wei, D. B. Laks, and A. Zunger, Appl. Phys. Lett. **62,** 1937 (1993).

[3]　Zhang, A. Mascarenhas, P. Ernst, F. A. J. M. Driessen, D. J. Friedman, C. Geng, F. Scholz, and H. Schweizer, J. Appl. Phys. **81,** 6365 (1997).

[4]　G. S. Horner, A. Mascarenhas, R. G. Alonso, S. Froyen, K. A. Bertness, and J. M. Olson, Phys. Rev. B **49,** 1727 (1994).

[5]　H. Lee, M. V. Klein, J. M. Olson, and K. C. Hsieh, Phys. Rev. B **53,** 4015 (1996).

[6]　S. R. Kurtz, J. M. Olson, and A. Kibbler, Appl. Phys. Lett. **57,** 1922 (1990).

[7]　S. L. Feng, J. Krynicki, V. Donchev, J. C. Bourgoin, M. Difortepoisson, C. Brylinski, S. Delage, H. Blanck, and S. Alaya, Semicond. Sci. Technol. **8,** 2092 (1993) and references therein.

8   J. J. O'Shea, C. M. Reaves, M. A. Chin, S. P. DenBaars, A. C. Gossard,
    V. Narayanamurti, and E. D. Jones, Mat. Res. Soc. Symp. Proc. **417,** 79 (1996).

9   E. Greger, P. Riel, M. Moser, T. Kippenberg, P. Kiesel, and G. H. Döhler, Appl. Phys.
    Lett. **71,** 3245 (1997).

10 R. Wirth, A. Moritz, F. Scholz, and A. Hangleiter, Appl. Phys. Lett. **69,** 2225 (1996).

11 Y. Zhang and A. Mascarenhas, Phys. Rev. B **55,** 13100 (1997).

12 M. K. Lee, R. H. Horng, and L. C. Haung, J. Appl. Phys. **72,** 5420 (1992).

13 K. Nakano, A. Toda, T. Yamamoto, and A. Ishibashi, Appl. Phys. Lett. **61,** 1959 (1992).

14 R. Wirth, A. Moritz, C. Geng, F. Scholz, and A. Hangleiter, Phys. Rev. B **55,** 1730
    (1997).

15 M. C. DeLong, D. J. Mowbray, R. A. Hogg, M. S. Skolnick, J. E. Williams, K. Meehan,
    S. R. Kurtz, J. M. Olson, R. P. Schneider, M. C. Wu, and M. Hopkinson, Appl. Phys. Lett.
    66, 3185 (1995).

16 P. Ernst, C. Geng, F. Scholz, H. Schweizer, Y. Zhang, and A. Mascarenhas, Appl. Phys.
    Lett. **67,** 2347 (1995).

17 B. T. McDermott, K. G. Reid, N. A. El-Masry, S. M. Bedair, W. M. Duncan, X. Yin, and
    F. H. Pollak, Appl. Phys. Lett. **56,** 1172 (1990).

18 M. H. Hecht, L. D. Bell, W. J. Kaiser, and L. C. Davis, Phys. Rev. B **42,** 7663 (1990).

19 L. D. Bell, W. J. Kaiser, M. H. Hecht, and L. C. Davis, in *Methods of Experimental
    Physics 27: Scanning Tunneling Microscopy*, edited by J. A. Stroscio and W. J. Kaiser
    (Academic Press, New York, 1993).

20 J. G. Simmons, J. Appl. Phys. **34,** 1793 (1963).

21 L. D. Bell and W. J. Kaiser, Phys. Rev. Lett. **61,** 2368 (1988).

22 R. Ludeke and M. Prietsch, J. Vac. Sci. Technol. **A9,** 885 (1991).

23 D. L. Smith and S. M. Kogan, Phys. Rev. B **54,** 10354 (1996).

24 L. J. Schowalter and E. Y. Lee, Phys. Rev. B **43,** 9308 (1991).

25 R. Ludeke, Phys. Rev. Lett. **70,** 214 (1993).

26 W. J. Kaiser and L. D. Bell, Phys. Rev. Lett. **60,** 1406 (1988).

27 J. J. O'Shea, E. G. Brazel, M. E. Rubin, S. Bhargava, M. A. Chin, and V. Narayanamurti,
    Phys. Rev. B **56,** 2026 (1997).

28 D. L. Smith, E. Y. Lee, and V. Narayanamurti, Phys. Rev. Lett. **80,** 2433 (1998).

29 D. L. Smith, M. Kozhevnikov, E. Y. Lee, and V. Narayanamurti, Phys. Rev. B **61,** 13914
    (2000).

30 M. Kozhevnikov, V. Narayanamurti, C. Zheng, Y.-J. Chiu, and D. L. Smith, Phys. Rev.
    Lett. **82,** 3677 (1999).

31 M. Prietsch, Phys. Rep. **253,** 163 (1995).

32 L. D. Bell and W. J. Kaiser, Annu. Rev. Mater. Sci. **26,** 189 (1996).

33 A. Bauer, M. T. Cuberes, M. Prietsch, and G. Kaindl, Phys. Rev. Lett. **71,** 149 (1993).

34 J. J. O'Shea, C. M. Reaves, S. P. DenBaars, M. A. Chin, and V. Narayanamurti, Appl.
    Phys. Lett. **69,** 3022 (1996).

35 M. Kozhevnikov, V. Narayanamurti, A. Mascarenhas, Y. Zhang, J.M. Olson, and
    D. L. Smith, Appl. Phys. Lett **75,** 1128 (1999).

36 L. D. Bell, R. P. Smith, B. T. McDermott, E. R. Gertner, R. Pittman, R. L. Pierson, and
    G. L. Sullivan, J. Vac. Sci. Technol. B **16,** 2286 (1998).

37 E. Brazel, M. A. Chin, and V. Narayanamurti, Appl. Phys. Lett. **74,** 2367 (1999).

38 L. D. Bell, R. P. Smith, B. T. McDermott, E. R. Gertner, R. Pittman, R. L. Pierson, and
    G. J. Sullivan, Appl. Phys. Lett. **76,** 1725 (2000).

39 M. Kozhevnikov, V. Narayanamurti, C. V. Reddy, H. P. Xin, C. W. Tu, A. Mascarenhas,
    and Y. Zhang, Phys. Rev. B **61,** R7861 (2000).

[40] H.-J. Im, B. Kaczer, J. P. Pelz, and W. J. Choyke, Appl. Phys. Lett. **72,** 839 (1998).

[41] B. Kaczer, H.-J. Im, J. P. Pelz, J. Chen, and W. J. Choyke, Phys. Rev. B **57,** 4027 (1998).

[42] For review, see, e.g., V. Narayanamurti, Sci. Rep. Res. Inst. Tokohu Univ. A, Phys. Chem. Metall. **44**, 165 (1997).

[43] G. B. Stringfellow and G. S. Chen, J. Vac. Sci. Technol. B **9,** 2182 (1991).

[44] D. J. Friedman, J. G. Zhu, A. E. Kibber, J. M. Olson, and J. Moreland, Appl. Phys. Lett. **63,** 1774 (1993).

[45] D. Biswas, N. Debbar, P. Bhattacharya, M. Razeghi, M. Defour, and F. Omnes, Appl. Phys. Lett. **56,** 833 (1990).

[46] S. R. Kurtz, J. Appl. Phys. **74,** 4130 (1993).

[47] S. H. Kwok, P. Y. Yu, and K. Uchida, Phys. Rev. B **58,** R13395 (1998).

[48] T. Kippenberg, J. Krauss, J. Spieler, P. Kiesel, G. H. Dohler, R. Stubner, R. Winkler, O. Pankratov, and M. Moser, Phys. Rev. B **60,** 4446 (1999).

[49] A. Zunger, private communications.

2[50] S.-H. Wei, A. Franceschetti, and A. Zunger, Phys. Rev. B **51,** 13 097 (1995).

Chapter 10

Cross-Sectional Scanning Tunneling Microscopy as a Probe of Local Order in Semiconductor Alloys

Jeremy D. Steinshnider and Michael B. Weimer
Texas A&M University, Department of Physics, College Station, TX, U.S.A.

Mark C. Hanna
National Renewable Energy Laboratory, Golden, CO, U.S.A.

Keywords: cross-sectional scanning tunneling microscopy, order parameter, GaInP, GaInAs

Abstract: Cross–sectional scanning tunneling microscopy (STM) is used to examine and characterize spontaneous ordering in metal-organic vapor phase epitaxy grown GaInP and GaInAs. Under appropriate growth conditions, these alloys exhibit $CuPt_B$ type order in which the group III cations form a monolayer superlattice of alternating Ga (III_A) and In (III_B) planes oriented along either $<111>-B$ direction. Atomic-resolution STM images of these ordered ternaries display III_A-III_B site discrimination allowing direct, real-space visualization of the $CuPt_B$ arrangement. The same STM images likewise permit a quantitative assessment of the magnitude and range of the local order throughout selected regions of either material by facilitating reconstruction of the In–In pair correlation function. This pair correlation function is used to define a local order parameter, analogous to the convential Bragg-Williams order parameter, that enables direct comparison of the cross-sectional STM data with optical or x-ray measurements. The spatial evolution of the order parameter near the alloy / buffer interface is studied in the case of GaInP on GaAs, where the STM images reveal antiphase boundaries that presumably play a role in the local onset of $CuPt_B$ order. The demonstrated capacity to visualize and quantify the degree of order in these alloy systems at near-atomic length scales with STM suggests the technique may prove an especially powerful tool for understanding and controlling the growth of ordered materials.

Cross-sectional scanning tunneling microscopy (STM) has been widely used to characterize atomic-scale composition fluctuations, isovalent intermixing, and interfacial structure in a variety of III–V semiconductor multilayers and heterostructures [1-9] for some time. Only recently, however, has the technique been employed to investigate ordered GaInP [10,11] and to image the

273

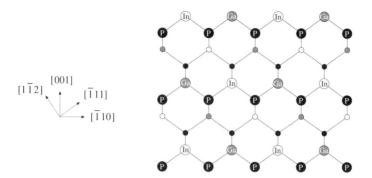

Figure 1. (110) cleavage surface of ideal, single-variant, $CuPt_B$ ordered GaInP. Alternating In and Ga planes with ordering vector $[\bar{1}11]$ intersect this surface along $<1\bar{1}2>$ directions.

natural $(InP)_1(GaP)_1$ superlattice associated with $CuPt_B$ ordering on an atomic scale. These early exploratory studies uncovered evidence of a new ordering arrangement in GaInP — $(InP)_2(GaP)_1$ [10] — as well as of long–period modulations in the electronic structure of $(InP)_1(GaP)_1$ multilayers [11].

As described at greater length here, cross-sectional STM facilitates structural and compositional analysis of ordered epitaxial layers through III_A–III_B site discrimination on the atomic scale. This precision enables direct visualization and quantitative analysis of the local $CuPt_B$ order in GaInP and GaInAs films that, in turn, can shed unprecedented light on the respective roles played by interfaces, ordering defects, and growth conditions in controlling the evolution of cation sublattice order during vapor phase epitaxy.

The spatially varying order naturally present in lattice-matched GaInP or GaInAs films on (001)-oriented GaAs or InP substrates may be examined by cleaving a sample in ultra-high vacuum to expose a (110) crystal plane in cross section, and subsequently positioning an STM tip over the areas of interest. Since the {111}–*B* superlattice planes associated with both $CuPt_B$ variants are normal to this cleavage face, alternating <112> rows of either III_A (gallium) or III_B (indium) constituents will be revealed at the surface as shown in Fig. 1.

High-resolution STM contours can then be used to distinguish between individual III_A and III_B cations on the basis of their distinct valence orbitals and covalent radii. For example, because an In—P back bond is slightly longer than a Ga—P one, top-layer indium atoms will sit above the surface plane defined by surrounding gallium atoms, as indicated with the schematic cation contour shown in Fig. 2, and thus appear somewhat brighter than their neighbors. The corresponding empty-state STM image should therefore display alternating <112> rows of III_A and III_B sites stacked along the $[\bar{1}11]$ ordering direction, as illustrated in Fig. 3.

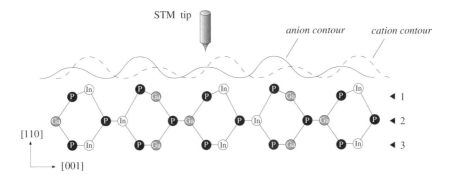

Figure 2. Schematic (110) STM contours for ideal CuPt$_B$ order.

Similar discrimination between geometrically inequivalent group–V sites — for example, phosphorous atoms back bonded to gallium versus those back bonded to indium (Fig. 2) — is likewise possible with high-resolution anion contours. The same logic dictates in this case that III$_B$–V bonded anions appear somewhat brighter than III$_A$–V ones (Fig. 3), mirroring the arrangement of subsurface cations and, thus, the CuPt$_B$ order, within a neighboring (110) plane directly beneath the cleavage surface. Since there is no *a priori* reason to prefer any one (110) plane over another, anion or cation images may be interchangeably relied on, as convenient.

Quantitative information concerning the extent, as well as the spatial dependence, of CuPt$_B$ order in non-ideal samples is available through the In–In pair correlation function [12],

$$g^{(2)}(\mathbf{R}) = \frac{1}{x_{In}^2}\left\{\frac{N_{In-In\,pairs}(\mathbf{R})}{N_{cation\,pairs}(\mathbf{R})}\right\}, \tag{1}$$

which can be reconstructed with atomic-resolution STM data. This quantity provides an estimate of the magnitude, range, and anisotropy of the statistical correlations between selected pairs of cleavage-exposed III$_B$–V-like lattice sites by contrasting the actual distribution of these sites, as a function of their in-plane separation vector $\mathbf{R}$, with a random distribution of the same density. This comparison amounts to normalizing the number of In–In pairs observed (with a given $\mathbf{R}$) to the total number of cation pairs surveyed (with the same $\mathbf{R}$) in such a way that a random distribution produces a pair correlation equal to unity for all possible lattice vectors in the plane (Fig. 3). Given a random distribution comprised of equal numbers of III$_A$ and III$_B$ atoms ($x_{In} = 0.5$), only

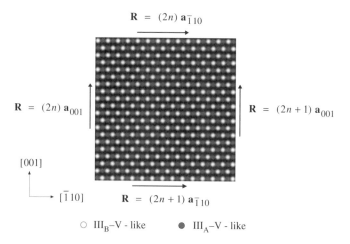

Figure 3. Schematic (110) STM image for ideal CuPt$_B$ order.

one quarter of the sites connected by a specified vector will yield In–In as opposed to In–Ga or Ga–Ga pairs.

The In–In pair probability associated with perfect CuPt order, on the other hand, oscillates between one-half and zero in both the $[\bar{1}10]$ and [001] directions, depending upon whether the lattice vector $\mathbf{R}$ is an even or odd multiple of the corresponding primitive translation (Fig. 3); the ideal In–In pair correlation likewise oscillates between zero and two (Fig. 4) as a consequence of the normalization to a random distribution.

The Bragg-Williams order parameter, S, serves as a convenient metric for the long-range correlations between lattice sites exhibited by ordered alloys [13]. This parameter is unity in the case of perfect order but vanishes with a random arrangement; for partially ordered alloys, S lies between zero and one. Analysis of the pair correlation function reveals that the magnitude of its oscillatory deviation from unity is to be identified with the square of S under the assumption of long-range order. Schematic plots of the In–In pair correlation versus $[\bar{1}10]$ separation for ideal CuPt order ($S = 1$), for a realistic alloy ($S = 0.7$), and for a random alloy, are contrasted in Fig. 4.

Due to the finite sampling of III$_A$ and III$_B$ sites imposed by fixed-area STM images, there will also be uncertainties at each lattice separation arising from statistical error. This statistical error naturally limits one's ability to distinguish long–range order, with constant oscillation amplitude, from short–range order, with exponentially-decaying oscillation amplitude. Should $S(\mathbf{R})$ remain independent of $\mathbf{R}$ (within statistical uncertainty) up to the largest separations available, however, one can improve this local estimate of the Bragg-Williams

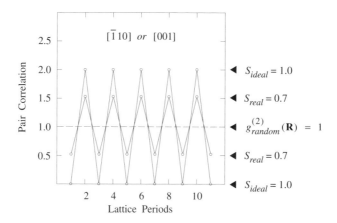

Figure 4. Schematic III_B–III_B pair correlation functions with ideal CuPt order, realistic CuPt order, and random III_B site occupancy.

order parameter by treating the measurements at distinct lattice vectors as independent observations that may be averaged together.

Atomic-resolution, cross-sectional STM images were used as just described to investigate the local order in GaInP and GaInAs films grown by low-pressure, metal-organic vapor phase epitaxy (MOVPE) on vicinal substrates. Triethylgallium, trimethylindium, phosphine and arsine were employed as precursors. The GaInP overlayer was deposited at 670°C on a *p*-type (001) GaAs wafer, miscut 4° toward $[\bar{1}11]$–*B*, with a V/III ratio of 260 and growth rate of 5 Å/s. The GaInAs overlayer was likewise deposited at 550°C on a *p*-type (001) InP wafer, miscut 6° toward $[\bar{1}11]$–*B*, with a V/III ratio of 240 and growth rate of 3.3 Å/s. X-ray analysis confirmed each sample was lattice-matched to its respective substrate. These conditions lead to reasonably strong $CuPt_B$ order as inferred from Raman and photoluminescence measurements.

Figure 5 shows a typical anion sublattice image across the GaInP–on–GaAs interface revealed through (110) cleavage of the GaInP sample. The GaAs buffer layer is the homogeneous region in the lower right portion of the scan, and the GaInP alloy film the more complex looking region on the upper left. Three successive $[\bar{1}11]$–*B* steps are resolved along the alloy/buffer interface and a number of phosphorous vacancies (dark sites) are evident within the alloy film. From this image and others one finds that the $[\bar{1}11]$–*B* inclination is 4.1° from (001), in good agreement with the 4° miscut specified for this substrate. The precision with which the geometrically inequivalent phosphorous atoms — those back bonded to indium and those back bonded to gallium — may be distinguished is illustrated in the accompanying surface section, which

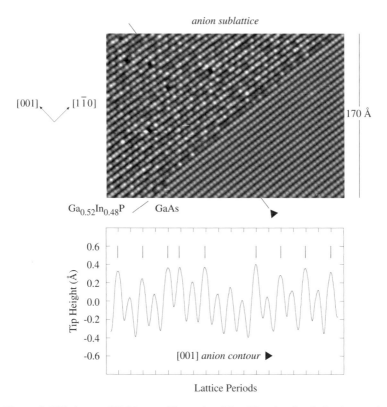

Figure 5. Filled-state STM image illustrating III_A–III_B site discrimination at the GaInP–on–GaAs interface.

begins within a few lattice periods of the GaAs buffer and moves outward from there along the (001) growth direction; tick marks indicate the III_B–V bonded anion sites identified with underlying indium atoms. The anticipated $CuPt_B$ pattern is recognizably regular, with the height difference between III_B–V-like and III_A–V-like sites amounting to about 0.2 Å. There are also occasional defects in this ordering sequence — for example, where two III_B–V-like phosphorous atoms follow one another in succession, or where one that is expected appears to be missing.

The development of local $CuPt_B$ order during vapor phase epitaxy may be traced by following the evolution in the In–In pair correlation function, and hence the order parameter, as a function of distance from the GaAs buffer. Examining the first 20 monolayers of the alloy film (outlined by the black box nearest the substrate in Fig. 6) one sees the faint emergence of short-range order with several, small-amplitude oscillations in the In–In pair correlation (left panel) that quickly decay toward a random distribution within a few lattice

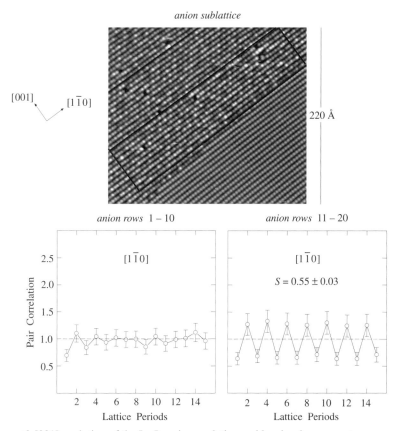

Figure 6. [001] evolution of the In–In pair correlation and local order parameter.

periods. It is important to note that this surprising behavior of the order parameter in the near-interfacial region does not appear to be caused by a change in local stoichiometry, since the indium fraction detected with STM here agrees, within experimental error, with the overall stoichiometry of the alloy layer concluded from x-ray rocking curves. The In–In pair correlation function apparently stabilizes over the next 20 monolayers (right panel), as the oscillations in $g^{(2)}$ approach their asymptotic amplitude deeper into the alloy film and the local order parameter converges on 0.55.

The evolution in the In–In pair correlation tracked with STM suggests at least 20 monolayers are needed to establish measurable $CuPt_B$ order across the GaInP–on–GaAs interface. This kind of insight into the emergence of order on the atomic scale is very difficult to obtain with more conventional optical (photoluminescence, Raman, far-infrared spectroscopy) and structural (TEM, x-ray diffraction) characterization techniques.

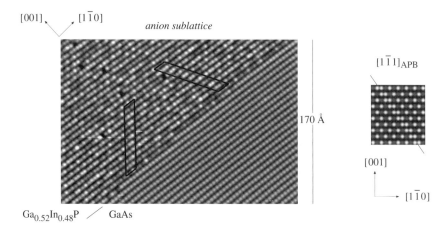

Figure 7. Atomically-abrupt antiphase boundaries near the GaInP–on–GaAs interface.

One may likewise use STM to correlate the onset of order with the step morphology and defect structure of the alloy film in the near-interfacial region over which the order parameter is evolving. Fig. 7, for example, outlines a potentially important ordering defect directly above a step in the GaAs buffer layer, where two adjacent [$\bar{1}$12]-oriented rows of phosphorous atoms back bonded to indium interrupt the expected CuPt$_B$ pattern. This structure suggests an antiphase boundary aligned with the (1$\bar{1}$1) plane, as illustrated in the conventionally-oriented schematic image to the right of the experimental one. The actual structure is not a perfect match, with two gallium- rather than indium-like sites in the third row, and another gallium–for–indium substitution further up, but this is entirely natural absent perfect order to begin with. There is also suggestion of a second, complimentary antiphase boundary, aligned with the ($\bar{1}$11) plane, that appears in (110) cross-section as the mirror image of the first. These ordering defects represent either a local slip along the cation ordering planes in the first instance, or, in the second, a stacking fault in the cation ordering sequence.

Fig. 8 shows a typical filled-state image of the ordered GaInAs sample in (110) cross-section together with the corresponding pair correlation analysis; the [$\bar{1}$12]- and [1$\bar{1}$2]-oriented rows of III$_B$–V-like anion sites are easily recognized and the indium fraction detected with STM again agrees with the lattice-matched stoichiometry indicated by x-ray diffraction. Although the number of sites sampled with this particular image is comparatively small, the pair correlation function nevertheless oscillates in the expected way, yielding a well-defined order parameter, $S = 0.45 \pm 0.06$. This local order parameter is considerably greater than the corresponding global one ($S = 0.27$) determined from

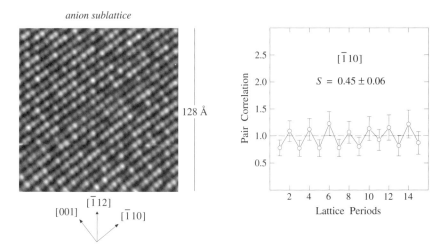

Figure 8. Filled-state STM image of ordered $Ga_{0.47}In_{0.53}As$ and corresponding pair correlation.

high-resolution x-ray diffraction measurements [14] of the [111] CuPt super-structure peak intensities over the entire sample. This difference presumably stems from the unavoidable averaging over domains of different size, orientational variant, and degree of order that such macroscopic measurements necessarily entail, again highlighting the potentially detailed structural insights that can only be obtained with cross-sectional STM.

These examples illustrate the enormous promise for advancing our understanding of spontaneous ordering in semiconductor alloys that springs from III_A–III_B site discrimination on the atomic-scale. And although not explicitly described, this discrimation may be straightforwardly extended to encompass single variant identification on the nanometer scale as well. The quantitative analysis of order facilitated by the introduction of a local order parameter based on the III_B–III_B pair correlation function likewise emphasizes the importance of atomic-resolution STM data. The utility of this construct has been demonstrated with a quantitative examination of the spatial evolution of order in the vicinity of the GaInP-on-GaAs interface revealed in (110) cross section. The evidence for atomically-abrupt antiphase boundaries at this heterojunction likewise suggests STM may prove equally useful for resolving the similar defect structures that delineate bulk ordered domains. Finally, careful correlation of the local order parameter with the atomic-scale morphology and composition of the alloy / buffer interface may make possible more stringent tests of current theories [15] concerning the onset of ordering in these materials, thereby guiding the future growth of ordered films, and related structures, of greatly improved quality.

ACKNOWLEDGMENTS

The authors are pleased to acknowledge the valuable assistance and kind encouragement of A.G. Norman (NREL), A. Mascarenhas (NREL), R.L. Forrest (University of Houston), S.C. Moss (University of Houston), and T.D. Golding (University of North Texas) throughout this work.

STM characterization at Texas A&M University is supported by the National Science Foundation, Division of Materials Research (DMR–9633011 and DMR–007316). MOVPE growth at NREL is supported by the U.S. Department of Energy, Office of Basic Energy Sciences, Division of Materials Sciences (DE-AC36-83CH100093).

REFERENCES

[1]   H.W.M. Salemink and O. Albrektsen, *Phys. Rev. B*. **47**, 16044 (1993).

[2]   J.F. Zheng, J.D. Walker, M.B. Salmeron, and E.R. Weber, *Phys. Rev. Lett.* **72**, 2414 (1994).

[3]   R.M. Feenstra, D.A. Collins, D.Z.-Y. Ting, M.W. Wang, and T.C. McGill, *Phys. Rev. Lett.* **72**, 2749 (1994).

[4]   M. Pfister, M.B. Johnson, S.F. Alvarado, H.W.M. Salemink, U. Marti, D. Martin, F. Morier-Genoud, and F.K. Reinhart, *Appl. Phys. Lett.* **67**, 1459 (1995).

[5]   K.-J. Chao, C.-K. Shih, D.W. Gotthold, and B.G. Streetman, *Phys. Rev. Lett.* **79**, 4822 (1997).

[6]   A.Y. Lew, S.L. Zuo, E.T. Yu, and R.H. Miles, *Phys. Rev. B*. **57**, 6534 (1998).

[7]   J. Harper, M. Weimer, D. Zhang, C.-H. Lin, and S.S. Pei, *Appl. Phys. Lett.* **73**, 2805 (1998).

[8]   J. Steinshnider, M. Weimer, R. Kaspi, and G.W. Turner, *Phys. Rev. Lett.* **85**, 2953 (2000).

[9]   J. Steinshnider, J. Harper, M. Weimer, C.-H. Lin, S.S. Pei, and D.H. Chow, *Phys. Rev. Lett.* **85**, 4562 (2000).

[10] N. Liu, C.-K. Shih, J. Geisz, A. Mascarenhas, and J.M. Olson, *Appl. Phys. Lett.* **73**, 1979 (1998).

[11] A.J. Heinrich, M. Wenderoth, M.A. Rosentreter, K. Engel, M.A. Schneider, R.G. Ulbrich, E.R. Weber, and K. Uchida, *Appl. Phys. A*. **66**, S959 (1998).

[12] L. Guttman, *Solid State Physics* **3**, 146 (1956).

[13] B.E. Warren, *X-ray Diffraction* (Addison Wesley, Reading, 1969).

[14] R.L. Forrest, E.D. Meserole, R.T. Nielsen, M.S. Goorsky, Y. Zhang, A. Mascarenhas, M. Hanna, and S. Francoeur, *Mat. Res. Soc. Symp. Proc.* **583**, 249 (2000).

[15] See, for example, the companion chapters in this volume by A.G. Norman, G.B. Stringfellow, and T. Suzuki, respectively.

Chapter 11

The Physics of Tunable Disorder in Semiconductor Alloys

Angelo Mascarenhas and Yong Zhang
National Renewable Energy Laboratory, Golden, Colorado, U.S.A.

Key words: ordering, alloy statistics, band structure, optical anisotropy, band offset, GaInP

Abstract: A review of the key changes in electronic properties that result from spontaneous ordering in III-V semiconductor alloy is presented. The intrinsic as well as extrinsic effects of the phenomenon are reviewed. The band structure changes and resulting optical anisotropy, the order parameter and the effects of controllable alloy statistical fluctuations on optical properties, orientational domain boundaries and the formation of orientational superlattices, the band-offsets between ordered GaInP and GaAs, and the effects of microstructural features such as anti-phase boundary defects on optical spectra are discussed. Wherever applicable, both the experimental and theoretical aspects of the phenomenon are examined to illustrate the current status of the field.

1. INTRODUCTION

During the past three decades the explosive progress in the area of electronic and photonic devices has been possible largely because of technological advances in the growth and physical understanding of semiconductor alloys. The silicon-germanium heterojunction bipolar transistor, high electron mobility transistor, light emitting diode, semiconductor diode laser and solar cell are illustrative examples of the pivotal role alloy semiconductors have played as precursors of whole new technologies. The relentless pursuit towards miniaturization of electronic devices and the recent excitement in nanoscience and nanotechnologies have generated a demand for a much better understanding of semiconductor alloys and their properties at ultra-short length scales. Although there have been

283

significant advances made in understanding the physical properties of alloys on a macroscopic scale, it is anticipated that on a submicroscopic scale certain peculiarities will be manifested. Although the conventional approach to treating semiconductor alloys as substitutional solid solutions has yielded satisfactory results for macroscopic physical properties, it is unclear how this will breakdown due to the more violent effects of statistical fluctuations at the nanoscale. With regard to theoretical modelling, there has been a need for experimental guidance but little progress made due to the paucity of techniques for exploring this realm. The phenomenon of spontaneous ordering appears well suited for exploring the physical consequences of disorder because it provides a convenient avenue for controlling statistical fluctuations. The ability to controllably achieve desired order parameters using this process allows for the possibility of tailored disorder in a lattice and there has been a great deal of excitement towards understanding how the consequences of this tunable disorder are manifested on the electronic and optical properties of spontaneously ordered alloys. For example, one can choose to tailor disorder on the cation or anion sublattice and study the consequences this has on the scattering of Bloch states, phonons, and excitons. Spontaneous ordering changes the alloys symmetry which in turn brings about changes to the electronic and optical properties. Bandgap lowering, valence band splitting, effective mass anisotropy, birefringence, electron spin polarization, second harmonic generation and spontaneously generated electric fields are examples of such intrinsic symmetry induced changes that have been investigated in the past. The intrinsic effects that are the results of the alternation in statistical fluctuations in partially ordered alloys have received much less attention. These include the effects of alloy scattering on mobility, exciton linewidth, and lattice dynamics. The process of spontaneous ordering is inevitably associated with the formation of structural defects comprised of anti-phase boundaries, orientational domain boundaries, and spatial variations in the order parameter. These inherent microstructural changes that are extrinsic to the phenomenon of spontaneous ordering have been associated with peculiarities in the low temperature photoluminescence spectra, the formation of orientational superlattices, and the cancellation of spontaneous electric fields that have been predicted to exist in ordered alloys. Finally, the effect of spontaneous ordering on the band alignment between GaInP and GaAs has become the subject of recent investigation because of the manner in which this effects the performance of GaInP/GaAs heterojunction bipolar transistors.

Spontaneous ordering is a result of a short wavelength instability that results in a special point at a Brillouin zone boundary along the ordering axis collapsing onto the Brillouin zone centre. The process as observed in epitaxially grown semiconductor alloys is irreversible but results in group-subgroup relations characteristic of the Landau theory of structural phase transformations. Here, as opposed to ordering in bulk grown crystals, ordering is initiated at the surface of the epitaxially growing layer and as such is essentially controlled by kinetics and thermodynamics at the growth surface. It is the two dimensional structural transformation at the surface that evolves into the final observed three-dimensional structural transformation. In–situ optical characterization techniques such as Reflectance Difference Spectroscopy and Surface Photoabsorption Spectroscopy that probe changes occurring at the growth surface have been used to investigate details of the ordering mechanism. In this chapter we will provide a brief review of the various studies that have been done as regards the intrinsic and extrinsic changes to the electronic and optical properties that result from spontaneous ordering in epitaxially grown semiconductor alloys. The focus will primarily be on the alloy GaInP because this system has proven to be most amenable to experimental investigation

2.   ORDERING INDUCED BAND GAP REDUCTION AND VALENCE BAND SPLITTING

2.1   Band Gap Reduction

Long before CuPt ordering was actually observed in III-V semiconductors[1-13], Pikhtin[14] had already noticed the scatter in the values for the band gap of GaInP in the literature, and pointed out that it was possible that the disagreement between the experimental results obtained by different research groups for the $Ga_xIn_{1-x}P$ system were due to some ordering components in these solid solutions resulting from specific growth techniques. For ordered GaInP alloy, a band gap reduction was first inferred from the red shift of the photoluminescence (PL) peak[9], and later confirmed by an absorption measurement[15] by Gomyo et al. However, it was not clear whether the GaInP samples were fully or partially ordered. The fact that CuPt ordered samples were found to have different band gaps[16-18] and valence band splittings[19] logically led to the recognition of the partial ordering in GaInP alloys[19]. The experimentally measured band gap reduction (< 100 meV typically) was found to be significantly smaller than the band gap bowing obtained from the earliest band structure calculations for a fully CuPt ordered $Ga_{0.5}In_{0.5}P$: 455 meV of Wei and Zunger[20], and 330 meV of Kurimoto and Hamada[21]. The band gap bowing was defined as the band gap difference between the average value of the binaries and the value for the actual structure, where the band gaps were calculated using a self-consistent general-potential linear augmented-plane-wave (LAPW) method within the local-density-functional approximation (LDA)[20,21]. A direct comparison of these theoretical results with the experimental data was not possible without knowing the bowing for the disordered structure. However, the random structure was too difficult to calculate using this technique. Kurimoto and Hamada[21] believed that their result was in good agreement with the experimental data of Ref.[9], by assuming a large bowing already existing for the random structure. Wei and Zunger[20] pointed out a large discrepancy between the experimental result of Ref.[9] and their calculation. A few hypothesis were given for the discrepancy[20]: (1) partial ordering, (2) coexistence of different types of ordering, and (3) antiphase boundaries. Later, the band gap of the random structure, simulated by a so-called quasi-random structure (SQS)[22], was calculated using the LAPW method[23,24]. The band gap reduction for the fully ordered structure was calculated by Wei et al to be $\delta E_g(x = 0.5, \eta = 1) = E_g(x = 0.5, \eta = 0) - E_g(x = 0.5, \eta = 1) = 320$ meV[24], where η is the order parameter that can vary from 0 to 1. Note that the accuracy of this value relies on three assumptions: (1) the quasi-random structure can adequately simulate the random structure, (2) the LDA error is negligible for the energy separations among different conduction band critical points, and (3) the LDA error is approximately the same for different structures (either the random or ordered). Since the samples are only partially ordered (i.e., ($\eta < 1$), to compare experimental data with the above theoretical calculation one needs to (1) determine the order parameter experimentally and (2) know the functional form for $\delta E_g(x,\eta)$. Two approaches have been used to obtain the functional form of $\delta E_g(x,\eta)$: one is to directly calculate $\delta E_g(x,\eta)$ for different values of x and η[25-27], and the other is to seek an interpolation function between the end points $\eta = 0$ and 1[23]. Note that if one only knows the functional form but not the order parameter of the sample, the comparison is still impossible, although one may find such comparisons with claimed good agreements in the literature. Although the results of direct calculations by Capaz and Koiller[25] and by Mäder and Zunger[26] have been available for quite some time, the most widely used functional form for $\delta E_g(x,\eta)$ has been the so-called η^2 rule proposed by Laks et al[23]: $P(x, \eta) = P(x,0) + \eta^2 [P(X_\sigma,1) - P(X_\sigma,0)]$, where X_σ is the composition of the ordered structure. The combination of this scaling rule and the end-point value of $\delta E_g(0.5, 1) = 320$ meV has been extensively used for determining the order parameter of partially ordered samples. An indirect but meaningful method of making the comparison between experiment and theory was suggested by Zhang and Mascarenhas[28]: assuming the validity of the η^2 rule, the ratio $r = \delta E_g(\eta)/\Delta_{CF}(\eta) = \delta E_g(\eta = 1)/\Delta_{CF}(\eta = 1)$ should be a constant, and can be compared with experimental data. Here $\Delta_{CF}(\eta = 1)$ is the crystal-field splitting parameter for the fully order structure at x = 0.5. Utilizing the piezo-reflectance data of Alonso et al[29], they found r = 2.0 ± 0.1[28]. Subsequently, more accurate experimental data of Ernst et al[30] (measured by PL excitation, PLE, spectroscopy) yielded r = 2.36 ± 0.06, and of Fluegel et al[31] (measured by differential absorption using a time-resolved pump-probe technique) yielded 2.66 ± 0.15. Comparing to the theoretical value of r = 1.6 (with $\Delta_{CF}(1) = 0.20$ eV)[24], these experimental data indicated that the theory of Ref. [24] either underestimated $\delta E_g(1)$ or overestimated $\Delta_{CF}(1)$. A revised calculation of Wei and Zunger[32] has given new values of $\delta E_g(1) = 430$ meV and $\Delta_{CF}(1) = 160$ meV, resulting in r = 2.69, which is in good agreement with the experimental result.

Despite the good agreement that has been achieved in the above-mentioned comparison, it is still not possible to make a direct comparison of any individual physical property for a partially ordered sample. Another concern relates to how accurate the η^2 rule really is. Two techniques, NMR[33,34] and x-ray diffraction[35,36], have been used for determining the order parameter experimentally. For the NMR technique, two approaches have been adopted. One used by Tycko et al[33] was to analyze the relative areas of the ^{31}P NMR lines (there are five lines corresponding to five possible $Ga_nIn_{1-n}P$ clusters with n = 0 - 4). They found $\eta \leq 0.6$ for the ordered samples studied, but no explicit η dependence was given. The other used by Mao et al[34] was to analyze the NMR spin echo of ^{71}Ga with the help of a point-charge model. Only the result of one sample was given with relatively large error bars. Wei and Zunger[37] later pointed out that the point-charge model used in Ref.[34] was inadequate for modelling the experiment. The x-ray technique is perhaps the most traditional technique that has been used for the measurement of order parameters. Forrest et al[35] successfully applied this technique, in conjunction with optical measurements, to obtain the dependence of the band gap reduction and valence band splitting as a function of order parameter. Using the η^2 rule, they were able to get by extrapolation the end-point values of $\delta E_g(1) = 498 \pm 27$ meV and $\Delta_{CF}(1) = 189 \pm 11$ meV. These values are in fact the first experimentally obtained band structure parameters for the currently unachievable fully ordered GaInP. However, one has to remember that the validity of these values relies on the validity of the η^2 rule. In addition to this concern, a very recent x-ray diffraction study of Li et al[36] indicates that the domain size affects the order parameter derived from the modeling method used in Ref.[35] for samples with small domain sizes. A systematic investigation of this issue has not yet been accomplished.

The results of two early direct calculations[25,26] of the band gap for partially ordered structures (0 < η < 1) had largely been ignored, because of the convenient use of the η^2 rule proposed by Laks et al[23]. The result of

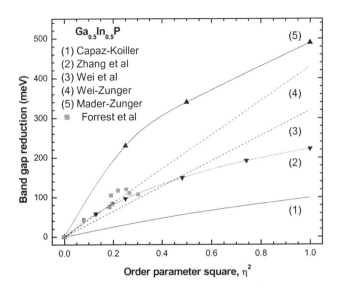

Figure 1. A comparison of the band gap reduction vs. order parameter between theoretical calculations and the experimental data. Theoretical results are from Ref.[25] (Capaz and Koiller), Ref.[27] (Zhang et al), Ref.[24] (Wei et al), Ref.[32] (Wei and Zunger), and Ref.[26] (Mäder and A. Zunger). Experimental data are from Ref.[35] (Forrest et al).

Ref.[25], $\delta E_g(\eta) = 130 \, \eta^2 - 30 \, \eta^4$ (meV), appears to have underestimated the band gap reduction, but this was the first attempt to directly calculate $\delta E_g(\eta)$. The result of Ref.[26], in fact, shows a strong deviation from the η^2 rule for $\delta E_g(\eta)$, but its end point value $\delta E_g(1) = 490$ meV appears to be in very good agreement with the extrapolated value of Forrest et al[35] using the η^2 rule, which presents an apparent paradox. The most recent calculation of Zhang et al[27] has yielded a value for $\delta E_g(\eta)$ that is in very good agreement with the experimental data of Forrest et al[35] available for $\eta <$ 0.55, but severely deviates from the η^2 rule for $\eta > 0.5$, with an end point value $\delta E_g(1) = 223$ meV.

Fig.1 compares $\delta E_g(\eta)$ obtained from all the three direct calculations[25-27], the η^2 rule[24,32], and the experimental data[35]. Among the three direct calculations, the most recently calculation of Zhang et al[27] appears to best match the experimental data. In this empirical pseudopotential calculation, the partial ordering has been more realistically modeled by using a ~3,500 atoms size supercell and averaging over 100 configurations, compared to Capaz and Koiller's tight binding calculation[25] using a 64 atom size unit cell and averaging over 400 structures or Mäder and Zunger's

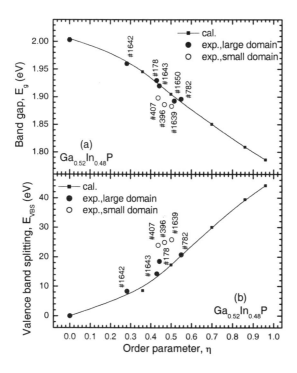

Figure 2. Band gap (Eg) and valence band splitting (EVBS) as functions of order parameter η for partially ordered $Ga_{0.52}In_{0.52}P$ alloys. Theoretical curves are from Ref.[27] (Zhang et al). Experimental data are from Ref.[35] (Forrest et al), except for data points for #178 and #782 (J. H. Li et al, unpublished).

pseudopotential calculation[26] using 32 atom size quasi-random structures. The empirical pseudopotential method of Ref. [27] so far is the only method capable of calculating the absolute band gap energy as a function of order parameter in good agreement with experimental data.

Fig.2 shows a comparison for the absolute band gap energy between the experimental data and the theoretical results for $x_{Ga} = 0.52$ (at which $Ga_xIn_{1-x}P$ is lattice matched to GaAs). One can see in Fig.2 that the data for samples with small crystalline domains do not agree with the theoretical curve as well as the data for samples with large domains. It is worth mentioning that the direct calculation of Ref.[27] confirms that the crystal field splitting parameter, $\Delta_{CF}(\eta)$, does approximately obey the η^2 rule. Although without experimental data for higher ordered samples, it is hard to judge which results of Ref. [27] and Ref.[32] for the band gap reduction is more accurate for $\eta > 0.55$, there is a logical difficulty in believing that the η^2 rule should be valid for the large η region[27]. If one views ordering as a perturbation of the random alloy, this perturbation causes a folding of the Brillouin-zone L point to the Γ point, and the repulsion between the folded L state and the original conduction band edge state may be considered as the primary contribution to the band gap reduction. A perturbation scheme proposed by Wei and Zunger[20] gives $\delta E_c \propto |<c,L|\Delta V|c, \Gamma>|^2/(E_{cL} - E_{c\Gamma})$. If the matrix element $|M_{L\Gamma}| = |<c,L|\Delta V|c, \Gamma>| \propto \eta$, one does have $\delta E_c \propto \eta^2$. However, to achieve the large band gap reduction of ~ 400 meV[32], the coupling would be too strong for this perturbation scheme to be valid, considering the fact that $\delta E_{L\Gamma} = E_{cL} - E_{c\Gamma}$ is ~ 350 meV for the random alloy. Thus, a higher order theory would naturally be expected to bring in higher order terms beyond the η^2 term. In general, without actually performing the calculation, it is not trivial to make a judgment as to whether or not a physical quantity $P(x,\eta)$ should follow the η^2 rule. An obvious reason for not taking the validity of η^2 rule for granted is that physical properties are not always linearly related to each other. Thus, their relationship to η^2 is not guaranteed to be linear, unless the ordering effect is very weak. In fact, as will be discussed in the next subsection, the dependence of the valence band splitting on η is a good example of how a more complicated η dependence emerges from the strong interaction amongst the valence bands, even though the crystal field splitting parameter $\Delta_{CF}(\eta)$ involved follows the η^2 rule reasonably well.

2.2   Valence Band Splitting

The ordering induced valence band splitting was first observed in a polarized PL measurement at room temperature by Mascarenhas et al[19]. Kanata et al obtained the valence band splittings for a set of samples with varying degree of order through temperature dependent PL measurements[38]. Usually, because of the involvement of the Boltzmann occupation factor in the emission process, PL is a less accurate technique for determining the critical points, comparing to other techniques like PLE spectroscopy[30,39] and modulation spectroscopy (electroreflectance[40], piezoreflactance[29], and differential absorption[31]). PLE could resolve the split valence band quite accurately, but could not access the spin orbit band. All the modulation techniques could resolve all the three valence band states near the band edge, but electroreflectance as well as piezoreflactance generally require a complex fitting procedure. The time-resolved pump-probe differential absorption technique used by Fluegel et al[31] appears to be the most accurate technique for this purpose. Fig.3 shows the experimental data obtained by PL[38], piezoreflactance[29], PLE[30], and differential absorption[31]. Fluegel et al[31] also found the spin-orbit splitting parameter $\Delta_{SO} = 103$ meV to be independent of the order parameter (up to $\eta \sim 0.6$).

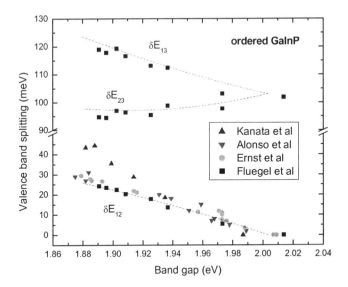

Figure 3. A comparison of the valence band splitting vs. the band gap reduction determined by using different experimental techniques: photoluminescence (Ref.[38] of Kanata et al), piezoreflectance (Ref.[29] of Alonso et al), photoluminescence excitation (Ref.[30] of Ernst et al), and differential absorption (Ref.[31] of Fluegel et al).

The valence band splitting can be approximately but conveniently described by the quasicubic model which was originally proposed by Hopfield for treating simultaneous perturbations of a uniaxial crystalline field and spin-orbit coupling to the triply degenerate Γ_{15} valence band[41]. This model had been used by Shay et al[42] for describing the valence band splitting and optical polarization in chalcopyrite ordered ZnSiAs$_2$ crystals. Not only had they pointed out that the valence band splitting and polarization dependence agreed with observations in stressed zinc-blende crystals, but also they used the concept of zone folding by stating "much additional structure is observed in ZnSiAs due to pseudodirect band gaps which result from the doubling of unit cell in chalcopyrite relative to zinc-blende. This change in the unit cell causes the Brillouin zone of zinc-blende to be imbedded into the smaller Brillouin zone of chalcopyrite". Wei and Zunger[43] and others[28,44] applied this model to CuPt ordered III-V alloys. The three valence band edge states are given as follows:[41-44]

$$E_1 = \frac{\Delta_{CF}}{3}, \tag{1}$$

$$E_{2,3} = -\frac{1}{2}(\Delta_{SO} + \frac{\Delta_{CF}}{3}) \pm \frac{1}{2}[(\Delta_{SO} + \Delta_{CF})^2 - \frac{8}{3}\Delta_{SO}\Delta_{CF}]^{1/2}, \qquad (2)$$

where the energy reference is the valence band maximum with spin-orbit interaction taken into account. Based on the calculations of Ref.[43] or Ref.[24], it was unclear whether the ordering caused any shift in the center of gravity of the valence band. The recent empirical pseudopotential calculation of Zhang et al[27] has yielded $\Delta_{CF}(1)$ = 135 meV and a net valence band upward shift of 30 meV for η = 1. Thus, the center of gravity actually moves downward by 15 meV. Although for most optical measurements the relevant parameter is the band gap change, the absolute shift of the band edge, which determines the band offset, is important for understanding phenomena involving heterostructures (e.g., ordered GaInP/GaAs). Issues related to the ordering induced change in band offsets will be discussed later. Note that in Eq.(1) and (2) the spin-orbit splitting is assumed unchanged with ordering. In general, the spin-orbit interaction may change due to ordering. Then, Eq.(1) and (2) should be modified accordingly[28], in analogy to the situation for the effect of strain.

3.   ORDERING INDUCED OPTICAL ANISOTROPY OR POLARIZATION

3.1   Linear Polarization

Since ordering reduces the crystal symmetry from T_d for the random alloy to C_{3v} for the CuPt structure, selection rules for various types of optical transitions are expected to change. The symmetry effect was first demonstration in a polarized PL measurement[19], and later in various polarized spectroscopic studies which include PLE[19,30,39,45], cleaved edge PL[46,47], piezo-reflectance[29], reflectance difference[48], electroreflectance[40], electroabsorption[49], photocurrent[50], elliposometric measurement[51], second harmonic generation[52], and birefringence[53-55]. The ordering induced optical anisotropy has also been used advantageously for making various polarization selective or sensitive devices such as polarization rotators[53,55], lasers[56], LEDs[57], and optical switches[58]. As mentioned above, the ordering induced perturbation to the random alloy is closely analogous to that of uniaxial strain to a zinc-blende semiconductor. Thus, the well developed perturbation theory[59] for strain can be readily used for calculating the ordering induced valence band splitting and the interband optical transition probability[28,56,60-62]. The three most frequently encountered optical transitions are those from the three valence bands to the conduction band. If there is no epitaxial strain in the ordered layer (i.e., the epilayer is lattice matched to the substrate), the transition intensity, which is proportional to the square of the interband transition matrix element, are given in the following analytical forms[61]:

$$I_1 = e_1^2 + e_2^2 \tag{3}$$

for the transition between the topmost valence band (the heavy hole like) and the conduction band,

$$I_2 = \frac{2a_2^2}{3} + \frac{(a_1^2 + 2\sqrt{2}a_1 a_2)}{3}(e_1^2 + e_2^2) + \frac{4(a_1^2 - \sqrt{2}a_1 a_2)}{3}e_3^2 \tag{4}$$

for the transition between the second valence band (the light hole like) and the conduction band, and

$$I_3 = \frac{2b_2^2}{3} + \frac{(b_1^2 + 2\sqrt{2}b_1 b_2)}{3}(e_1^2 + e_2^2) + \frac{4(b_1^2 - \sqrt{2}b_1 b_2)}{3}e_3^2 \tag{5}$$

for the transition between the third valence band (the spin-orbit split-off band) and the conduction band, where $e = (e_1, e_2, e_3)$ is a unit vector in the direction of the light polarization in a coordinate system (x', y', z') with z' along the ordering direction [111], x' and y' in the plane perpendicular to the ordering direction ($x' \sim [11\overline{2}]$ and $y' \sim [\overline{1}10]$). The four coefficients in the above equations are $a_1 = (E_3 + d)/\sqrt{(E_3 + d)^2 + 2d^2}$, $a_2 = -\sqrt{2}d / \sqrt{(E_3 + d)^2 + 2d^2}$, $b_1 = (E_2 + d)/\sqrt{(E_2 + d)^2 + 2d^2}$, $b_2 = -\sqrt{2}d/\sqrt{(E_2 + d)^2 + 2d^2}$, where $d = -\Delta_{CF}/3$. Fig.4 shows how the transition intensity varies with the strength of ordering, measured by the crystal field splitting parameter, for the two frequently encountered polarizations in the growth plane[61].

There had been a few attempts[40,60] to make quantitative analyses of the experimental data by using the theoretical results like that of Fig.4. However, it was later shown[61] that various other effects could result in significant deviations from the theoretical curves of Fig.4. For instance, the substrate tilt angle is a critical factor for quantitatively evaluating the optical anisotropy in two important aspects. Firstly, the crystalline structure strongly depends on the tilt angle. On an exact (001) substrate, two equally probable ordered variants tend to form quasi-periodic micro-domain twins[63,64]. Such a more complex form of ordering, termed an orientational superlattice by Mascarenhas et al[65,66], in fact has distinctly different electronic and optical properties from the simple single-variant CuPt structure. It is, thus, quite inappropriate to apply the theory which is only meant for the CuPt structure to this special category of superlattices[67,68]. Secondly, the single-variant CuPt ordered structure can

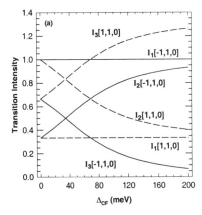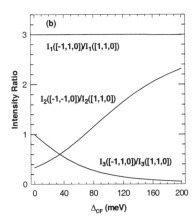

Figure 4. Calculated band-to-band transition intensities at k = 0 from the three valence bands to the conduction, respectively, with light polarized along the [-110] and [110] direction (the ordering is assumed to be along the [111] direction): (a) for the intensities and (b) for the intensity ratios. Curves are re-plotted from Fig. 4 of Ref.[61] (Zhang et al).

be obtained by using substrates tilted toward one of the $[111]_B$ directions, but the substrate tilt introduces a substantial effect on the polarization anisotropy[61]. For instance, the anisotropy ratio for the interband transition involving the topmost valence band is calculated to be $R_1 = 3$ between the two orthogonal [110] directions[60], as shown in Fig.4(b). However, this ratio is expected to reduce to 2.3 for a sample grown on a $6°B$ tilt substrate, when measured on the growth surface[61]. Also, the transition intensity shown in Fig.4 is calculated only for the electron-hole direct transition at k = 0. In reality, the excitonic effect is involved in the band edge transition, which requires knowledge of the transition matrix element at $k \neq 0$[61]. Fig.5 shows typical polarized PL spectra for a pair of (nearly) random and ordered GaInP samples. The polarization ratio for the band edge excitonic transition is found to be ~ 2.0, instead of 3, for the ordered sample, almost independent of the order parameter for samples with reasonably large order parameters[61]. Such a result indicates that to quantitatively analyze the optical anisotropy one has to take into account the factors of substrate tilt (both orientation and angle), excitonic effect, and even epitaxial strain. It is worth pointing out that in all the above-mentioned theoretical considerations for the interband transition intensity, the conduction band wavefunction has been assumed unchanged with the occurrence of ordering. Since CuPt ordering causes a mixing of the Γ and folded L point, it is expected that transition intensity should decrease on increasing the degree of order. Such

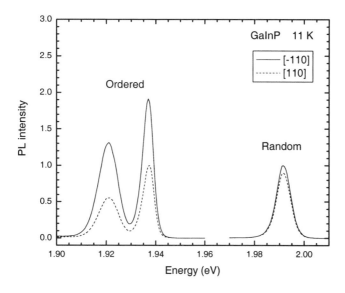

Figure 5. Typical polarized photoluminescence spectra for a random and a partially ordered GaInP alloy. Spectra are re-plotted from Fig.1 of Ref.[61] (Zhang et al).

an effect can be estimated by using the coupling matrix element given in Ref.[27].

3.2 Circular (Spin) Polarization

It is well known that for a zinc-blende semiconductor near-band-edge interband optical pumping by circularly polarized light can produce conduction band electrons with a maximum degree of polarization $P = (n\uparrow - n\downarrow)/(n\uparrow + n\downarrow) = 50\%$, owing to the opposite polarization for electrons transferred from the degenerate heavy and light hole state[69]. It was pointed out by Ciccacci et al[70] that for a CuAu ordered AlGaAs alloy (i.e., GaAs/AlAs monolayer superlattice) 100% polarization could be achieved for the electrons, because of the ordering induced valence band splitting. Wei and Zunger[60] extended this idea to the CuPt ordered alloy. Experimentally, Kita et al[71] studied spin polarization of the band edge excitonic luminescence pumped along the [001] direction. The maximum polarization, measured at nearly zero time delay, was found to be ~ 55%. The primary reason given for the maximum polarization being significantly less than 100% was that the direction of the optical pumping was not along the ordering direction. Indeed, the maximum electron polarization for [001] optical pumping P_{001} is only 50%, which can be evaluated using the following wave functions for the doublet degenerate topmost valence band

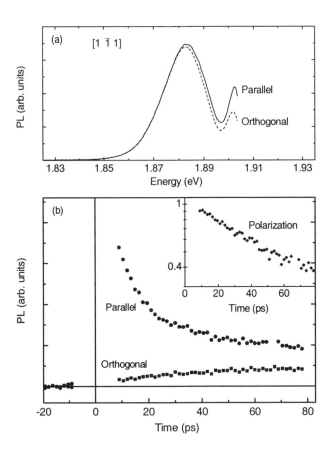

Figure 6. Spin polarization pumped and measured along the ordering direction. (a) Time-integrated photoluminescence (PL) spectra polarized parallel and orthogonal to the circularly polarized pump; and (b) time –resolved PL at the peak of the band edge excitonic transition, under identical conditions. Inset: log plot of degree of polarization (from Ref.[72] of Fluegel et al).

state:

$$\varphi_1 = \frac{1}{\sqrt{2}} \left| \frac{3}{2}, -\frac{3}{2} \right\rangle_{(001)} + \frac{1-i}{\sqrt{6}} \left| \frac{3}{2}, -\frac{1}{2} \right\rangle_{(001)} + \frac{i}{\sqrt{6}} \left| \frac{3}{2}, \frac{1}{2} \right\rangle_{(001)}, \tag{6}$$

$$\varphi_2 = \frac{i}{\sqrt{6}} \left| \frac{3}{2}, -\frac{1}{2} \right\rangle_{(001)} + \frac{1+i}{\sqrt{6}} \left| \frac{3}{2}, \frac{1}{2} \right\rangle_{(001)} + \frac{1}{\sqrt{2}} \left| \frac{3}{2}, \frac{3}{2} \right\rangle_{(001)}. \tag{7}$$

One should note than the PL polarization, which differs from the polarization of the electron populations, not only depends on the polarization of the electrons but also on that of the holes. If the spin orientation of holes is assumed unrelaxed for near band edge pumping, the maximum value for the PL polarization with an exact [001] pumping is expected to be 75%[69]. In fact, Fluegel et al[72] showed that even for pumping along the direction normal to the sample surface, at zero time delay, the PL polarization could approach values as high as 90%, which can be explained by taking into account the 6° substrate tilt angle as well as a finite collection angle of similar amount. Furthermore, Fluegel et al[72] actually performed the spin-polarized PL measurement pumped exactly along the ordering direction. In this case, as shown in Fig.6, a near 100% polarization at zero time delay was in fact observed.

4.   ORDERING INDUCED CHANGES IN EFFECTIVE MASS

The first attempt by Jones et al[73] to study the effect of ordering on the effective mass using magneto-photoluminescence indicated that the exciton reduced mass of an ordered sample was smaller than that of a disordered sample. The change of the reduced mass was attributed to the reduction of the conduction band effective mass. Emanuelsson et al[74] subsequently measured the electron effective mass of an ordered and a disordered sample by optically detected cyclotron resonance measurements, and found m_c = 0.088 ± 0.003 for the ordered sample and m_c = 0.092 ± 0.003 for the disordered sample. In both of these studies, the reduction of the conduction band mass was explained as being a result of ordering induced band gap reduction using a **k.p** model. It was later pointed out by Raikh and Tsiper[75] for the conduction band and Zhang and Mascarenhas[28] for the valence band that ordering not only modifies the effective mass but also makes it anisotropic. Thus, any magneto-measurement should be sensitive to the direction of the field. The conduction band effective mass m_c was derived by Raikh and Tsiper as follows, with only the repulsion between the conduction band and the folded L-band considered[75]:

$$\frac{1}{m_{c\|,\perp}} = \frac{1}{2m_\Gamma}\left(1 + \frac{\delta E_{\Gamma L}}{\sqrt{\delta E_{\Gamma L}^2 + 4M_{\Gamma L}^2}}\right) + \frac{1}{2m_{L\|,\perp}}\left(1 - \frac{\delta E_{\Gamma L}}{\sqrt{\delta E_{\Gamma L}^2 + 4M_{\Gamma L}^2}}\right), \quad (8)$$

where m_Γ (m_L) is the conduction band effective mass at Γ (L) point for the disordered alloy, and $m_\|$ and $m_\perp$ represent the effective mass parallel and perpendicular to the ordering direction, respectively. Assuming $m_{L\|} > m_{L\perp} > m_\Gamma$, the authors found $m_{c\|} > m_{c\perp} > m_\Gamma$, i.e., the conduction band effective mass of an ordered structure would always be heavier than that of a disordered structure, and it is more heavier along the ordering direction. However, Zhang and Mascarenhas pointed out[28] that the coupling to the valence would reduce m_c and, in the process, cause m_c to be anisotropic. The following results were derived[28]:

$$\frac{1}{m_{c\parallel}} = \frac{1}{m_\Gamma} + \frac{E_p}{3}\left(\frac{2a_0}{E_g^2} + \frac{a_0}{E_d^2}\right) - \frac{2E_p}{9}\left(\frac{\Delta_{CF}}{E_g^2} + \frac{2\Delta_{CF}}{E_g E_d}\right), \qquad (9)$$

$$\frac{1}{m_{c\perp}} = \frac{1}{m_\Gamma} + \frac{E_p}{3}\left(\frac{2a_0}{E_g^2} + \frac{a_0}{E_d^2}\right) + \frac{E_p}{9}\left(\frac{\Delta_{CF}}{E_g^2} + \frac{2\Delta_{CF}}{E_g E_d}\right), \qquad (10)$$

where $a_0 = \delta E_v - \delta E_c = \delta E_g - \Delta_{CF}/3$ (here $\delta E_g > 0$) is the band gap reduction caused by the shift of the conduction band and the shift of the center of gravity of the valence band, and $E_d = E_g + \Delta_{SO}$. In principle, one could combine the two effects described by Eq.(8) – (10), but a few key parameters (e.g., m_L, $\delta E_{\Gamma L}$ and $M_{\Gamma L}$) are not so well known. Nevertheless, both effects considered above would make $m_{c\parallel} > m_{c\perp}$. Later, Franceschetti et al[76] used a first-principles method to calculate the effective mass for the fully ordered structure. They found that $m_{c\parallel}$ increased significantly while $m_{c\perp}$ slightly decreased from m_Γ. However, ambiguities arose when the authors attempted to generate interpolation curves between $\eta = 0$ and 1, because the validity of the η^2 rule was uncertain. It now appears unlikely to be true[27]. Even if the η^2 rule was valid, there was an ambiguity as to whether the rule should have been applied to m_c itself or to the energy $E_c(k) \propto 1/m_c$.

The valence band effective masses were derived analytically in Ref.[28]. For the topmost (heavy-hole like) valence band,

$$\frac{1}{m_{hh\perp}} = \gamma_1 + \gamma_3 + \frac{\delta E_g E_p}{2E_g^2}, \qquad (11)$$

$$\frac{m_e}{m_{hh\parallel}} = \gamma_1 - 2\gamma_3 ; \qquad (12)$$

for the second (light-hole like) valence band,

$$\frac{1}{m_{lh\perp}} = \alpha_1(\gamma_1 - \gamma_3) - \alpha_2\gamma_1' + \alpha_3\gamma_3', \qquad (13)$$

$$\frac{1}{m_{lh\parallel}} = \alpha_1(\gamma_1 + 2\gamma_3) - \alpha_2\gamma_1' - 2\alpha_3\gamma_3' ; \qquad (14)$$

and for the spin-orbital split-off band,

$$\frac{1}{m_{sh\perp}} = \beta_1\gamma_1' - \beta_2\gamma_3' + \beta_3(\gamma_1 - \gamma_3),$$ (15)

$$\frac{1}{m_{sh\parallel}} = \beta_1\gamma_1' + 2\beta_2\gamma_3' + \beta_3(\gamma_1 + 2\gamma_3);$$ (16)

where

$$\alpha_1 = \frac{(2dx + 2\Delta_{SO}d + 6d^2)}{x^2 - (\Delta_{SO} - 3d)x} \qquad \alpha_2 = \frac{(\Delta_{SO} - d)x - \Delta_{SO}^2 - 3d^2}{x^2 - (\Delta_{SO} - 3d)x}$$

$$\alpha_3 = \frac{4\Delta_{SO}d - 4dx - 12d^2}{x^2 - (\Delta_{SO} - 3d)x} \qquad \beta_1 = \frac{\Delta_{SO}^2 + 3d^2 + (\Delta_{SO} - d)x}{x^2 + (\Delta_{SO} - 3d)x}$$

$$\beta_2 = \frac{-4\Delta_{SO}d + 12d^2 - 4dx}{x^2 + (\Delta_{SO} - 3d)x} \qquad \beta_3 = \frac{2\Delta_{SO}d + 6d^2 - 2dx}{x^2 + (\Delta_{SO} - 3d)x},$$

and x = $[(\Delta_{SO} + d)^2 + 8 \ d^2]^{1/2}$. γ_1, γ_3, γ_1', and γ_3' are Luttinger parameters. Qualitatively, one can conclude that (1) The "heavy-hole" mass m_{hh} along the ordering direction is indeed heavy, and independent of the degree of order; but in the ordering plane, m_{hh} is actually light and has a weak dependence on the order parameter. (2) Both the "light hole" mass m_{lh} and split-off band mass m_{sh} strongly depend on the order parameter, and show strong anisotropy. These valence band effective masses have later also been calculated by Yeo et al with the spin-orbit coupling ignored[77], and by Tsitsishvili with the coupling to the conduction band ignored[62].

The effective mass anisotropy was demonstrated by Ernst et al[78] using magneto-luminescence with the magnetic field oriented along either the [$\bar{1}$11] ordering direction or the [001] growth direction. As shown in Fig.7, they found that the diamagnetic shift (the shift of the excitonic emission energy) is larger with the field along the [$\bar{1}$11] direction than when along the [001] direction, as well as that the diamagnetic shift is larger for an ordered as compared to a disordered sample. The diamagnetic shift in the low field region can be described by the following perturbation formulae:

$$\delta E_{[\bar{1}11]} = \frac{e^2 B^2}{8\mu_\perp c^2} \langle f | x^2 + y^2 | f \rangle,$$ (17)

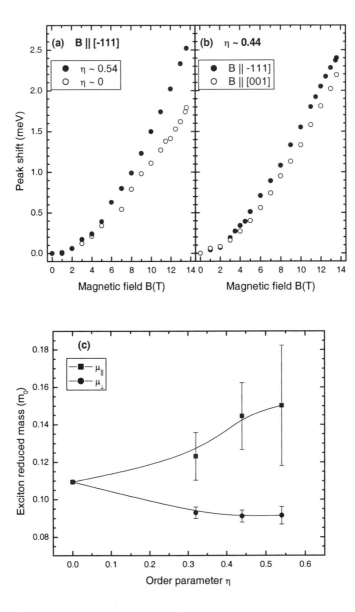

Figure 7. Results of a magneto-PL study on partially ordered GaInP alloys. (a) The effect of ordering indicated by the difference between the disordered and ordered sample in their energy shifts. (b) The effect of the effective mass anisotropy indicated by the difference in energy shift between two field directions. (c) Exciton reduced mass vs. order parameter, derived from the magnetic field induced peak shift (from Ref.[78] of Ernst et al).

$$\delta E_{[001]} = \frac{e^2 B^2}{8\mu_\perp c^2} \frac{1}{3} \langle f | x^2 + y^2 + 2z^2 | f \rangle + \frac{e^2 B^2}{8\mu_\parallel c^2} \frac{1}{3} \langle f | x^2 + y^2 | f \rangle, \quad (18)$$

where $\mu_{\parallel}$ and $\mu_{\perp}$ are exciton reduced masses parallel and perpendicular to the ordering direction, and $f(x,y,z)$ is the exciton wavefunction at zero field. Applying these equations to the experimental data, exciton reduced masses were extracted, as shown in Fig.7(c). Zhang et al[79] later demonstrated that both $\mu_{\parallel}$ and $\mu_{\perp}$ could in fact be obtained from the experimental data with the field along the ordering direction alone, if the data for the entire range of field were modeled using a generalized theory. Given the fact that the effective masses of the topmost valence band are independent of or weakly dependent on the order parameter[28], the results for the reduced masses shown in Fig.7(c) should represent the trend of the variation of the conduction band masses. Thus, the variation of the conduction band masses appear to agree qualitatively, but not quantitatively, with the theoretical results of Ref.[76].

5. REFLECTANCE DIFFERENCE SPECTROSCOPY STUDY OF ORDERED STRUCTURE

Reflectance difference spectroscopy (RDS) has been used to investigate the anisotropic surface reconstruction of III-V materials with isotropic bulk properties[80]. This technique has been shown to also be a sensitive technique for detecting the presence of ordering in III-V alloys and can easily be adapted to *in situ* measurements during and after growth[48,81,82]. For the CuPt order structure, typically one measures the reflectance difference between the $[\bar{1}10]$ and $[110]$ direction (the ordering direction is assumed to be $[\bar{1}11]$), that is

$$\frac{\Delta R}{R} = \frac{R_{\bar{1}10} - R_{110}}{(R_{\bar{1}10} + R_{110})/2},$$ (19)

where $R_{\bar{1}10}$ and R_{110} are the reflectances of light polarized along $[\bar{1}10]$ and $[110]$, respectively. The RD signal originates from the ordering induced anisotropy in the dielectric function ε which exhibits a uniaxial symmetry:

$$\varepsilon = \begin{pmatrix} \varepsilon_{\perp} & 0 & 0 \\ 0 & \varepsilon_{\perp} & 0 \\ 0 & 0 & \varepsilon_{\parallel} \end{pmatrix},$$ (20)

where $\varepsilon_{\parallel}$ and $\varepsilon_{\perp}$ are dielectric functions for light polarized parallel and perpendicular to the ordering direction. For light incident normal to the (001) plane, the reflectances for the two orthogonal directions are[55]

$$R_{110} = \left| \frac{\sqrt{\varepsilon_\perp} - 1}{\sqrt{\varepsilon_\perp} + 1} \right|^2 , \tag{21}$$

$$R_{\bar{1}10} = \left| \frac{\sqrt{\dfrac{\varepsilon_\perp \varepsilon_\parallel}{\bar{\varepsilon}}} - 1}{\sqrt{\dfrac{\varepsilon_\perp \varepsilon_\parallel}{\bar{\varepsilon}}} + 1} \right|^2 , \tag{22}$$

where $\bar{\varepsilon} = (2\varepsilon_\perp + \varepsilon_\parallel)/3$. Assuming $\delta\varepsilon = (\varepsilon_\perp - \varepsilon_\parallel)/3 \ll \bar{\varepsilon}$, one approximately has $\Delta R / R \propto -2\delta\varepsilon \approx \varepsilon_{\bar{1}10} - \varepsilon_{110}$. Another commonly adopted approximation is to neglect the contribution of the imaginary part of the dielectric function near the fundamental band gap. The real part ε_1 takes the general form of[83]

$$\varepsilon_1 \propto E^{-2}[2\sqrt{E} - \sqrt{E_0 + E} - \sqrt{E_0 - E}\theta(E_0 - E)] . \tag{23}$$

There have been three iteratively improved models for modeling the near band-edge RD spectrum of a CuPt ordered GaInP alloy. Luo et al[48,81] first used a four-band model, Wei and Zunger[84] then used a six-band model. In all these approaches, the first two terms in Eq.23 were not included. It was later found out by Luo et al[85] that these two terms are critical for getting the sign correct for the calculated lineshape function above the band gap. It was also showed[85] that it was necessary to take into account the k-dependence of the interband transition matrix element in order to achieve an overall good agreement between the experimental data and the theoretical lineshape function. Fig.8 shows the comparison between an experimental RD spectrum and the calculated lineshape functions.

Since RDS is sensitive to the bulk anisotropy due to the ordering as well as the surface anisotropy that may exist in a zinc-blende structure, Luo et al[85] has managed to suppress the surface induced features in order to reveal the effect of bulk ordering near the band gap. However, the surface effect may contain information related to the formation mechanism of ordering. Zorn et al[86] have in turn applied RDS to investigate the correlation between CuPt ordering and surface reconstruction. They concluded that there was a clear and unambiguous correlation between the occurrence of ordering and the presence of P-dimers with a (2x1) surface reconstruction during growth, which would agree with the theoretical model of Zhang et al[87]. It is worth pointing out that the validity of this conclusion critically depends on the reliability of the procedure and assumption for separating the bulk (band structure) effect and the surface

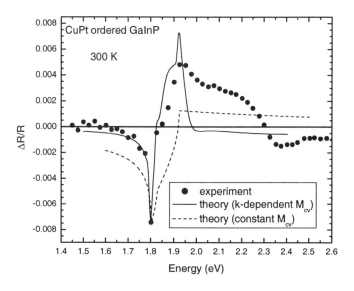

Figure 8. A comparison of an experimental reflectance difference spectrum and calculated lineshape functions with a constant or k-dependent interband transition matrix element. Curves are re-plotted from Fig.2 of Ref.[85] (Luo et al).

effect. The major assumption was that the anisotropic contribution from the oxidized GaInP surface was only small and independent from the bulk ordering, which allowed them to obtain the contribution of the bulk ordering by subtracting the spectrum of a disordered sample from that of the ordered sample for the temperature range T < 775 K. The oxide layer was assumed to desorb at T > 800 K. Thus, the RD signal at T > 800 K only had the ordering contribution due to either the surface or bulk effect. After the bulk contribution, obtained by extrapolating the lower temperature data, was subtracted, they were in principle left with the surface contribution of ordering. The focus was on a spectroscopy feature near 3 eV that was assumed to be associated with the (2x1) reconstruction, because the (2x1) reconstruction was simultaneously observed by a RHEED measurement for CBE (chemical beam epitaxy) grown samples. No independent theoretical justification was given for whether or not the (2x1) reconstruction would give rise to the ~ 3 eV peak. Furthermore, even though MOVPE samples did have the similar feature at ~ 3 eV, there was no guarantee that the same reconstruction existed in the MOVPE growth; since the ordering was observed only for the MOVPE but not for the CBE grown samples, logically there was no guarantee that the ordering was correlated to the (2x1) reconstruction, even if the reconstruction did exist for both growths. Thus, the conclusion of Zorn et al needs further verification.

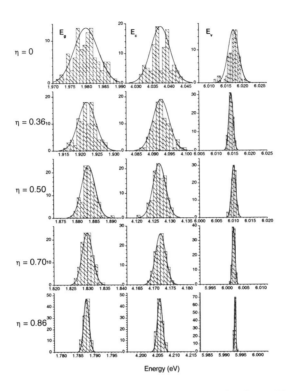

Figure 9. Histogram plots of the energy distributions of the band gap (Eg), conduction band edge (E$_c$) and valence band edge (E$_v$) for partially ordered Ga$_{0.5}$In$_{0.5}$P alloys with order parameter η = 0, 0.36, 0.50, 0.70, and 0.86 (from Ref.[94] of Zhang et al).

Another technique – surface photoabsorption (SPA), has been used by Murata et al[88,89] for studying the ordering induced surface effect. In fact, this technique is very similar to RDS, when the difference of two polarizations is evaluated. Murata et al have suggested that (2x4) reconstruction was necessary for ordering. According to the theoretical model of Zhang et al[87], the (2x4) reconstruction can indeed generate CuPt ordering, but its efficiency is expected to be weaker that that of the (2x1) reconstruction. We would like to note that the possible bulk effect was not taken into account in Murata et al's analysis.

6.   STATISTICAL ASPECTS OF SPONTANEOUS ORDERING

A great deal of experimental and theoretical studies on spontaneous ordering have focused on the dependence of the ensemble average properties of the alloy on the order parameter: e.g., the band-gap reduction, valence band splitting, and optical anisotropy, as has been discussed in previous sections. In contrast, a fundamental aspect of spontaneous ordering relating to the statistical nature of the phenomenon has largely been ignored. Only until very recently, the most anticipated statistical effect of ordering, a reduction of alloy fluctuations, has been observed experimentally by Zhang et al[90] through a continuous reduction of the exciton linewidth with increasing order parameter. The statistical effects referred to here are those intrinsic to ordering and not related to imperfections in sample growth. For example, phenomena associated with the macroscopic spatial variations of the alloy composition, the order parameter, and antiphase domain boundaries[45,91-93] are considered to be related to growth imperfections which can in principle be eliminated or minimized by improving the growth technique. Evidently, the investigation of statistical effects for partially ordered structures is more difficult in terms of sample quality (experimentally) or computation effort (theoretically) than that of ensemble average properties. It is well known that the influence of alloy statistical fluctuations on many physical properties is a function of the alloy composition x. The first order effect can frequently be described by a simple function $x(1-x)$. For a spontaneously ordered alloy, the effects of alloy fluctuations will not only be a function of the average composition x but also of the order parameter η. Two statistical aspects of the influence of ordering are particularly interesting. One is the effect of alloy fluctuations on the band structure parameters (band gap and etc.), and the other is the effect on the crystal structural parameters (bond length etc.).

The first study of the statistical effect on the band structure was performed by Capaz and Koiller[25]. In their study the band gap fluctuation, obtained by averaging over 400 configurations of a 64 atom unit cell, was found to decrease on increasing the order parameter. A very recent study by Zhang et al[94] has attempted to simulate the partially ordered structure more realistically by using a rather large unit cell of ~ 3500 atoms and averaging over 100 configurations. The energy fluctuation not only for the band gap but also for the band edge of the conduction and valence band have been calculated. Fig.9 shows the histogram plots for these energy fluctuations, and Fig.10 shows the full width at half maximum (FWHM) of the histogram plots with a comparison of he experimentally measured excitonic linewidth[90]. The information for the band gap fluctuation is most relevant for various optical measurements (e.g., emission and absorption), but that for the individual band edge is most valuable for transport measurements related to either electrons or holes. In comparison with the results of Ref.[25], the new results indicate a somewhat stronger dependence of the alloy fluctuation on order parameter. Fig.9 also reveals

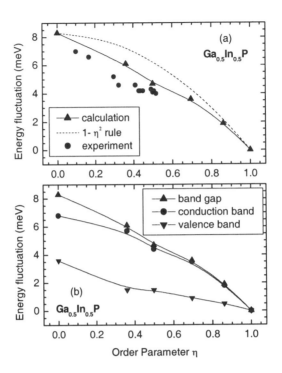

Figure 10. The energy fluctuation of the band gap $W_{gap}(\eta)$, of the conduction band edge $W_c(\eta)$, and of the valence band edge $W_v(\eta)$ for a partially ordered $Ga_{0.5}In_{0.5}P$ alloy, measured by the full width at half maximums (FWHM) of the histogram plots (shown in Fig.9), as a function of the order parameter η. (a) A comparison of the calculated $W_{gap}(\eta)$ with the low temperature photoluminescence linewidth $W_{ex}(\eta)$ of Ref.[90] and a curve predicted by a simple η^2 dependence. (b) Calculated $W_c(\eta)$, $W_v(\eta)$ as well as $W_{gap}(\eta)$ (from Ref.[94] of Zhang et al).

that the band gap fluctuation does not obey the simple η^2 rule, although it was argued in Ref.[23] that the majority of physical properties should follow this rule.

As regards the effect of ordering on the crystal structure parameters, Capaz and Koiller[25] pointed out that if all the Ga-P and In-P bonds are divided into four groups: Ga-P and In-P bonds along the ordering direction ("O") and in the lateral direction ("L"), the average bond length for each group would follow a η^2 dependence. However, a recent x-ray absorption fine- structure (XAFS) study by Meyer et al[95] on a partially ordered GaInP sample only resolved a single average Ga-P or In-P bond length for the entire sample (except for possible bond length modifications in certain localized regions), in agreement with the typical bimodal behavior of conventional alloys [96,97]. Zhang et al[94] pointed out that there are two

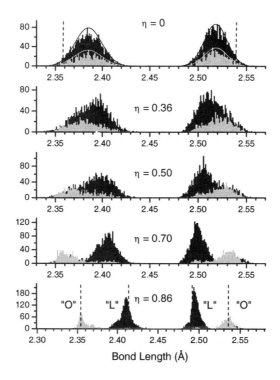

Figure 11. Histogram plots of the bond length distributions of partially ordered $Ga_{0.5}In_{0.5}P$ alloys with order parameter $\eta = 0, 0.36, 0.50, 0.70$, and 0.86. Black – for the "L" type bonds along the lateral directions; Gray - for the "O" type bonds along the ordering direction. The dashed vertical lines denote the Ga-P and In-P bond length in the binaries on the $\eta = 0$ panel (the up most), and in the fully order structure ($\eta = 1$) on the $\eta = 0.86$ panel (the bottom) (from Ref.[94] of Zhang et al).

factors which have prevented the observation of the ordering effect in the XAFS measurement: (1) the unpolarized nature of the measurement could not distinguish the O- and L-type bonds, and (2) the order parameter for the sample investigated was too small. Fig.11 shows the evolution of the distribution of the Ga-P and In-P bonds in partially ordered GaInP with order parameter. Fig.12 shows the average bond lengths and their statistical fluctuations as functions of η^2. In agreement with the results of Ref.[25], the average bond length follows the η^2 dependence very well. One can see that (1) from Fig.12(a) for η up to 0.5, the "O" –"L" splitting is smaller than 0.2 Å which is the typical experimental uncertainty of any EXAFS measurements[95-97]; (2) the strong overlap between the distributions of O- and L-bonds and the 1: 3 ratio for the numbers of the types of bonds make it unfeasible to distinguish them by using any unpolarized EXAFS techniques, unless the sample is very highly ordered. Note that a

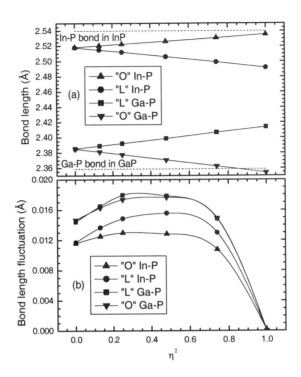

Figure 12. (a) Average bond lengths for the four types of bonds in partially ordered $Ga_{0.5}In_{0.5}P$ alloys versus η^2 ("O" – along the ordering direction, "L" – along the lateral directions). (b) Bond length fluctuations versus η^2 for the four types of bonds (from Ref.[94] of Zhang et al).

superpositionof the distribution of the O- and L-bond will result in a mixed distribution that shows two peaks. An unpolarized XAFS measurement in principle should be able to resolve two average bond lengths for a highly ordered sample, but the average bound lengths so obtained will not follow the η^2 rule.

7.   BAND OFFSET BETWEEN ORDERED GaInP AND GaAs

The GaInP/GaAs heterojunction bipolar transistor has emerged as a frontrunner for high-speed power transistors used for cellular communications. Because of the very practical device interest, the band alignment for the $Ga_xIn_{1-x}P$ (x ~ 0.5)/GaAs heterostructure has attracted great

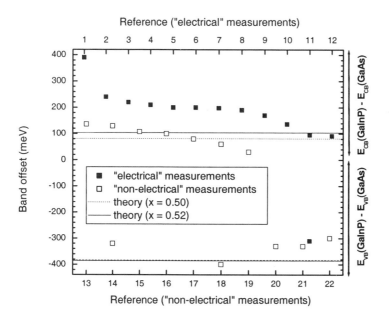

Figure 13. Comparison between the calculated values of Ref.[128] and experimental data for the conduction and valence band offset. Tick labels 1-22 correspond to the reference numbers 112-133 (from Ref.[128] of Zhang et al).

deal of attention. A number of different approaches have yielded largely scattered values for the conduction band offset $\Delta E_c = E_c(\text{GaInP}) - E_c(\text{GaAs})$, ranging from 30 to 390 meV[98-119]. It has recently been noticed that the existence of spontaneous ordering in the GaInP layer could significantly alter the band offset[110,120-124]. It thus appears that besides the possible intrinsic limitation of each technique[98-119], the ordering effect has contributed to the large scatter in the reported values at least to some extent. The scatter for theoretically calculated values is actually smaller. For the random-GaInP/GaAs heterostructure, Harrison found $\Delta E_c = 160$ meV[125]; Foulon et al found a valence band offset $\Delta E_v = 320$ to 390 meV[126]; Froyen et al found $\Delta E_c = 120$ meV and $\Delta E_v = 370$ meV[127]. A most recent calculation of Zhang et al gave $\Delta E_c = 81$ (104) mV and $\Delta E_v = 383$ (385) meV for x = 0.50 (0.52)[27,128]. Fig.13 shows the values of ΔE_c and/or ΔE_v for the GaInP/GaAs heterostructure found in the literature. It appears that the conduction band offsets derived from various "electrical measurements" (e.g., capacitance-voltage, current-voltage) [98-111] show a larger scatter (ranging from 390 to 91 meV) as compared to those obtained from other techniques: 108 ± 6 meV from internal photoemission[112], 137 meV or 100 meV from BEEM spectroscopy[110,113], 80 meV from photoluminescence[114], 160 meV[125], 120 meV[127], and 104 meV (x

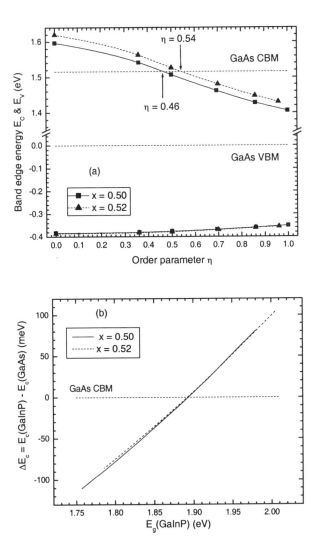

Figure 14. Band edge energies of partially CuPt ordered $Ga_xIn_{1-x}P$ alloys, varying with (a) order parameter η and (b) band gap Eg (from Ref.[128] of Zhang et al).

= 0.52) or 81 meV (x = 0.50)[128] from theoretical calculations. Note that for the conduction band offset the results of the more recent "electrical" measurements (Refs.[107-109]), except for Ref.[103], have approached those of "non-electrical" measurements (Refs.[110-115]) and the theoretical results.

The conduction band offset between ordered-GaInP and GaAs has also been a critical issue on the debate of the mechanism for the up-converted PL observed in GaInP/GaAs heterostructure[122,129-131]. Up-converted PL for an ordered-GaInP/GaAs heterostructure was first reported by Driessen[129]. He observed PL from the GaInP layer for excitation at the GaAs band gap, and explained the phenomenon by the so-called cold Auger process[129]. Su et al later attributed the same phenomenon to a two-step two-photon absorption (TS-TPA) process[122]. Driessen subsequently showed that the up-conversion could even be observed for (AlGa)InP/GaAs interfaces[130]. Since the band offset was likely to increase significantly with the incorporation of Al, they believed that the two-step model of Su et al was invalid. However, this later result of Driessen et al actually only proved that the nearly zero band offset was not a necessity for the up-conversion, and did not exclude the possible role of the two-step mechanism in the observed up-conversion at the ordered-GaInP/GaAs interface. Kita et al performed a time-resolved up-conversion study on the ordered-GaInP/GaAs interface[131]. They pointed out that there were two channels for the up-conversion: (1) direct TS-TPA processes for both electrons and holes localized near the interface; and (2) TS-TPA and Auger processes caused by the GaAs PL, but they excluded the need of a flat conduction band alignment (since they believed that the sample should have a type I alignment). Although based on the existing data it is inconclusive as to what is the dominant mechanism for the up-conversion and whether the band offset is the single dominant factor for generating the up-conversion, an accurate knowledge of how the band offset changes with the ordering is important for answering these questions. Froyen et al's calculation indicated that the conduction band alignment between GaAs and perfectly CuPt ordered GaInP would become type II with ΔE_c = - 130 meV, but that the valence band alignment would remain type I with a reduced value of ΔE_v = 270 meV. Applying the η^2 rule to the band edge energies, Froyen et al obtained a crossover point at $\eta_0 = 0.70$, implying that all the samples used for the up-conversion studies were most likely in the type I region[35]. However, a recent band structure calculation by Zhang et al[27] has shown that the η^2 rule is generally invalid. Thus, a direct calculation of the band edge energy with varying order parameter is highly desirable. Fig.14 shows the results of the first such attempt, where the band offsets are given both as a function of the order parameter and band gap[128]. According to Fig. 14, the crossover point is $\eta_0 = 0.54$ for x = 0.52. Therefore, on one hand, for the sample used by Kita et al[131] with E_g ~ 1.89 eV, the order parameter actually should have been very close to η_0 instead of $\eta = 0.44$ as estimated by the authors, implying that a nearly flat conduction band alignment could have played a role in the observed up-conversion. On the other hand, in order to justify the existence of a type II band alignment, Kwork et al[132] claimed that their sample had a value of η as high as 0.75 to conform with the crossover point $\eta_0 = 0.7$ estimated by Froyen et al[127]. Since the PL peak was taken as a measure of the band gap and the η^2 rule was used for converting the "band gap reduction" to the order parameter, Kwok et al likely overestimated the order parameter for their sample. Zeman et al[123,124] found that a set of ordered-GaInP/GaAs samples that exhibited the up-conversion had ΔE_c in a range of $\pm$ 3 meV and PL peak energies around 1.9 eV. If one takes the PL peak as a measure of band gap, one finds that Zeman et al's results agree quite well with that of Fig.14. However, one has to keep in mind two factors: (1) a PL peak is frequently not an accurate measure of the band gap; (2) the samples they used were grown on exact (001) substrates, and thus, were not expected to have the simple CuPt ordered structure, but rather to have the double variant structure[65-68]. Also, we notice that the result of Ref.[116] (i.e., $\Delta E_c = 30$ meV for an ordered GaInP sample with 60 meV band gap reduction) appears to agree quite well with the theoretical result of Ref.[128], as shown in Fig.14. We would like to point out that the finding of Ref.[103], $\Delta E_c = 200$ meV not changing with ordering, is contradictory with either the experimental data of Refs.[120-124] or the theoretical results of Refs.[127,128].

8.   NOVEL SUPERLATTICES–ORIENTATIONAL SUPERLATTICES

It is well-known that when an exact (001) substrate or [111]$_A$ tilt substrate is used, two [111]$_B$ CuPt ordered variants are usually present simultaneously. However, the domain size, the stacking direction as well as the regularity of the domain distribution strongly depends on the growth conditions (e.g., the exact substrate tilt angle and orientation, growth rate, epilayer thickness and etc.)[67,68]. Quasi-periodic structures of domain twins of the two ordered variants have been observed either along the [001] growth direction[63,64,68] or along the [$\bar{1}$10] direction[133] (where we define the [$\bar{1}$11] and [1$\bar{1}$1] as the ordering direction for the two variants). The [001] structures are found to have a periodicity of typically less than 5 nm, as observed in the TEM study (e.g., Morita et al[63], Baxter et al[64], Ahrenkiel et al[68]) and recently in a x-ray study by Li et al[36]. The [$\bar{1}$10] structures usually have a large periodicity of the order of 1 μm and show facets on the sample surface, as reported by Friedman et al[133]. Ahrenkiel found that during growth the double variant structure gradually evolved from the [001] stacking (roughly below 2 μm thickness) to the [$\bar{1}$10] stacking arrangement for the upper part of the epilayer for file thickness up to 10 μm[134]. In fact, domain twin structure is a frequently seen phenomenon

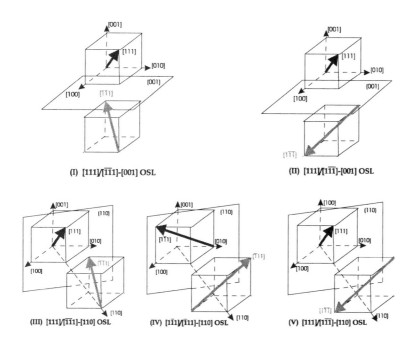

Figure 15. Five polytype orientational superlattices based on CuPt-ordered GaInP alloys (from Ref.[65] of Mascarenhas et al).

in different kinds of crystals, but typically the domain size is of the order of the wavelength of light or bigger. Thus, when viewed under polarized light, one expects to see bright and dark contrast between domains. In deed, optical effects have been observed in ordered GaInP with the large $[\bar{1}10]$ type double variant structure by Alsina et al[135] and Sapriel and Hassine[136]. However, quantum mechanical effects which are expected to occur when the domain size is reduced to less than a few hundred Å have only been discussed recently, first by Ikonic et al[137] for a hypothetic twinning superlattice formed by periodically stacked [111] twin defects in diamond and zinc-blende type semiconductors, and then by Mascarenhas et al[65,66] for orientational superlattices (OSLs) formed by domain twins of CuPt ordered III-V alloys. What distinguishes the OSL from the conventional superlattice where the superlattice effect results from a discontinuity in the band edge energy (a scalar) is that the superlattice effect for the OSL originates from a discontinuity (a rotation) in the orientation of the symmetry axis (a vector) for the constituent layers. Fig.15 schematically depicts five simple polytypes which can be formed by two ordered variants, where polytype I, III, and IV can be viewed as comprised of twin boundaries of two $[111]_A$ or $[111]_B$ ordered domains, and polytype II and V as comprised of twin boundariesy of a $[111]_A$ and a $[111]_B$ ordered domain. Since the ordering induced sphase transition is ferroelastic[65], Sapriel and

(a)   (b)

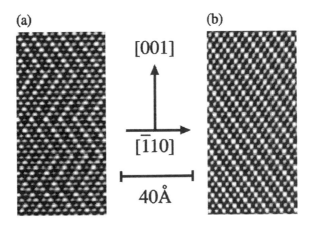

[001]

[$\bar{1}$10]

40Å

Figure 16. High-resolution cross-sectional TEM pictures of (a) a double-variant ordered and (b) a single-variant ordered GaInP alloys (from Ref. [68] of Zhang et al).

Hassine[136] have shown that only polytype I, III and IV are strain-compatible (for free standing CuPt ordered layers).

As demonstrated by Zhang et al[67,68,138], the CuPt ordered micro-domain twin structures that closely resemble the proposed OSL have distinctly different electronic and optical properties from those of the single variant CuPt ordered structure. Fig.16 shows a comparison of high resolution TEM pictures of an OSL like sample and a simple CuPt ordered sample[68]. Fig.17 shows polarized PL and PLE spectra for a pair of ordered samples both having order parameters near $\eta = 0.50$[138]: one grown on a 6°B tilted substrate, and thus, a typical single variant ordered sample; whilst the other grown on a 6°A tilted substrate, and thus having an OSL like structure along the [001] direction. As is evident that the second sample not only has a 60 meV larger band gap than the first one but also shows much stronger optical anisotropy between the [110] and [$\bar{1}$ 10] polarizations. The enhanced optical anisotropy can be readily explained by the symmetry change for the topmost valence band state upon the formation of an OSL[68], and was the first observed signature of the OSL effect[66]. One might attempt to ascribe the band gap increase for the double variant ordered sample shown in Fig.17 as being due to a lower degree of order, but the enhanced optical anisotropy cannot possibly be explained by this effect. To further rule out this possibility, the order parameter as well as the structural parameters of each individual variant have been determined experimentally[36]. In addition, Fig.18 further reveals that the correlation between the band gap reduction and valence band splitting are clearly different for the single and double variant order samples[67]. Mascarenhas et al[65,66] first attempted to model the band structure of OSLs using a 6-band **k.p** envelope function

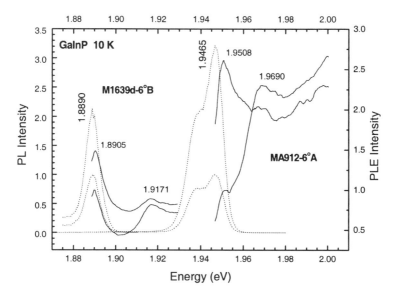

Figure 17. A comparison of polarized PL and PLE spectra between a single-variant CuPt ordered and a double-variant CuPt ordered GaInP with the same measured order parameter (from Ref.[138] of Zhang et al).

approach. They showed that even without band offsets and effective mass discontinuities, the OSL was capable of giving rise to quantum confinement effects that are normally achieved via the band offsets in conventional superlattices. The optical anisotropy due to the symmetry change of the wavefunction is associated with the "orientational" nature of this superlattice, and this becomes the dominant effect, because the confinement energy for the hole (the downward shift of the valence band edge) is relatively small. A comparison with a later first principle calculation of Wei and Zunger[139] indicates that the simple envelope function can indeed describe the valence band very well[66]. However, the first principle calculations of both Munzar et al (where the authors referred to the ordered domain twins as antiphase domain boundaries or APBs)[140] and Wei and Zunger[139] revealed a larger shift in the conduction band edge than that of the valence band. Even though both the domain size and the order parameter for the double variant ordered sample can now been determined experimentally[36], a quantitative comparison with theory is still impossible because the full band structure calculation has only been performed for the fully ordered structure with small number of layers[139,140].

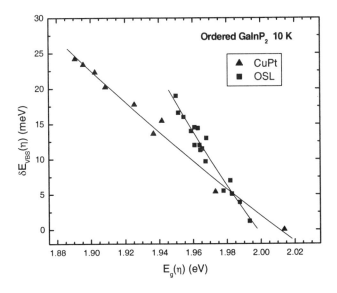

Figure 18. A comparison of the functional dependence of the valence-band splitting versus the band-gap between singe-variant and double-variant (small domain) ordered $GaInP_2$ samples (from Ref.[67] of Zhang et al).

9.   EXTRINSIC EFFECTS IN ORDERED GaInP

For CuPt ordered GaInP samples used in the early stage of the ordering study, it was plausible that there existed considerable structural non-uniformity, which was evidenced by the observed relatively large PL linewidth (typically > 10 meV for partially ordered samples[16,39,45,141,142], compared to < 7 meV for disordered samples[90,141]. Although these of samples were useful for the purpose of demonstrating some important effects of ordering (e.g., the ordering features in TEM, the band gap reduction, the optical anisotropy), they also gave rise to certain peculiar properties which attracted a great deal of attention for a period of time as though these properties were inherent to ordering. The so-called "moving emission"[142] has been perhaps the most extensively discussed issue among these extrinsic phenomena observed in ordered GaInP alloys. A broad PL band, either as the only major emission band for some samples[141] or the lower energy band for samples having two PL bands[142], was found to shift to high energy with increasing excitation density at a rate (> 10 meV/decade) significantly faster than that for the band edge emission of a typical semiconductor alloy, or for the typical donor-acceptor pair emission. DeLong et al[142] attributed this moving emission, which exhibited a long decay time > 1 µs, to a spatially indirect transition between the disordered and ordered domain, and suggested that the ordered domains were distributed throughout a disordered matrix. However, Hahn et al[143] shown later, by means of stereo images, that domain boundaries were microscopically thin, and the reason that they generally appeared as broad dark lines in TEM dark-field images was due to their projection onto the image plane. This study supports the conclusion based on spectroscopy studies that there is no disordered phase surrounding ordered domains, because not has only no optical transition associated with the disordered phase ever been observed for partially ordered samples[29-31] but also the spatial variation of the order parameter has been found to be minimal in the sub-micron scale[91]. DeLong et al[144] also observed that the moving emission only appeared in double-variant ordered samples, thus, suggesting that the moving emission was related to the interfaces between domains with different ordering directions. Ernst et al[145,146] investigated the correlation between optical properties and the ordered domain size. They found that samples with small domain sizes typically showed a single broad PL peak (as in the case of Ref.[141]), while samples with large domain sizes typically showed two PL peaks of which the higher energy peak (HE) was due to the band edge excitonic transition and the lower energy peak (LE) behaved similarly to the "moving emission" of Ref.[142]. However, the two peaks in Ref.[145] were well separated under low excitation density (the separation is larger than 30 meV, see Fig.5 for typical spectra of a sample of this type) and considerably sharper than those observed in the earlier study of Ref.[142]. Since the large domain samples used by Ernst et al were grown on substrates tilted 6° toward the $[111]_B$ direction, thus, being single-variant ordered, the correlation made by DeLong et al[144] between the "moving emission" and the double-variant ordering appears not to be valid in general. Indeed, in Ref. [144], samples on 6^OB tilted substrates did not show the "moving peak"; instead, a peak, behaving like a normal impurity transition, appeared at ~ 20 meV below the band edge transition. Similar spectra to those reported by Delong et al have also been reported for single-variant samples grown under similar conditions (e.g., Ref.[47] and Ref.[72]). Thus, the "moving emission" does not always appear with partial ordering, and the below band gap emission could have different origins for different samples. Given these factors, any attempt to provide a generalized model or theory for the extrinsic below band gap emission is unrealistic.

There have been quite a few efforts to investigate the below band gap emission in partially ordered GaInP by using sub-micron spatially resolved PL and PLE[91-93,147,148]. Gregor et al[147] reported an anti-correlation for the intensity of the band edge and the below band gap emission, thus

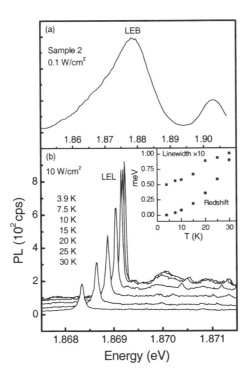

Figure 19. (a) Macro PL spectra cw excited at 1.908 eV. (b) Micro PL from the apertured sample area at lattice temperatures from 3.9 K (largest) to 30 K (smallest). Curves are plotted on the same origin, i.e., not displaced. Inset: the linewidth and peak position relative to the 3.9 K peak are plotted as a function of temperature (from Ref.[148] of Fluegel et al).

drawing a conclusion that the HE and LE emission originated from distinct spatial locations, and indicated that their results supported the interpretation of the LE emission being spatially indirect. Although Smith et al[92] also observed the anti-correlation for the PL intensities, they found that the LE emission coexisted with the HE emission in regions whose scales were much larger than the average domain size of 0.9 μm, which contradicts the idea that the LE emission originates solely from domain boundaries of either the ordered/disordered or antiphase domains, and suggests that other defects within the ordered domain might also contribute to the LE emission. Although the LE emission has been shown to contain some sharp emission lines in several studies by Cheong et al[91], Smith et al[[92], and Kops et al[93], only Kops et al have attempted to correlated them to a specific type of crystalline defect, namely, antiphase domain boundaries (APBs). Kops et al investigated the LE emission in partially ordered samples with different

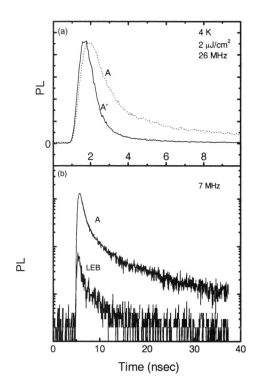

Figure 20. (a) Normalized time-resolved PL at two LEL peaks (A and A' for a single exciton and a multi-exciton state, respectively). (b) Log plot of the PL decay at the ground state peak and at the LE band (from Ref.[148] of Fluegel et al).

order parameters by varying temperature, external magnetic field, and excitation density. They indicated that the LE emission could be decomposed into two distinctly different types of transitions: a broad emission band (LEB) and superimposed sharp emission lines (LEL) (with a width of 0.3 – 1.0 meV). LEB and LEL were found to have different temperature and magnetic field dependence: the linewidth of the LEB showed a large temperature broadening while that of the LEL was temperature independent within an accuracy of ± 0.1meV; the LEL showed an excitonic behavior under magnetic field while the LEB's behavior was typical of free carriers (similar to the behavior of the moving peak observed in macro-PL by Jones et al[73]). While the origin for the LEB was not explicitly discussed except that the LEB might be related to a recombination between a strongly localized electron and a free hole, Kops et al associated the LEL to disks at APBs of two monolayers of InP aligned in the $[111]_B$ direction, and pointed out a correlation between the number of the LELs and the density of APBs as shown in TEM pictures. In an effort to give a unified picture for the LE below band gap emission, Matilla et al[149] performed an empirical pseudopotential calculation for the electronic states of the APB suggested by Kops et al, and claimed that their results explained the experimentally observed type II behavior for the below band gap emission. The calculation indicated that that for the APB with two adjacent InP layers (the so-called V_2 structure) the first hole state (h_1) would be localized at the interface, whereas the higher hole states (h_2 and etc.) and electron states (e_1 and etc.) would be delocalized in the bulk CuPt ordered region. Matilla et al assigned the indirect $e_1 - h_1$ transition to the LE emission, but it was not clear which of the LEB or LEL they meant. If they meant the LEB, they actually disagreed with Kops et al who believed that the LEB was related to a strongly localized electron state rather than a localized hole state; if they meant the LEL, the sharp emission line and the excitonic behavior under magnetic field seem to disprove this transition being type II. Not only has no long lifetime ever been measured for the LEL, but it was also shown by Smith et al[92] that the sharp lines could appear even within an ordered domain. Besides these points, we would like to point out that the specific kind of APB's suggested by Kops et al and modeled by Matilla et al have not been clearly observed in any microscopic study, although it has been found that the APB's may exist with almost any orientations[64,93,146]. Thus, even though this particular type of APB could perhaps generate a hole bound state, there is no direct experimental evidence for relating the theoretical results to the experimentally observed LEB or LEL. However, it is clear that the LEL are related to some type of defects in partially ordered alloys, but the exact nature of these defects is yet unknown.

The data of Kops et al[93] indicated that the behavior of LEL was much like that of quantum dots (QDs). However, unlike those QDs which normally distribute in a two dimensional plane (either by "self-assembling"[150] or formed unintentionally due to lateral layer width fluctuations in a quantum well[151]), these defects may be view as three dimensionally distributed QDs and thus LELs resemble the recombination of bound excitons associated with impurities in a bulk crystal[152]. One could in fact consider a single impurity analogous to a smallest QD. Using a higher spectral resolution than that of Kops et al, Fluegel et al[148] have resolved LELs with linewidth as small as 50 μeV (still limited by the spectral resolution), and found that the LEL linewidth increased by a factor of two on increasing the temperature to 30 K, just like the temperature dependence of the linewidth for the impurity transition[152]. They also observed that the transition energy decreased with increasing temperature, which merely reflects the band structure change of the host semiconductor with temperature. The results are shown in Fig.19. Through time resolved measurements, it has now been revealed that for a LEL due either to a single exciton or to a multi-exciton bound state, the radiative decay time is of the order of ~ 1 ns, typical of a direct transition. Even the LEB has been found to have a decay time similar to that of the LEL. Fig.20 shows typical PL decay curves for the LELs (one for a single exciton state and one for a multiexciton state) and the LEB. Another interesting finding is that by performing selective excitation at individual LELs, an energy transfer process amongst different LELs has been demonstrated, which again shows a close similarity between the LELs in ordered GaInP and impurity bound excitons in a bulk semiconductor. For instance, this type of energy transfer has long been observed amongst different nitrogen related trap centers in GaP:N[153].

10. CONCLUSIONS

Significant progress has been made during the past two decades in understanding and controlling the phenomenon of spontaneous ordering during epitaxial growth of semiconductor alloys. A variety of experimental techniques have been successfully utilized for probing the structural and electronic changes that result from this phenomenon and there now appears to be reasonable agreement between some of the experimental and theoretical results. However, even for the GaInP system there still remains a great deal of work to be done with respect to clarifying the ordering mechanism, obtaining higher order parameters, elucidating the effect of spontaneously generated electric fields on the properties of the ordered GaInP/GaAs heterojunction and understanding the role played by the microstructure on some of the peculiar optical properties. The more interesting research on the effects of statistical fluctuations that are tunable by control of the order parameter has only just begun and the studies that have been discussed should provide a solid foundation for further advances in the field.

11. ACKNOWLEDGEMENT

We would like to acknowledge the invaluable contributions of Brian Fluegel, Steve Smith, Hyeonsik M. Cheong, S. Phil Ahrenkiel, Andrew Norman, M. Al-Jassim, Jerry Olson, John Geisz, Daniel Friedman, Mark Hanna, Michael Kozhevnikov, Venky Narayanamuti, Lin-Wang Wang, Sverre Froyen, Eric D. Jones, Rebecca Forrest, Jianhua Li, and Simon Moss to the research presented in this chapter. We thank the Office of

Science/Basic Energy Sciences/Division of Materials Sciences for their generous support of this research at NREL.

REFERENCES

1. I. J. Murgatroyd, A. G. Norman, G. R. Booker and T. M. Kerr, J. Electron Microsc., **35**, 1497 (1986).
2. A. G. Norman (1987). Ph.D thesis, University of Oxford.
3. I. J. Murgatroyd (1987). Ph. D Thesis, University of Oxford.
4. A. G. Norman, R. E. Mallard, I. J. Murgatroyd, G. R. Booker, A. H. Moore and M. D. Scott. *Proc. Microsc. Semicond. Mater. Conf., Oxford, 6–8 April 1987 (IOP Conf. Ser. No. 87)*. A. C. Cullis and P. D. Augustus. (IOP Publishing, Bristol, UK, 1987), 77.
5. R. Hull, K. W. Carey, J. E. Fouquet, G. A. Reid, S. J. Rosner, D. Bimberg and D. Oertel. *GaAs and Related Compounds 1986*. (Inst. of Phys., Bristol, 1987), 209.
6. M.A. Shahid, S. Mahajan, D.E. Laughlin and H.M. Cox, Phys. Rev. Lett., **58**, 2567 (1987).
7. Y.-E. Ihm, N. Otsuka, J. Klem and H. Morkoc, Appl. Phys. Lett., **51**, 2013 (1987).
8. O. Ueda, M. Takikawa, J. Komeno and I. Umebu, Jpn. J. Appl. Phys., **26**, L1824 (1987).
9. A. Gomyo, T. Suzuki and S. Iijima, Phys. Rev. Lett., **60**, 2645 (1988).
10. J. P. Goral, M. M. Al-Jassim, J. M. Olson and A. Kibler. *Epitaxy of Semiconductor Layered Structures*. 1988), 583.
11. S. McKernan, B. C. D. Cooman, C. B. Carter, D. P. Bour and J. R. Shealy, J. Mater. Res., **3**, 406 (1988).
12. P. Bellon, J. P. Chevalier, G. P. Martin, E. Dupont-Nivet, C. Thiebaut and J. P. André, Appl. Phys. Lett., **52**, 567 (1988).
13. M. Kondow, H. Kakibayashi and S. Minagawa, J. Cryst. Growth, **88**, 291 (1988).
14. A. N. Pikhtin, Sov. Phys. Semicond., **11**, 245.
15. T. Suzuki, A. Gomyo, S. Iijima, K. Kobayashi, S. Kawata, I. Hino and T. Yuasa, Jpn. J. Appl. Phys., **27**, 2098 (1988).
16. A. Gomyo, K. Kobayashi, S. Kawata, I. Hino, T. Suzuki and T. Yuasa, J. Cryst. Growth, **77**, 367 (1986).
17. Y. Ohba, M. Ishikawa, H. Sugawara, M. Yamamoto and T. Nakanishi, J. Cryst. Growth, **77**, 374 (1986).
18. S. R. Kurtz, J. M. Olson and A. Kibbler, Solar Cells, **24**, 307 (1988).
19. A. Mascarenhas, S.R. Kurtz, A. Kibbler and J.M. Olson, Phys. Rev. Lett, **63**, 2108 (1989).
20. S.-H. Wei and A. Zunger, Phys. Rev. B, **39**, 3279 (1989).
21. T. Kurimoto and N. Hamada, Phys. Rev. B, **40**, 3889 (1989).
22. S.-H. Wei, L. G. Ferreira, J. E. Bernard and A. Zunger, Phys. Rev. B, **42**, 9622 (1990).
23. D. B. Laks, S.-H. Wei and A. Zunger, Phys. Rev. Lett, **69**, 3766 (1992).
24. S.-H. Wei, D. B. Laks and A. Zunger, Appl. Phys. Lett., **62**, 1937 (1993).
25. R. B. Capaz and B. Koiller, Phys. Rev. B, **47**, R4044 (1993).
26. K. A. Mäder and A. Zunger, Phys. Rev. B, **51**, 10462 (1995).
27. Y. Zhang, A. Mascarenhas and L.-W. Wang, Phys. Rev. B, **63**, R201312 (2001).
28. Y. Zhang and A. Mascarenhas, Phys. Rev. B, **51**, 13162 (1995).
29. R. G. Alonso, A. Mascarenhas, G. S. Horner, K. A. Bertness, S. R. Kurtz and J. M. Olson, Phys. Rev. B, **48**, 11833 (1993).
30. P. Ernst, C. Geng, F. Scholz, H. Schweizer, Y. Zhang and A. Mascarenhas, Appl. Phys. Lett., **67**, 2347 (1995).

31. B. Fluegel, Y. Zhang, H. M. Cheong, A. Mascarenhas, J. F. Geisz, J. M. Olson and A. Duda, Phys. Rev. B, **55**, 13647 (1997).
32. S.-H. Wei and A. Zunger, Phys. Rev. B, **57**, 8983 (1998).
33. R. Tycko, G. Dabbagh, S. R. Kurtz and J. P. Goral, Phys. Rev. B, **45**, 13452 (1992).
34. D. Mao, P. C. Tylor, S. R. Kurtz, M. C. Wu and W. A. Harrison, Phys. Rev. Lett, **76**, 4769 (1996).
35. R. L. Forrest, T. D. Golding, S. C. Moss, Y. Zhang, J. F. Geisz, J. M. Olson, A. Mascarenhas, E. Ernst and C. Geng, Phys. Rev. B, **58**, 15 355 (1998).
36. J. H. Li, J. Kulik, V. Holy, Z. Zhong, S. C. Moss, Y. Zhang, S. P. Ahrenkiel, A. Mascarenhas and J. Bai, Phys. Rev. B, **63**, 155310 (2001).
37. S.-H. Wei and A. Zunger, J. Chem. Phys., **107**, 1931 (1997).
38. T. Kanata, M. Nishimoto, H. Nakayama and T. Nishino, Phys. Rev. B, **45**, 6637 (1992).
39. D. J. Mowbray, R. A. Hogg, M. S. Skolnick, M. C. DeLong, S. R. Kurtz and J. M. Olson, Phys. Rev. B, **46**, 7232 (1992).
40. T. Kanata, M. Nishimoto, H. Nakayama and T. Nishino, Appl. Phys. Lett., **63**, 512 (1993).
41. J. J. Hopfield, J. Phys. Chem. Solids, **15**, 97 (1960).
42. J. L. Shay, E. Buehler and J. H. Wernick, Phys. Rev. B, **3**, 2004 (1971).
43. S.-H. Wei and A. Zunger, Appl. Phys. Lett., **56**, 662 (1990).
44. D. Teng, J. Shen, K. E. Newman and B.-L. Gu, J. Phys. Chem. Solids, **52**, 1109 (1991).
45. G. S. Horner, A. Mascarenhas, S. Froyen, R. G. Alonso, K. Bertness and J. M. Olson, Phys. Rev. B, **47**, 4041 (1993).
46. G. S. Horner, A. Mascarenhas, R. G. Alonso, D. J. Friedman, K. Sinha, K. A. Bertness, J. G. Zhu and J. M. Olson, Phys. Rev. B, **48**, 4944 (1993).
47. H. M. Cheong, Y. Zhang, A. Mascarenhas, J. F. Geisz and J. M. Olson, J. Appl. Phys., **83**, 1773 (1998).
48. J. S. Luo, J. M. Olson, K. A. Bertness, M. E. Raikh and T. E. V., J. Vac. Sci. Technol. B, **12**, 2552 (1994).
49. E. Greger, K. H. Gulden, M. Moser, G. Schmiedel, P. Kiesel and G. H. Dohler, Appl. Phys. Lett., **70**, 1459 (1997).
50. T. Kita, A. Fujiwara, H. Nakayama and T. Nishino, Appl. Phys. Lett., **66**, 1794 (1995).
51. F. Alsina, M. Garriga, M. I. Alonso, J. Pascual, J. Camassel and R. W. Glew (1995). Proceedings of the 22nd International Conference in Physics of Semiconductors. D. J. Lockwood. Singapore, World Scientific: 253.
52. B. Fluegel, A. Mascarenhas, J. F. Geisz and J. M. Olson, Phys. Rev. B, **57**, R6787 (1998).
53. M. Moser, R. Winterhoff, C. Geng, I. Queisser, F. Scholz and A. Dornen, Appl. Phys. Lett., **64**, 235 (1994).
54. R. Wirth, A. Moritz, C. Geng, F. Scholz and A. Hangleiter, Phys. Rev. B, **55**, 1730 (1997).
55. Y. Zhang, B. Fluegel, A. Mascarenhas, J. F. Geisz, J. M. Olson, F. Alsina and A. Duda, Solid State Commun., **104**, 577 (1997).
56. G. G. Forstmann, F. Barth, H. Schweizer, M. Moser, C. Geng, F. Scholz and E. P. O'Reilly, Semicond. Sci. Technol., **9**, 1268 (1994).
57. E. Greger, K. H. Gulden, P. Riel, H. P. Schweizer, M. Moser, G. Schmiedel, P. Kiesel and G. H. Dohler, Appl. Phys. Lett., **68**, 2383 (1996).
58. E. Greger, K. H. Gulden, P. Riel, M. Moser, T. Kippenberg and G. H. Dohler, Appl. Phys. Lett., **71**, 3245 (1997).
59. G. L. Bir and G. E. Pikus.*Symmetry and Strain-Induced Effects in Semiconductors* (Wiley, New York, 1974).
60. S.-H. Wei and A. Zunger, Phys. Rev. B, **49**, 14337 (1994).
61. Y. Zhang, A. Mascarenhas, P. Ernst, F. A. J. M. Driessen, D. J. Friedman, K. A. Bertness, J. M. Olson, C. Geng, F. Scholz and H. Schweizer, J. Appl. Phys., **81**, 6365 (1997).

62. E. G. Tsitsishvili, Phys. Rev. B, **59**, 10044 (1999).
63. E. Morita, M. Ikeda, O. Kumagai and K. Kaneko, Appl. Phys. Lett., **53**, 2164 (1988).
64. C. S. Baxter, W. M. Stobbs and J. H. Wilkie, Journal of Crystal Growth, **112**, 373 (1991).
65. A. Mascarenhas, Y. Zhang, R. G. Alonso and S. Froyen, "Orientational superlattices in ordered GaInP$_2$," Solid State Commun., **100**, 47 (1996), Fig.1 used with permission of Elsevier Science.
66. Y. Zhang and A. Mascarenhas, Phys. Rev. B, **56**, 9975 (1997).
67. Y. Zhang, A. Mascarenhas, S. P. Ahrenkiel, D. J. Friedman, J. Geisz and J. M. Olson, "Electronic and optical properties of periodically stacked orientational domains in CuPt-ordered GaInP$_2$," Solid State Commun., **109**, 99 (1998), Fig.3 used with permission of Elsevier Science.
68. Y. Zhang, B. Fluegel, S. P. Ahrenkiel, D. J. Friedman, J. F. Geisz, J. M. Olson and A. Mascarenhas, Mat. Res. Soc. Symp. Proc., **583**, 255 (2000).
69. M. I. Dyakonov and V. I. Perel. *Optical Orientation*. F. Meier and B. P. Zakharchenya. (North-Holland, Amsterdam, 1984), 11.
70. F. Ciccacci, E. Molinari and N. E. Christensen, Solid State Commun., **62**, 1 (1987).
71. T. Kita, M. Sakurai, K. Bhattacharya, K. Yamashita, T. Nishino, C. Geng, F. Scholz and H. Schweizer, Phys. Rev. B, **57**, R15044 (1998).
72. B. Fluegel, Y. Zhang, A. Mascarenhas, J. F. Geisz, J. M. Olson and A. Duda, Phys. Rev. B, **60**, R11261 (1999).
73. E. D. Jones, D. M. Follstaedt, S. K. Lyo and J. R. P. Schneider, Mat. Res. Soc. Symp. Proc., **281**, 61 (1993).
74. P. Emanuelsson, M. Drechsler, D. M. Hofmann, B. K. Meyer, M. Moser and F. Scholz, Appl. Phys. Lett., **64**, 2849 (1994).
75. M. E. Raikh and E. V. Tsiper, Phys. Rev. B, **49**, 2509 (1994).
76. A. Franceschetti, S.-H. Wei and A. Zunger, Phys. Rev. B, **53**, 13992 (1995).
77. Y. C. Yeo, M. F. Li, T. C. Chong and P. Y. Yu, Phys. Rev. B, **55**, 16414 (1997).
78. P. Ernst, Y. Zhang, F. A. J. M. Driessen, A. Mascarenhas, E. D. Jones, C. Geng, F. Scholz and H. Schweizer, J. Appl. Phys., **81**, 2814 (1997).
79. Y. Zhang, A. Mascarenhas and E. D. Jones, J. Appl. Phys., **83**, 448 (1998).
80. D. E. Aspnes, J. P. Harbison, A. A. Studna and L. T. Florez, J. Vac. Sci. Technol A, **6**, 1327 (1988).
81. J. S. Luo, J. M. Olson, S. R. Kurtz, D. J. Arent, K. A. Bertness, M. E. Raikh and E. V. Tsiper, Phys. Rev. B, **51**, 7603 (1995).
82. B. A. Philips, I. Kamiya, K. Hingerl, L. T. Florez, D. E. Aspnes, S. Mahajan and J. P. Harbison, Phys. Rev. Lett., **74**, 3640 (1995).
83. J. Callaway. *Quantum Theory of the Solid State* (Academic Press, Boston, 1991).
84. S. H. Wei and A. Zunger, Phys. Rev. B, **51**, 14110 (1995).
85. J. S. Luo, J. M. Olson, Y. Zhang and A. Mascarenhas, Phys. Rev. B, **55**, 16385 (1997).
86. M. Zorn, P. Kurpas, A. I. Shkrebtii, B. Junno, A. Bhattacharya, K. Knorr, M. Weyers, L. Samuelson, J. T. Zettler and W. Richter, Phys. Rev. B, **60**, 8185 (1999).
87. S. B. Zhang, S. Froyen and A. Zunger, Appl. Phys. Lett., **67**, 3141 (1995).
88. H. Murata, I. H. Ho, T. C. Hsu and G. B. Stringfellow, Appl. Phys. Lett., **67**, 3747 (1995).
89. H. Murata, I. H. Ho, L. C. Su, Y. Hosokawa and G. B. Stringfellow, J. Appl. Phys., **79**, 6895 (1996).
90. Y. Zhang, A. Mascarenhas, S. Smith, J. F. Geisz, J. M. Olson and M. Hanna, Phys. Rev. B, **61**, 9910 (2000).
91. H. M. Cheong, A. Mascarenhas, J. F. Geisz, J. M. Olson, M. W. Keller and J. R. Wendt, Phys. Rev. B, **57**, R9400 (1998).

92. S. Smith, H. M. Cheong, B. D. Fluegel, J. F. Geisz, J. M. Olson, L. L. Kazmerski and A. Mascarenhas, Appl. Phys. Lett., **74**, 706 (1999).

93. U. Kops, P. G. Blome, M. Wenderoth, R. G. Ulbrich, C. Geng and F. Scholz, Phys. Rev. B, **61**, 1992 (2000).

94. Y. Zhang, A. Mascarenhas and L.-W. Wang, Phys. Rev. B, **64**, 125207 (2001).

95. D. C. Meyer, K. Richter, P. Paufler and G. Wagner, Phys. Rev. B, **59**, 15253 (1999).

96. J. C. Mikkelsen(Jr.) and J. B. Boyce, Phys. Rev. Lett., **49**, 1412 (1982).

97. J. B. Boyce and J. C. Mikkelsen(Jr.), J. Cryst. Growth, **98**, 37 (1989).

98. K. Kodama, M. Hoshino, K. Kilahara, M. Takikawa and M. Ozeki, Jap. J. Appl. Phys., **25**, L127 (1986).

99. S. L. Feng, J. Krynicki, V. Donchev, J. C. Bourgoin, M. D. Forte-Poisson, C. Brylinski, S. Delage, H. Blanck and S. Alaya, Semicon. Sci. Technol., **8**, 2092 (1993).

100. M.A. Rao, E.J. Caine, H. Kroemer, S.I. Long and D.I. Babic, J.Appl. Phys.,**61**,643(1987).

101. T. W. Lee, P. A. Houston, R. Kumar, X. F. Yang, G. Hill, M. Hopkinson and P. A. Claxton, Appl. Phys. Lett., **60**, 474 (1992).

102. W. Liu, S. K. Fan, T. S. Kim, E. A. Beam and D. B. Davito, IEEE Electron Devices, **40**, 1378 (1993).

103. P. Krispin, M. Asghar, A. Knauer and H. Kostial, J. Cryst. Growth, **220**, 220 (2000).

104. D. Biswas, N. Debbar, P. Bhattacharya, M. Razeghi, M. Defour and F. Omnes, Appl. Phys. Lett., **56**, 833 (1990).

105. M. O. Watanabe and Y. Ohba, Appl. Phys. Lett., **50**, 906 (1987).

106. M. S. Faleh, J. Tasselli, J. P. Bailbe and A. Marty, Appl. Phys. Lett., **69**, 1288 (1996).

107. S. Lan, C. Q. Yang, W. J. Wu and H. D. Liu, J. Appl. Phys., **79**, 2162 (1996).

108. A. Lindell, M. Pessa, A. Salokatve, F. Bernardini, R. M. Nieminen and M. Paalanen, J. Appl. Phys., **82**, 3374 (1997).

109. Y.-H. Cho, K.-S. Kim, S.-W. Ryu, S.-K. Kim, B.-D. Choe and H. Lim, Appl. Phys. Lett., **66**, 1785 (1995).

110. J. J. O'Shea, C. M. Reaves, S. P. DenBaars, M. A. Chin and V. Narayanamurti, Appl. Phys. Lett., **69**, 3022 (1996).

111. G. Arnaud, P. Boring, B. Gil, J.-C. Garcia, J.-P. Landesman and M. Leroux, Phys. Rev. B, **46**, 1886 (1992).

112. M. A. Haase, M. J. Hafich and G. Y. Robinson, Appl. Phys. Lett., **58**, 616 (1981).

113. M. Kozhevnikov, V. Narayanamurti, A. Mascarenhas, Y. Zhang, J. M. Olson and D. L. Smith, Appl. Phys. Lett., **75**, 1128 (1999).

114. K.-S. Kim, Y.-H. Cho, B.-D. Choe, W. G. Jeong and H. Lim, Appl. Phys. Lett., **67**, 1718 (1995).

115. J. Chen, J. R. Sites, I. L. Spain, M. J. Hafich and J. Y. Robinson, Appl. Phys. Lett., **58**, 744 (1991).

116. T. Kobayashi, K. Taira, F. Nakamura and H. Kawai, J. Appl. Phys., **65**, 4898 (1989).

117. M. Leroux, M. L. Fille, B. Gil, J. P. Landesman and J. C. Garcia, Phys. Rev. B, **47**, 6465 (1993).

118. M. F. Whitaker, D. J. Dunstan, M. Missous and L. Gonzalez, Phys. Stat. Sol. B, **198**, 349 (1996).

119. O. Dehaese, X. Wallart, O. Schuler and F. Mollot, J. Appl. Phys., **84**, 2127 (1998).

120. Q. Liu, S. Derksen, A. Lindner, F. Scheffer, W. Prost and F.-J. Tegude, J. Appl. Phys., **77**, 1154 (1995).

121. Q. Liu, S. Derksen, W. Prost, A. Lindner and F.-J. Tegude, J. Appl. Phys., **79**, 305 (1996).

122. Z. P. Su, K. L. Teo, P. Y. Yu and K. Uchida, Solid State Commun., **99**, 933 (1996).

123. J. Zeman, G. Martinez, P. Y. Yu, S. H. Kwok and K. Uchida, Phys. Stat. Sol. (b), **211**, 239 (1999).
124. J. Zeman, G. Martinez, P. Y. Yu and K. Uchida, Phys. Rev. B, **55**, R13 428 (1997).
125. W. A. Harrison, J. Vac. Sci. Technol. B, **1**, 126 (1983).
126. Y. Foulon, C. Priester, G. Allan, J. C. Garcia and J. P. Landesman, J. Vac. Sci. Technol. B, **10**, 1754 (1992).
127. S. Froyen, A. Zunger and A. Mascarenhas, Appl. Phys. Lett., **68**, 2852 (1996).
128. Y. Zhang, A. Mascarenhas and L.-W. Wang, unpublished.
129. F. A. J. M. Driessen, Appl. Phys. Lett., **67**, 2813 (1995).
130. F. A. J. M. Driessen, H. M. Cheong, A. Mascarenhas, S. K. Deb, P. R. Hageman, G. J. Bauhuis and L. J. Giling, Phys. Rev. B, **54**, R5263 (1996).
131. T. Kita, T. Nishino, C. Geng, F. Scholz and H. Schweizer, Phys.Rev. B, **59**, 15358(1999).
132. S. H. Kwok, P. Y. Yu, K. Uchida and T. Arai, Appl. Phys. Lett., **71**, 1110 (1997).
133. D. J. Friedman, G. S. Horner, S. R. Kurtz, K. A. Bertness, J. M. Olson and J. Moreland, Appl. Phys. Lett., **65**, 878 (1994).
134. S. P. Ahrenkiel, unpublished.
135. F. Alsina, J. Pascual, M. Garriga, M. I. Alonso, S. Tortosa, C. Geng, F. Scholz and R. W. Glew, Solid State Commun., **101**, 10 (1997).
136. J. Sapriel and A. Hassine, Phys. Rev. B, **56**, R7112 (1997).
137. Z. Ikonic, G. P. Srivastava and C. J. Inkson, Phys. Rev. B, **48**, 17181 (1993).
138. Y. Zhang and A. Mascarenhas, "Effect of the orientational superlattice on the electronic and vibrational properties of CuPt ordered GaInP alloys", J. Raman Spec., **32**, 831, ©2001, [Wiley-Liss, Inc., a subsidiary of John Wiley & Sons, Inc.].
139. S.-H. Wei, S. B. Zhang and A. Zunger, Phys. Rev. B, **59**, R2478 (1998).
140. D. Munzar, E. Dobrocka, I. Vavra, R. Kudela, M. Harvanka and N. E. Christensen, Phys. Rev. B, **57**, 4642 (1998).
141. J.E. Fouquet, V.M. Robbins, S.J. Rosner and O. Blum, Appl. Phys. Lett., **57**, 1566 (1990).
142. M. C. DeLong, W. D. Ohlsen, I. Viohl, P. C. Taylor and J. M. Olson, J. Appl. Phys., **70**, 2780 (1991).
143. G. Hahn, C. Geng, P. Ernst, H. Schweizer, F. Scholz and F. Phillipp, Superlattices and Microstructures, **22**, 301 (1997).
144. M. C. DeLong, C. E. Ingefield, P. C. Taylor, L. C. Su, I. H. Ho, T. C. Hsu, G. B. Stringfellow, K. A. Bertness and J. M. Olson. *Proc. 21th Int. Symp. Comp. Semicond.* H. Goronkin and U. Mishra. (IOP Publishing, Bristol, 1995), 207.
145. P. Ernst, C. Geng, F. Scholz and H. Schweizer, Phys. Status Solidi B, **193**, 213 (1996).
146. P. Ernst, C. Geng, G. Hahn, F. Scholz, H. Schweizer, F. Phillipp and A. Mascarenhas, J. Appl. Phys., **79**, 2633 (1996).
147. M. J. Gregor, P. G. Blome, R. G. Ulbrich, P. Grossmann, S. Grosse, J. Feldmann, W. Stolz, E. O. Göbel, D. J. Arent, M. Bode, K. A. Bertness and J. M. Olson, Appl. Phys. Lett., **67**, 3572 (1995).
148. B. Fluegel, S. Smith, Y. Zhang, A. Mascarenhas, J. F. Geisz and J. M. Olson, unpublished.
149. T. Matilla, S.-H. Wei and A. Zunger, Phys. Rev. Lett., **83**, 2010 (1999).
150. D. Leonard, M. Krishnamurthy, C. M. Reaves, S. P. Denbaars and P. M. Petroff, App. Phys. Lett., **63**, 3203 (1993).
151. A. Zrenner, L. V. Butov, M. Hagn, G. Abstreiter, G. Böhm and G. Weimann, Phys. Rev. Lett, **72**, 3382 (1994).
152. D. E. McCumber and M. D. Sturge, J. Appl. Phys., **34**, 1682 (1963).
153. P. J. Wiesner, R. A. Street and H. D. Wolf, Phys. Rev. Lett., **35**, 1366 (1975).

Chapter 12

Spectroscopic Study of the Interface and Band Alignment at the GaInP(Partially Ordered)/GaAs Heterojunction under High Pressure and High Magnetic Field

P. Y. Yu[1], G. Martinez[2], J. Zeman,[2*] and K. Uchida[3]

1) Department of Physics, University of California, Berkeley and Materials Science Division, Lawrence Berkeley National Laboratory, Berkeley CA 94720, USA;
2) Grenoble High Magnetic Field Laboratory MPI/CNRS, 25, Av. des Martyrs, 38042 Grenoble Cedex 9, France (on leave from the Institute of Physics, Academy of Sciences, Prague, Czech Republic);*
3) Department of Electrical Engineering, The University of Electro-Communications, 1-5-1, Choufugaoka, Choufu, Tokyo 182, Japan

Key words: GaInP, PL

Abstract: The semiconducting alloy $Ga_{0.5}In_{0.5}P$ grown epitaxially on a GaAs substrate forms a rather unique system in which not only the band gap is tunable by varying the degree of CuPt ordering in the $Ga_{0.5}In_{0.5}P$ but also the band alignment can be varied from type I to type II. The band alignment has been found to have an important influence on the optoelectronic properties of the $Ga_{0.5}In_{0.5}P$ layer in terms of the efficiency of photoluminescence upconversion at the $Ga_{0.5}In_{0.5}P$ /GaAs interface and the observation of quantum well emission from $Ga_{0.5}In_{0.5}P/GaAs/-Ga_{0.5}In_{0.5}P$ structure. In this paper we shall demonstrate how the band offset and alignment in such heterostructures can be manipulated and measured with meV resolution by combining spectroscopic measurements with high pressure and high magnetic field. In addition, the magnetic length can be used as a "nanometer-stick" in probing the size of type II domains in samples where most of the interface is of type I except for tiny islands of more ordered domains.

1.   INTRODUCTION

Studies of the *random alloy* $Ga_xIn_{1-x}P$ (where $x \sim 0.5$) grown pseudomorphically on GaAs (to be abbreviated as GaInP in the rest of this article) were initiated in the 1970s. These alloys were found to make good materials for visible lasers since their band gap is about 1.99 eV in the red [1]. They also have potential applications in forming heterojunction bipolar transistors [2]. For the latter applications it is important to know the band alignment and offset between GaInP and GaAs (whose band gap is ~ 1.5 eV and therefore smaller than GaInP). This was determined by high pressure photoluminescence (PL) and found to be type I with the conduction band offset equal to ~ 0.1 eV [3]. Thus most of the difference between the band gaps of GaInP and GaAs shows up in the valence band offset. Self-organization of GaInP grown on GaAs into a superlattice along one of the equivalent [111] crystallographic axes was discovered by Gomyo *et al.* in 1987 [4]. Such ordering is usually known as CuPt ordering. The band gap in the ordered samples was found to be reduced by as much as 0.1 eV when compared with the random alloy [5]. It was generally assumed at that time that the band alignment remains always type I and most of the band gap reduction is absorbed by the large valence band offset (for a definition of type I and II band alignments see, for example, [6]). Although later Liu *et al.* [7] have proposed that the band alignment between the ordered GaInP and the GaAs substrate is type II, the evidence on which they made this band assignment was not considered to be conclusive.

In 1995 Driessen [8] in the Netherlands and in 1996 Su *et al.* [9] at Berkeley reported that partially ordered GaInP/GaAs samples exhibited highly efficient photoluminescence upconversion. In this process photons with energy *below* the band gap energy (E_g) of GaInP can generate band gap emission from GaInP. In other words, *the emission has higher photon energy than the excitation.* By tuning the excitation photon energy around the band gap of GaAs both Driessen [8] and Su *et al.* [9] found that the upconversion process occurred efficiently only if the excitation photon energy was larger than the band gap of the GaAs substrate. Much weaker upconversion was, however, still observable when the photon energy could excite electrons from neutral acceptors (typically carbon which are incorporated unintentionally into GaAs during growth via metal-organic chemical vapor deposition or MOCVD) to the conduction band of GaAs [9]. These results demonstrated conclusively that the upconversion process involved the creation of electron and hole pairs first in GaAs, which subsequently were excited across the band edge discontinuities into GaInP where recombination occurred. Driessen [8] assumed that the band offset between GaAs and GaInP was of type I and proposed that the upconversion process resulted from the excitation of both electrons and holes via Auger processes occurring near the interface. However, given the low excitation power of the experiment, Su *et al.* [9] concluded that the probability for exciting *both* the electron and hole via Auger processes would be too small to explain the observed upconversion efficiency. They suggested instead that Auger process may account for at most the excitation of the holes across the valence barrier. For the electrons, Su *et al.* [9] suggested that they probably transferred spontaneously from GaInP to GaAs as a result of a type II band alignment. Again this idea was not generally accepted since there was no direct evidence of a type II band alignment at the GaInP/GaAs junction. Subsequently, theoretical calculation of the band offset between GaAs and both disordered GaInP and fully ordered GaInP were reported by Froyen *et al.* [10]. These authors found that the band alignment was type I and the band offset ~ 0.12 eV for the disordered GaInP/GaAs interface in agreement with experiment. However, the band alignment between GaAs and *fully ordered* GaInP was predicted to be type II with a band offset of ~ 0.2 eV in magnitude. To estimate the band offset in partially ordered GaInP, it is customary to use the model proposed by Laks *et al.* [11] in which the electronic properties of GaInP varies quadratically with an "order parameter" η reflecting the degree of CuPt ordering in the sample. With the assumption that the band offset also varies quadratically with the order parameter one can deduce from the calculation of Froyen *et al.* [10] that the conduction band offset between GaAs and partially ordered GaInP with order parameter of around 0.5 should be almost zero.

Although theoretical calculations suggest that it is possible to have a type II band alignment between the GaAs substrate and the partially ordered GaInP epilayer, it is not easy to verify this experimentally nor to determine the value of the band offset since the magnitude of this offset can be of the order of meV. Experimental techniques for measuring band offsets like photoemission have relatively poor energy resolutions with typical uncertainties larger than 20 meV [12,13]. Other techniques such as ballistic-electron-emission-microscopy [14,15] cannot probe an interface too far below the surface and hence are susceptible to surface effects. One technique which is well suited to measuring a small type II band offset is photoluminescence (PL) at low temperature. When the type II band offset is small it is often possible to observe recombination of both spatially direct and indirect transitions occurring in the vicinity of the interface. Figure 1 shows a situation appropriate for a type II band alignment at the GaAs/partially ordered GaInP where most of the electrons are located in GaInP while the holes are confined in GaAs.

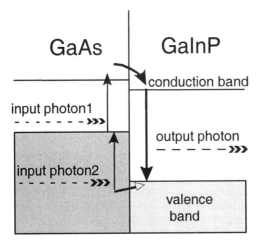

Figure 1. Processes responsible for upconversion of photoluminescence at a type II alignment between the GaAs and partially ordered GaInP band structure where the photo-excited electrons are mostly located in GaInP while the holes are confined in GaAs.

At non-zero temperatures a small amount of electrons are also present in GaAs and hence a spatially direct transition arising from their recombination with holes in GaAs is also observable. Since both the spatially direct and indirect transitions involve the same holes in GaAs, the energy difference of these two transitions is a precise measurement of the conduction band offset. The identification of these transitions can be facilitated by applying a magnetic field to the sample and measuring the effective masses of the carriers via their cyclotron frequencies. The magnetic field can be complemented by the application of hydrostatic pressure, which can modify the band offset. In this article we shall review various photoluminescence experiments carried out under high magnetic field and/or high pressure to determine the band conduction band offset in partially ordered GaInP/GaAs. In the special case, where both high magnetic field and high pressure are applied simultaneously, we successfully determine the size of type II domains in partially ordered GaInP near the interface.

2.   BACKGROUND

In this section we shall review some background materials necessary for understanding the rest of this paper. Although some of these materials are covered in much more detail in the literature including chapters in the rest of this book, we have nevertheless included them here in order that this article can be self-contained and be understood with minimal references to other sources. The topics we shall cover here include the effects of: ordering, pressure and magnetic field on the electronic band structure and optical properties of partially CuPt-ordered GaInP.

2.1   Effect of Ordering on Band Structure of GaInP and Its Band Offset with GaAs

For the purpose of understanding the band alignment between GaAs and GaInP we shall concentrate in this article mainly on the conduction and valence band extrema at the Brillouin zone center or Γ point. In both materials the band gap is direct (for a review of the band structure of GaAs see, for example, Chapter 2 of [6]). While the conduction band is s-like and non-degenerate, the valence band is p-like and doubly degenerate (the spin-orbit splitting in both GaAs and GaInP are large enough to allow us to neglect the split-off band). The band structures of disordered GaInP (or d-GaInP in short) are compared with those of CuPt ordered GaInP (or o-GaInP in short) in Fig. 2.

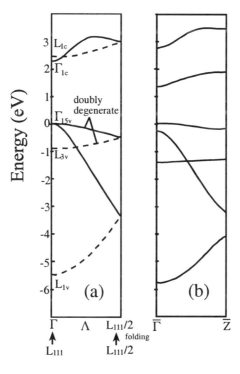

Figure 2. The band structures of a) disordered GaInP is compared with that of b) CuPt ordered GaInP. The broken curves in a) represent the bands along the [111] directions which would have been folded back into the zone center by the CuPt ordering [16].

At low temperatures the values of the band gaps in GaAs and d-GaInP are 1.54 and 2.01 eV, respectively. This means that the sum of the band offsets in the conduction and valence bands at the GaAs/d-GaInP interface is > 0.4 eV. According to the calculation of Froyen *et al.* [10] the conduction band offset is smaller and is only ~ 0.12 eV while the valence band offset is about 3 times larger. These values are in good agreement with experimental values obtained by a number of different techniques [3].

In perfectly CuPt-ordered GaInP the GaP and InP layers form a monolayer superlattice along one of the equivalent [111] directions. The reduction in the band gap at Γ as a result of this ordering can be understood in terms of two effects. The formation of the superlattice halves the size of the Brillouin zone along the [111] or L direction and produces the effect generally known as "zone folding." In other words, the bands of o-GaInP are obtained by folding the bands of d-GaInP along a line bisecting the Γ to L axis (also known as the Λ axis in group theory notation) [16]. As a result of this folding, the states originally at the L point on the surface of the Brillouin zone are folded into the zone center (Γ point). This is shown more clearly in Fig. 2(a). Zone folding can affect bands along other directions in addition to the L direction. For example, bands along the [110] (or Σ) direction are also folded [16]. The second effect of the superlattice is to introduce a crystal potential which mixes the states at Γ with the ones folded into the zone center by the zone-folding effect. Since the zone edge conduction band minima in d-GaInP are higher in energy than the conduction minimum at Γ, this mixing tends to *decrease* the energy of the conduction band minimum at Γ while raising the energies of the higher folded states. Similar coupling between the top of the valence band at Γ and the folded valence band states *increases* the energy of the valence band maxima at the zone center. Thus the net effect of ordering on the band gap is to *decrease* the band gap. According to Wei and Zunger [17] the band gap of o-GaInP is smaller than that of d-GaInP by as much as 212 meV. The corresponding change in band alignment results in a type II band offset between o-GaInP and GaAs as shown in Fig. 3.

However, in most GaInP samples the degree of CuPt ordering is not complete. This is manifested by a smaller band gap reduction in partially ordered GaInP (to be abbreviated as po-GaInP). Laks *et al.* [11] have proposed that the reduced CuPt ordering in po-GaInP can be characterized by an "average order parameter" η. Furthermore, the band gap reduction varies quadratically with η. So far there is no theoretical calculation or systematic experimental investigation of the relation between band offset and η. As a consequence one usually does not know what should be the band alignment and band offset between po-GaInP and GaAs. In this article we shall demonstrate how to utilize high magnetic field or high pressure

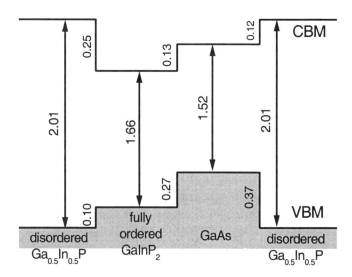

Figure 3. The theoretical band offsets between GaAs and GaInP (both fully ordered and disordered). Reproduced from [10].

experiments to determine the type and magnitude of the band offset at the po-GaInP/GaAs interface.

In addition to changing the band gap of GaInP, ordering also changes other electronic properties such as the effective masses of the electrons and holes [16]. Since ordering mixes the Γ states with smaller effective masses with zone-edge states with larger masses, the effective masses of both electrons and holes in o-GaInP are larger than those in d-GaInP. We shall not be interested in this particular effect of ordering in this article.

2.2   Effect of Pressure on Band Structure of GaAs and GaInP

The effect of hydrostatic pressure (P) on a medium is to decrease the distance between the atoms. Its influence on the electronic band structure of both zincblende- and chalcopyrite-type semiconductors has recently been computed by first principle band structure calculations [18]. However, it has been known for many years that in zincblende-type semiconductors the effect of pressure on the conduction band at different extrema of the Brillouin zone depends more on the symmetry of the extrema than on the chemical nature of the semiconductor (this is often referred to as the Paul's empirical rule [19]). This simple result can be understood in terms of the pseudopotential model [20]. In this model the pressure coefficient $\alpha = dE/dP$ of the electronic energy E is expressed in terms of a volume deformation potential $V(dE/dV)$. For each energy level dE/dV can be further decomposed into the sum of two terms: a "kinetic energy" term which is always negative and varies slowly from one level to another as long as these levels lie in the vicinity of the fundamental gap; and a "potential energy" term which varies with the symmetry of the level and can be either positive or negative. The magnitude of the kinetic energy contribution is always larger and hence dE/dV is always negative and the corresponding pressure coefficient is always positive. For p-like states (such as the top of the valence band or the conduction band minima at the X point or (001) of the Brillouin zone), the two contributions happen to have opposite signs and partially cancel each other. As a result, the pressure coefficient of the indirect band gap ($E_{\Gamma-X}$) between the conduction band minima at X and the valence band maximum at Γ is very small and negative ($dE_{\Gamma-X}/dP \approx -5$ meV/GPa). For the s-like states (such as the conduction band minimum at Γ) both contributions add together and as a result the pressure coefficient of the direct band gap ($E_{\Gamma-\Gamma}$) is rather large ($dE_{\Gamma-\Gamma}/dP \approx 100$ meV/GPa). For mixed s-p states (such as the conduction band minima at the L point or (111) of the Brillouin zone) the pressure coefficient of the indirect band gap ($E_{\Gamma-L}$) has intermediate values with $dE_{\Gamma-L}/dP \approx 50$ meV/GPa. Since the conduction band of po-GaInP is a mixture of the Γ and L conduction band extrema in d-GaInP, one expects the pressure coefficient of the band gap of po-GaInP to be *smaller* than that of d-GaInP. The experimentally determined pressure coefficients of the direct band gap at Γ (dE_g/dP) in GaAs, d-GaInP and po-GaInP as reported by various groups are: 105, 82 and 65 meV/GPa, respectively [21,22,3,23,24].

The band gap pressure coefficient dE_g/dP is related to the absolute pressure coefficients of the corresponding conduction and valence bands (dE_c/dP and dE_v/dP) by the equation:

$$\frac{dE_g}{dp} = \frac{dE_c}{dp} - \frac{dE_v}{dp} \tag{1}$$

Realizing that $dE_c/dP >> dE_v/dP$ for both GaAs and GaInP, one expects pressure to *decrease* the conduction band offset (which we shall define as: $\Delta E_c = E_c$ (GaInP)-E_c (GaAs)) although dE_v/dP is different for GaAs and GaInP. This means that if the band alignment at the GaAs/po-GaInP interface is originally type I with $\Delta E_c > 0$ then pressure can convert the band alignment into type II (provided ΔE_c is << 100 meV). From the above pressure coefficients one expects that a pressure of 1 GPa will decrease the band gap difference between po-GaInP and GaAs by about 40 meV. The corresponding decrease in the conduction band offset should be a sizable fraction of this value. So far there is no theoretical prediction of the pressure coefficient of ΔE_c. Still if one assumes that $d\Delta E_c /dP$ is given by more than 50% of the band gap pressure coefficient difference between GaAs and GaInP one expects that a pressure >1 GPa may be sufficient to convert a type I po-GaInP/GaAs interface to type II. This rough estimate, however, does not take into account the possible existence of confinement of electrons by band bending and other effects, which can affect the band alignment. As we shall see later, confinement can be an important consideration in some samples which contain nanometer-scale po-GaInP regions surrounded by disordered GaInP regions with larger band gap.

2.3   Effect of Magnetic Field on Free and Bound Carriers

From the above discussion on the effect of pressure on the band alignment between po-GaInP and the GaAs substrate, one expects that optical spectra such as photoluminescence in po-GaInP samples may show significant changes as a function of pressure. For example, when the band offset of type I optical spectra are dominated by strong spatially "direct" transitions (i.e. both electron and hole reside within the same semiconductor). On the other hand, when the interface is converted to type II under pressure, spatially "indirect" transitions (in which the electron and hole come from different semiconductors) may become stronger than direct ones. The appearance of such spatially indirect transitions is obviously an important signature of the conversion of an interface from type I to type II. However, from the optical spectra alone, one usually cannot determine definitively the origin of the carriers and hence the band offset. In principle, this can be deduced from the pressure coefficient of the spatially indirect transition provided one knows separately the pressure coefficients of the conduction and valence bands in the two semiconductors forming the interface. From the discussion in section 2.2 on the pressure effect on band gaps it is clear that the pressure coefficients of the *individual band edges*, especially those of the valence bands, are not as well known as those of the *band gaps*. In practice this technique may not be sensitive enough for samples with smaller band offsets. Having said that we shall demonstrate in this article that the pressure dependence can be used to determine the band offset in samples where the band offsets are of the order of *tens* of meV (the V-series of samples to be described in the next section). A much more precise way of identifying the electrons and holes in an optical transition is the measurement of the magnetic field dependence of optical transitions.

It is well established that under a magnetic field B, the energy of a charged free carrier of mass m^*, g-factor $g*$ and spin S undergoes a shift ΔE_n (see, for example, Chapter 9 of [6]):

$$\Delta E_n = \hbar\omega_c(n+1/2) + g^*\mu_0 \boldsymbol{S}\cdot\boldsymbol{B} \tag{2}$$

where

$$\mu_0 = \hbar e/(2m_0 c) \tag{3}$$

is the Bohr magneton and

$$\omega_c = eB/(m^* c) \tag{4}$$

is the cyclotron frequency and n is an integer ≥ 0. In Eqs. (2) to (4) $\hbar$ is the Planck's constant, e is the electronic charge, m_o is the mass of electron in free space and c is the speed of light (CGS units are used in this article). On the right hand side of Eq. (2), the first term gives rise to the Landau quantization of the cyclotron orbits (and is known as the Landau diamagnetic contribution) while the second term represents the Zeeman splitting ΔE_Z . For electrons in the lowest conduction band of GaAs, $g^* = -0.44$, $m^* = 0.067\, m_o$ and $S = \frac{1}{2}$ results in a ratio of $\hbar\omega_c/\Delta E_Z$ equal to ~ 16. Thus for the purpose of this article we expect the Landau term to be dominant. We note that the *Landau contribution is linear in B* with a slope which depends on the reciprocal of the mass of the carrier and therefore can be used, in principle, to identify the origin of carriers involved in optical transitions. For example, conduction electrons in GaAs have a field-induced energy shift of 0.86 meV/T in the lowest ($n = 0$) Landau level. The effective masses of the electron and hole in po-GaInP have been measured using a magnetic field to be $\sim (0.088\div0.10)\, m_o$ and $\sim 0.25\, m_o$ respectively [25]. The electron mass in d-GaInP measured in the same experiment is $0.092\, m_o$ [25]. A decrease in the conduction band effective mass in GaInP with ordering can be understood in terms of the shrinkage in band gap using the $\boldsymbol{k}\bullet\boldsymbol{p}$ approximation [6]. In principle, the conduction band offset ΔE_c can be changed with magnetic field since the effective masses of the electron are different in GaAs and GaInP. However, the change achieved with the application of a magnetic field is much smaller than that achievable with pressure so we shall neglect the effect of a magnetic field on band alignment.

The quantization of the Landau level (LL) energies implies that the radii of the cyclotron orbits are also quantized. For the $n= 0$ LL, this orbit has a radius l_H given by:

$$l_H = \sqrt{(\hbar c/eB)} = \sqrt{\hbar/(m^*\omega_c)} \tag{5}$$

where l_H is known as the *magnetic field length* and is *material-independent*. It sets *a size scale* for the charge particle orbit in the field B. The magnetic length for electrons has the value of ~ 25nm for a field of 1 Tesla. Most laboratory magnets nowadays can generate a field of the order of 10T so magnetic field measurements can probe sizes (equal to $2l_H$) varying from about ~ 14 to ~ 50 nm. With special high field magnets available at laboratories such as the Grenoble High Magnetic Field Laboratory (GHMFL) in France, it is possible to routinely achieve fields as high as 24T. With these magnets one can achieve spatial resolutions as small as 10 nm. In this article we shall demonstrate the application of this capability of high magnetic field to measure the size of type II domains at the po-GaInP/GaAs interface.

So far we have considered free charged carriers. In case these carriers are bound either to carriers of opposite signs to form excitons or to impurities by Coulomb attraction, the effect of the magnetic field would be different. The magnetic field dependence can be different from that described by Eq. (2) especially if the binding energy E_b involved is $>>\hbar\,\omega_c$. As an example, let us consider the case where the electron and hole form a hydrogen-like (or Wannier) exciton with the *1s* binding energy given by (see, for example, Chapter 6 of [6]):

$$E_b(1s) = \frac{m_0 e^4}{2\hbar^2 \varepsilon^2}\left(\frac{\mu}{m_0}\right) = 13.6 \frac{\mu\varepsilon^2}{m_0} \quad \text{[eV]} \qquad (6)$$

In Eq. (6) ε is the dielectric constant of the material and μ, the reduced mass of the exciton, is related to the electron (m_e) and hole effective masses (m_h) by:

$$\frac{1}{\mu} = \frac{1}{\mu_e} + \frac{1}{\mu_h} \qquad (7)$$

Under the applied field the exciton energy will *increase* by the amount:

$$\Delta E_b = \frac{\hbar^4 \varepsilon^2}{4\mu^3 e^2 c^2} B^2 \approx \frac{(\hbar\omega_c)^2}{E_b(1s)} \qquad (8)$$

The important point of Eq. (8) is that the energy of bound electron-hole pairs depends *quadratically* on the magnetic field when the binding energy is much larger than the cyclotron energy.

In summary, one can determine whether an optical transition involves free or bound carriers from the dependence of its energy on a magnetic field. In case of free carriers one can determine their effective masses and use this knowledge to identify their origins.

We shall now consider the special case where a confinement potential exists only in the direction perpendicular to B (such as in the case of quantum dots) and furthermore the confinement energy is *comparable* to $\hbar\omega_c$. In such cases, the effects of the magnetic field on the carrier energy are modified significantly. The simplest example, in terms of mathematics, is that of a radially symmetric parabolic confinement potential given by:

$$V(r) = \frac{1}{2}m^*\omega_0^2 r^2 \tag{9}$$

where r is a vector in the plane perpendicular to the applied field. This potential preserves the cylindrical symmetry around B. It can be shown that, neglecting the spin effects, the Landau contribution to Eq. (2) is modified as [26]:

$$\Delta E_{n,p}^{conf} = \left(\frac{2n + |p| + 1}{2}\right)\hbar\tilde{\omega} - p\left(\frac{\hbar\omega_c}{2}\right) \tag{10}$$

where p is now an integer, which can be either positive or negative while the new cyclotron frequency, $\tilde{\omega}$ is defined by:

$$\tilde{\omega} = \sqrt{\omega_c^2 + 4\omega_0^2} \tag{11}$$

Note that in this case the magnetic length defined previously by Eq. (5) has also to be replaced by a new one l_H defined by: $l_H = (\hbar/(m^*\tilde{\omega}))^{1/2}$. The field dependence of the LL's described by Eq. (10) is *no longer linear,* especially for low B. While the analysis of experimental data in such cases is more complicated, it can also be a powerful technique to extract information on the confinement potential $V(r)$ as demonstrated recently by Zeman *et al.* [24].

3.   SAMPLE PREPARATION AND CHARACTERIZATION

The GaInP samples studied in this article were all grown on GaAs substrates by low-pressure metal-organic vapor phase epitaxy (LP-MOVPE) and lattice-matched to GaAs with an accuracy of $\Delta a/a < 10^{-3}$ measured by Double Crystal X-ray Diffractometry. However, these samples have been grown under several different growth conditions, structures and types of substrate, so we have divided them into three series for convenience.

3.1   S Series

The first series of GaInP samples were investigated for the purpose of studying the effect of ordering on the pressure-induced Γ-X conduction band cross-over. These partially ordered GaInP samples have been labeled as S1 to S4 [23]. They were grown with a mass-production type vertical LP-MOVPE reactor in Shin-Etsu Handoutai Co., Japan.   The growth temperature was 700 °C with a (group V)/(group III) element ratio and a growth rate of 100 and 3 µm/hour, respectively. The sample structures are all the same: GaInP (1 µm) /GaAs buffer layer (0.1 µm) /GaAs substrate. However, the degree of ordering in the GaInP layers was controlled by cutting the GaAs substrates in such a way that the (100) surface has different degrees of mis-orientation towards the [011] direction. The degrees of mis-orientation are 0°, 5°, 10° and 15° for S1, S2, S3 and S4, respectively. As a result S1 is most ordered and shows the lowest PL peak energy of 1.86 eV at room temperature, while S4 is the least ordered and shows the highest PL peak energy of 1.92 eV. The degree of ordering in these samples has been analyzed qualitatively by electron diffraction using a transmission electron microscope (TEM). Evidence of CuPt type of ordering in GaInP is shown by electron diffraction from the planes of atoms perpendicular to the ordering directions, such as the (011) direction, which gives rise to satellite spots in the normal diffraction pattern. Figure 4 shows a comparison between the (011) electron diffraction patterns from the samples S1 and S4. From this figure one can see that satellite diffraction spots are found in the pattern from sample S1 but not in that of sample S4. This is clear evidence that CuPt ordering is present in S1 but not in S4.

3.2   T Series

A second series of three samples were grown for the study of the PL upconversion by LP-MOVPE in Nippon Sanso Co. Japan, with a single wafer handling type vertical reactor [9,27]. These samples were fabricated on a 1-mm-thick (100) GaAs substrate with no mis-orientations. Different growth temperatures were employed to control the degree of ordering, and these samples have been labeled as T650, T700 and T750 according to their growth temperatures which varies from 650 °C to 750 °C.

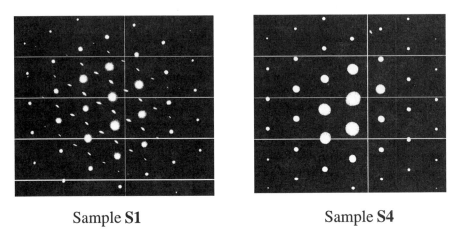

Sample **S1**   Sample **S4**

Figure 4. A comparison between the (011) electron diffraction patterns from the samples S1 and S4.

The thickness of the GaInP layers in these samples is also 1 μm and they were grown with a V/III ratio of 200 for the GaInP layer and a ratio of 50 for the GaAs buffer layer. The degree of ordering is found to be highest in T700 while lowest in T750 as indicated by the energy of their low temperature emission peaks (see Table 1).

Table 1. The low temperature PL emission peak energies and band offsets of the three series of po-GaInP samples studied in this article.

Sample	PL peak energy (eV)	$\Delta E_c = E_c(GaAs) - E_c(GaInP)$ at 1 bar (meV)	Upconversion at 1 bar
S1	1.906	2.4	No
S2	1.956		No
S3	1.983		No
S4	1.989		No
T650	1.895	−1.3	Yes
T700	1.888	−3	Yes
T750	1.928	−3	Yes
V1			
V2			
V3			
V4	1.81 (77K)	80	No

3.3 V Series

A series of four GaAs/GaInP single quantum well (QW) samples (labeled as the V samples) were grown on (100) GaAs substrates at 550 °C by a horizontal LP-MOVPE reactor at Nippon Sanso Co., Japan. This MOVPE system is capable of handling six 3-inch wafers in a run so that these QW samples were grown on 3-inch GaAs substrates. The sample V1 consists of 100-Å-thick GaAs well and 1100-Å-thick GaInP barriers. In the samples V2 and V3 there is a thin (nominal thickness ~ 14 Å) GaP layer between the 100-Å-thick GaAs QW and the 1100-Å-thick GaInP layers. In the case of sample V2 the GaP is below the GaAs QW while in sample V3 the GaP layer is grown on top of the GaAs QW. The sample V4 consists of 100-Å-thick GaAs well, 1100-Å-thick GaInP barriers and thin GaP layers sandwiching the GaAs well. A schematic diagram of the structure of sample V4 is shown in Fig. 5(c). Thus the main difference between these four samples lies in the presence or absence of thin GaP layers between the GaAs QW and the adjacent GaInP layers. We have also performed TEM measurements on both sample V4 and V2 and the resultant high resolution cross-sectional lattice images are shown, respectively, in Fig. 5 (a) and (b).

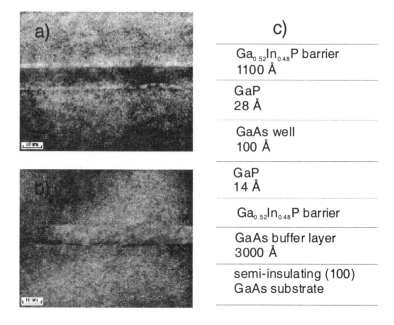

c)

| Ga$_{0.52}$In$_{0.48}$P barrier
1100 Å |
| GaP
28 Å |
| GaAs well
100 Å |
| GaP
14 Å |
| Ga$_{0.52}$In$_{0.48}$P barrier |
| GaAs buffer layer
3000 Å |
| semi-insulating (100)
GaAs substrate |

Figure 5. High resolution cross-sectional lattice images of sample V4 (a) and V2 (b); (c) schematic diagram of the structure of the sample V4.

These pictures were taken with an acceleration voltage of 200 kV. In both pictures we can see the CuPt type of ordering directed along the [111] direction in GaInP. This is confirmed also by satellite spots observed in the TEM diffraction pattern. First, we can see that the lower GaInP/GaAs interface in sample V4 is rougher than the top one. This rougher lower interface could be the result of three dimensional island growth of the GaP layer on GaInP since this sample has a nominal GaP layer on top of the lower GaInP/GaAs interface. We can also clearly see the existence of an intermediate layer of GaP in the upper GaAs/GaInP interface. The height of the GaP islands and the thickness of the GaP layer are 14 Å and 28 Å respectively. These values are in reasonable agreement with the values expected from the experimental growth rates of GaP layers. Similarly we can see a rough lower GaInP/GaAs interface in sample V2 as a result of GaP islands but a very smooth upper GaAs/GaInP interface without any intermediate GaP layer.

Finally, the expected free carrier concentrations for the GaAs and GaInP layers in all our samples are of the order of 10^{14} cm^{-3} for holes and 10^{16} cm^{-3} for electrons, respectively, based on considerations of the experimental V/III ratios during growth.

4.   EXPERIMENTAL DETAILS

Most of the experiments described in this article involve excitations of emission from samples via laser excitation. The samples are maintained at low temperature inside a cryostat using either liquid He or N_2. The excitation lasers are either visible lasers such as Ar$^+$ or Kr$^+$ ion laser or a tunable near-IR Ti:sapphire laser. The photoluminescence from the sample is analyzed by a Spex double monochromator and detected by a photon-counting technique. To vary the pressure on the samples, the sample is lapped to dimensions of <1 mm and loaded into a diamond anvil cell (DAC). This cell has been designed to allow it to be removed from the hydraulic press used to apply the pressure. The pressure is maintained inside the cell via a lock-ring. The sample is placed inside a hole drilled into a metal gasket and surrounded by a pressure media which is either a mixture of methanol/ethanol or liquid nitrogen. The pressure applied to the sample is measured by the standard ruby fluorescence technique. We estimate the pressure inhomogeneity to be slightly less than 10%. When the sample is located inside the high pressure cell the signal is reduced and for these experiments a triple spectrograph (Spex Triplemate) equipped with a cooled CCD detector (from Princeton Instruments) is used to analyze the signal. Our high magnetic field measurements were performed with the high static magnetic fields (up to

23T) produced by modified Bitter-type magnets at the Grenoble High Magnetic Field Laboratory (GHMFL). Finally, for experiments under both high pressure and high magnetic field, there are additional severe restrictions on the size of the DAC imposed by the diameter of the bore of the high field magnet. In these experiments we used a small DAC made from copper beryllium whose design has been described in detail already in [28] (see Fig. 6(a)). The cell was fixed at the bottom of a long insert filled with helium exchange gas. The entire assembly was immersed into a helium bath cryostat placed into the bore of the magnet. Radiation for exciting the sample and the signal collected from the sample were all transmitted into or out of the cryostat via optical fibers. In order to minimize the background signal generated in the optical fibers by the excitation laser, we designed a special optical splicing system consisting of a beam-splitter and three optical fibers shown schematically in Fig. 6 (b).

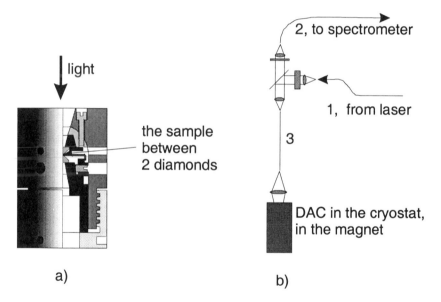

Figure 6. Schematic diagram of the diamond anvil high pressure cell (a) and optical system for performing measurements under high pressure and high magnetic fields (b).

The beam splitter was fixed to the assembly containing the DAC and is located on top of the cryostat. The three optical fibers consist of: (1) a fiber bringing in the excitation beam from Ar^+ or Kr^+ ion laser, 2) a fiber carrying the PL signal to the spectrometer and 3) a short fiber connecting the beam-splitter with the DAC. All coupling of light in and out of the optical fibers were achieved with microscope objectives except for the use of a small quartz lens placed between the DAC and the end of the short fiber for focusing the excitation beam onto the sample inside the DAC and at the same time collecting the PL signal from the sample back into this fiber. Band pass filters are used to reduce PL signal generated from the optical fibers. For example, a filter in front of the fiber 2 (see Fig. 6(b)) blocks the excitation laser from entering it and an interference filter placed at the exit of fiber 1 blocks the PL of fiber from being collected. In this arrangement any detected background PL signal from optical fibers originate only from the short fiber segment connecting the quartz lens and the DAC.

5.   EXPERIMENTAL RESULTS

5.1   Band Offsets in Samples Showing Upconversion (T Series)

As mentioned in the introduction the idea that the band alignment at the po-GaInP/GaAs may be type II and is responsible for the high PL upconversion efficiency in this system was proposed by Su *et al.* [29]. This was not generally accepted since the band alignment between GaAs and po-GaInP was assumed to be type I as at the d-GaInP/GaAs interface. Although subsequent theoretical calculation of the band alignment in both o-GaInP/GaAs and d-GaInP/GaAs by Froyen *et al.* [10] suggested that the band offset could be type II for the more ordered GaInP and GaAs interface, no strong correlation was found to exist between upconversion and the degree of order in GaInP as indicated by the band gap reduction. For example, Zeman *et al.* [27] observed upconversion in all three T-series samples but not in sample S1, although the PL peak energy of S1 is smaller than that of sample T750 (see Table 1). According to the theory of Wei *et al.* [30] S1 should be more ordered than T750 and therefore should have a stronger tendency to have a type II band offset. To determine whether there is any correlation between the band alignment in the T series and the existence of upconversion in their PL spectra we have studied their magneto-photoluminescence or magneto-PL in short.

When excited by a laser with photon energy above the band gap of GaAs but below the band gap of the GaInP layer, the PL spectra of all three samples in the T-series exhibit a broad structure located around 1.49 eV in addition to the exciton peak near the band gap of the GaInP layer as a result of upconversion. Although all three samples in the series show this PL upconversion the intensity of the up-converted PL (to be abbreviated UPL) is strongest in T700 so we shall concentrate on this sample first. Figure 7 shows both the "normal" (above-gap excited) PL and upconverted PL spectra in sample T700.

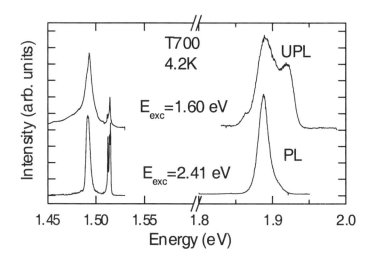

Figure 7. Photoluminescence of sample T700 for two different excitation laser energies showing both the normal and upconverted (labelled, respectively, as PL and UPL) from GaInP.

One can infer that the structure around 1.49 eV is related to the GaAs substrate from its energy. Since this energy is *lower* than that of the exciton in GaAs it may be due to either a spatially indirect transition which would imply that the band offset is type II or deep levels lying within the gap of GaAs. In order to distinguish between these two cases we have determined the effect of the magnetic field on its peak energy. The results are shown in Fig. 8(a). We note that this structure really contains two peaks and both of them exhibit a *linear* shift with B but with different slopes. As a result of this difference in slope the two peaks become readily resolvable at high fields. From these slopes effective masses of 0.07 m_0 and 0.084 m_0 [27] have been obtained for the high and low energy peaks, respectively.

The mass of 0.07 m_o is consistent with the electron mass in GaAs and suggests that the higher energy transition involves recombination of free electrons in GaAs with bound holes also in GaAs. By similar arguments, the mass of the lower energy peak suggests that it involves *free electrons in GaInP* and bound holes in GaAs and is, therefore, a *spatially indirect transition*. The holes in GaAs are presumably bound to carbon acceptors in GaAs since carbon is known to be usually present in samples grown by MOVPE. The crucial result obtained from this sample was that: the spatially indirect transition had *lower* energy than the spatially direct transition suggesting that *the band alignment in the sample T700 is type II.* From the separation of these PL peaks a conduction band offset $\Delta E_c = E_c$ (GaInP) $- E_c$ (GaAs) $= -3$ meV was obtained. Such a small band offset would have been

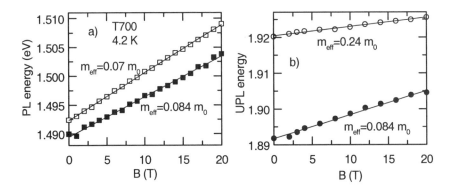

Figure 8. PL peak energies versus magnetic field B in the T700 sample for a) the two peaks around 1.49 eV and b) the two structures observed in the UPL spectrum of GaInP. The values of the effective masses (m_{eff}) deduced from their slopes are indicated.

difficult to determine using another technique. Similar spatially indirect transitions were identified in the other two samples T650 and T750 by magneto-photoluminescence. However, the band offsets were found to be also in the -2 to -3 meV range. Thus no significant difference was found between the band offsets among samples in the T-series, in spite of the fact that their PL peak energies are quite different. Since the PL peak energy is generally accepted to be an indicator of the degree of ordering, our result in the T-series suggests that either there is no strong correlation between the GaAs/GaInP band offset and the degree of ordering in GaInP or a strong correlation does not exist between ordering and PL peak energy.

For comparison we show also in Fig. 8 (b) the magnetic field dependence of the peaks in the UPL spectra from the GaInP in sample T700. Both peaks in the UPL spectra shift *linearly* with B up to 20T indicating that they correspond to recombination of either free carriers only or of free carriers with localized particles [31]. The slopes of the magnetic field dependence suggest that the higher energy peak involves the recombination of *free holes* in GaInP (with the effective mass $= 0.24\ m_0$) but *localized electrons* while the lower energy peak corresponds to the recombination of *free electrons* in GaInP (effective mass $= 0.084\ m_0$) with *localized holes* [27]. It was first pointed out by Driessen [8] that the UPL must involve localized carriers. Since Driessen had assumed a type I band alignment at the GaInP/GaAs interface, he noted that carriers in GaInP will have a strong tendency to return to GaAs unless they are trapped by defects in GaInP. Thus our results demonstrate that Driessen's proposal was basically correct. The nature of the traps responsible for localizing carriers in GaInP is not known. However, it is well established that defects associated with stacking faults or anti-phase domain boundaries are very abundant near the GaInP/GaAs interface. It is quite plausible that the abundance of such trapping centers near this interface is partially responsible for the strong upconversion efficiency in this system. Another interesting feature of the results in Fig. 8 (b) is that the energy of the transition (1.92 eV) involving localized electrons in the UPL has *higher* energy than that of the normal PL (1.888 eV as listed in Table 1) in the same sample when excited by above band gap excitation. Since the normal PL originates from the top region of the GaInP layer where the sample is more ordered and has fewer defects, one way to explain the higher energy of the upconverted PL peak is that they involve electrons which are localized by *less ordered* regions near the interface. This result suggests that near the interface of the T700 samples there may exist small regions of less ordered GaInP surrounded by more ordered domains. We shall later demonstrate that in the sample S1 the reverse is true in that small pockets of more ordered type II regions are surrounded by less ordered type I domains.

5.2   Band Offset in GaInP Sample which Shows No Upconversion at Ambient Pressure (S1)

We have pointed out earlier that PL peak emission energy in the sample S1 is higher than T700 but below that of T750 and that it does not exhibit any detectable UPL. The lack of UPL may be explained by the fact that S1 has a type I GaAs/GaInP interface. The S1 GaInP PL peak energy is not too far from that of T750 and this suggests that the magnitude of its band offset is probably small. Thus it may be possible to use hydrostatic pressure to convert its band alignment from type I to type II! In order to test this idea it is necessary to perform high pressure magneto-PL on this sample. For this purpose the high pressure magneto-PL setup described in Section 4 was developed. It turns out that high pressure magneto-PL in S1 not only provided us with direct proof that its band alignment is predominantly of type I at atmospheric pressure and can be converted into type II under pressure but also on the dimension of the type II domains induced by pressure. These complex and yet highly detailed results will be described in detail in this section.

We begin by examining the PL spectra of the GaAs substrate in S1 at atmospheric pressure and zero magnetic field (*B*=0). This spectrum is shown in Fig. 9.

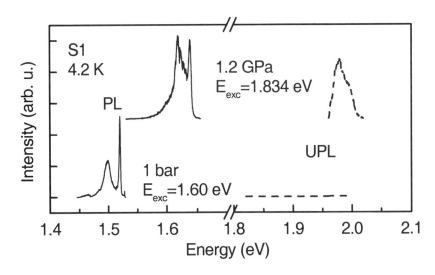

Figure 9. PL (continuous curves) and UPL (broken curves) spectra of the sample S1 measured at 1 bar and at 1.2 GPa.

One notices immediately that under relatively low spectral resolution this spectrum resembles the one from T700 (see Fig. 7) in that there is a structure around 1.49 eV. The only difference is that the sharper peak at higher energy near the band gap of GaAs is relatively stronger in S1 that in T700. The magnetic field dependence of both peaks at $B < 7$ T does not exhibit any unusual behavior. The field dependence of the lower energy peak is similar to the well known donor-acceptor transition in GaAs [32]. The field dependence of the higher energy peak indicates that free electrons and holes in GaAs are involved. Unlike the case of T700 no field induced splitting was observed in the 1.49 eV structure and therefore no peak attributable to spatially indirect transition can be identified. These results suggest that the GaAs/GaInP interface in sample S1 is indeed type I at atmospheric pressure as we have deduced from the absence of UPL in Fig. 9. Based on the arguments presented in Section 2.2 we expect that pressure may be able to convert its interface to type II. Using the appearance of UPL as a monitor of the band alignment conversion we find that this indeed happened in S1 at a pressure of >1.2 GPa as shown in Fig. 9. In addition to inducing UPL, pressure causes an additional peak to appear in between the donor-acceptor recombination peak and free electron-free hole structures in GaAs. Notice also that pressure has shifted both peaks in GaAs to higher energies. The low field ($B < 7$T) dependence of this new peak suggests that it involves a free electron in GaInP with effective mass = $0.086\, m_o$. This indicates that the new peak is a *spatially indirect* transition involving free electrons in GaInP and bound holes in GaAs similar to the peak observed in T700 at atmospheric pressure. The dependence of the intensity and peak position of this new peak at high field turns out to be quite unusual and will be discussed in more details in the next paragraph. At this point it suffices to note that the conduction band offset $\Delta E_c = E_c$ (GaInP) – E_c (GaAs) in the sample S1 is found to be *pressure dependent and it varies from –4 meV at 1.4 GPa to –11 meV at 2.6 GPa*. From these results in sample S1 we determined the pressure dependence of the band offset $|d(\Delta E_c)/dP|$ at the GaAs/GaInP heterojunction to be 4.6 ± 0.6 meV/GPa. As expected the pressure dependence of the band offset is determined mainly by the difference in the pressure coefficients of the conduction bands in GaAs and GaInP. Using these results we deduce that the band alignment in S1 at ambient pressure is indeed type I with $\Delta E_c = 2.4$ meV. Thus the high pressure magneto-PL results demonstrate conclusively that the interface between GaAs and po-GaInP in sample S1 can be converted from type I to type II with pressure. It is interesting to note that such a small type I band offset is already sufficient to suppress UPL in S1 at atmospheric pressure. Furthermore, efficient UPL becomes observable once the band alignment becomes type II. Our combined experiments in the T-series and S1 samples show that, at least in the GaAs/GaInP system, a type II band alignment is an important condition for efficient UPL and hence the mechanism responsible for the UPL process observed in this system is most likely the two-step two-photon absorption mechanism proposed by Su *et al.* [9].

Next we turn our attention to the effect of high magnetic field (B up to 22 T) on the PL of the GaAs substrate in S1 under pressure. In Fig. 10(a) we plot the PL spectra in S1 at $P = 1.4$ GPa as a function of magnetic field on an expanded energy scale when compared to Fig. 9. For convenience, we shall label the donor-acceptor (carbon) recombination peak in GaAs as C and the corresponding peak due to free electrons and free holes (excitonic recombination) in GaAs as H [24]. Note that compared to the $P = 1$ bar spectrum in Fig. 9 both peaks have shifted to higher energies as a result of the applied pressure. When B is increased, the intensities of both peaks decrease until they almost disappear at $B > 7$T. At higher fields ($B > 12$T) new structures at *lower* energies (denoted by C' and H') appear abruptly for B exceeding some threshold value B_c. The value of B_c *decreases* with increasing pressure until this threshold behavior disappears at $P = 3.2$ GPa. These results are shown more clearly in Fig. 10 (b). The peak H' has a *linear* field dependence whose slope indicates the involvement of a free carrier with mass = $0.086\,m_o$. Thus H' is a spatially indirect transition which becomes observable when the GaAs/GaInP interface in S1 is converted into type II by pressure. With increase in pressure the electron effective mass increases with P until it reaches a value of $0.094\,m_o$ at $P = 2.6$ GPa. At P above 3.2 GPa the conduction minimum in GaInP changes from the zone-center (Γ) to the X point in the Brillouin zone, in a phenomenon known as the Γ-X crossover [23].

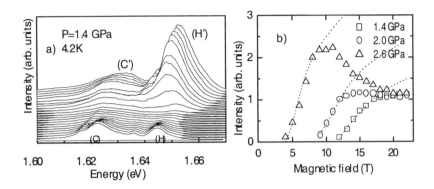

Figure 10. a) The PL spectra of the sample S1 for different magnetic fields *B* increasing by steps of 1T at a time from 0T (spectrum in front) to 22T (spectrum at the back); b) magnetic field dependence of the intensity of peak H' at *P* = 1.4, 2.0 and 2.6 GPa.

The fact that this threshold dependence on the magnetic field described above is observed for PL peaks involving both free and bound holes in GaAs suggests that it is most likely related to the property of the *electrons in GaInP* and not with the holes in GaAs. This behavior appears also to be sample dependent since it is not observed in sample T700. These observations led us to conclude that the threshold behavior is related to some *structural* rather than *electronic* properties of the GaAs/GaInP interface. The reason is because the threshold field B_c changes rather slowly with pressure. On one hand pressure will change the band offset at the rate of ~5 meV/GPa as shown above. On the other hand, the band offset depends on *B* with a coefficient of ~ 0.1 meV/T as a result of the difference in electron mass between GaAs and GaInP. If this phenomenon were to be determined entirely by the band offset at the GaInP/GaAs interface then we expect that the threshold field would depend very strongly on *P*. Alternately, the weak dependence of B_c on *P* may be understood by a change in the *size* of the type II domains at the GaAs/GaInP interface with pressure. If the magnetic length l_H of electrons in GaInP were somehow involved then one can understand the slow dependence of B_c on P since l_H varies with the *B* field only as $(1/B)^{1/2}$ (see Eq. (5)).

To relate domain size with magnetic lengths of electrons in GaInP we note that many studies exist on the structure of both disordered and partially ordered GaInP, of anti-phase domain boundaries and of the interface between GaInP and its GaAs substrate. TEM techniques especially have shown that near the interface there are usually many CuPt ordered domains of different sizes and degrees of ordering separated by anti-phase domain boundaries forming a zig-zag pattern. A rather typical pattern is reproduced from [33] in Fig. 11.

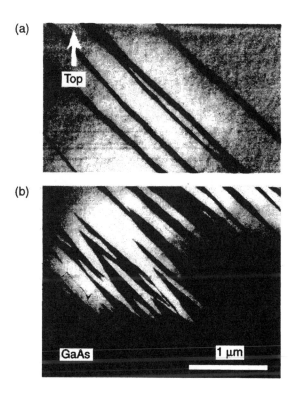

Figure 11. Dark field TEM picture of a po-GaInP sample taken from (a) an area near the top of the sample and farthest away from the GaAs/GaInP interface and (b) a region adjacent to the GaAs/GaInP interface. The dark regions in GaInP represent areas lacking in long range order. Reproduced from [33].

Pictures like these unfortunately provide no information about the electronic properties, such as the band alignment, within these domains. In this context magneto-PL is a very unique tool which can provide information on both the electronic and structural properties of these domains. The ability of this technique to measure nanometer domain sizes is associated with the magnetic length of photoexcited carriers. As noted in Section 2.3 the magnetic length is a variable "nanometer-stick." Using it one can measure the size of the type II domains near the GaAs/GaInP interface which are active in the UPL.

Since hydrostatic pressure can vary the band offset it can therefore vary the size of these domains which can, in turn, be probed by the magnetic length of the carriers. In the next paragraph we shall demonstrate how to apply these ideas to analyze *quantitatively* the pressure-dependent magneto-PL results including the occurrence of the threshold field B_c shown in Fig. 10.

We propose to explain the threshold behavior of the intensities of the spatially indirect transitions shown in Fig. 10 in terms of magnetic field induced *trapping* of electrons into "quantum boxes" in GaInP near the interface. The idea behind our model is based on the fact that electrons in GaInP are confined near the interface in type II domains as a result of their attraction to the positive holes left behind in GaAs. Within the plane of the interface these electrons are also confined by the surrounding regions of *lower degrees of CuPt ordering* and hence of larger band gaps as shown in Fig. 11(b). The combination of these two confinement potentials produce 3D "quantum boxes" containing the photo-excited electrons in GaInP. Note that type II band offset exists only in the more ordered region within the domains and these regions form our quantum boxes. While the size of the type II regions (and hence the "quantum boxes") depend on pressure, *the size of the CuPt domains does not depend on pressure.* For simplicity, we assume the potential across the GaInP/GaAs interface to be abrupt and is equal to the conduction band offset while the trapping potential parallel to the interface is a *parabola*. This model is sketched in Fig. 12. One of the motivations for assuming a parabolic confinement potential within the interface plane is to explain the pressure dependence of the electron mass. One expects a slight increase in the electron mass at zone center due to a pressure-induced increase in the band gap (see, for example, the discussion on the $k \bullet p$ theory in Chapter 2 of [6]). However, one can show easily that the observed increase of ~10% in the mass with $P = 2.6$ GPa is at least a factor of 2 larger than predicted by the $k \bullet p$ theory [6]. This means that the slope dE_H/dB of the peak energy $E_{H'}$ is not increasing with P as fast as would be predicted by the pressure dependence of the band gap of GaAs. By assuming that there is a parabolic confinement potential (of the form given in Eq. (9)) in the radial direction r perpendicular to the z-axis (see Fig. 12) one introduces an effective cyclotron frequency given by Eq. (11).

The intensity of the spatially indirect PL peak is then proportional to N_{capt}, the number of electrons captured into these quantum boxes, and to M_{opt}^2, the square of the optical transition matrix element, which can be assumed to be constant around B_c. In this model the probability of capturing an electron into a box of radius r_D is assumed to be given by:

$$P_{capt}(r_D) = \begin{cases} 0 & \text{when } B < B_c \\ \pi r_D^2 N_{dot}(r_D) & \text{when } B \geq B_c \end{cases} \qquad (12)$$

where $N_{dot}(r_D)$, the number of dots of radius r_D, is assumed to have a Gaussian distribution of width w around a mean value r_0. More details about this model can be found in [24].

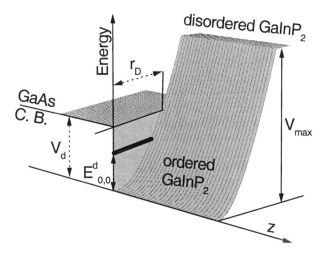

Figure 12. Schematic diagram of the confinement energies defining the "quantum box" model.

It follows from this model that the fraction of electrons trapped in type II domains is equal to:

$$N_{capt}(B) \propto N_0 \int_{r_D(B_c)}^{\infty} dr_D \left(\pi r_D^2\right) \exp\left[-\left(\frac{r_D - r_0}{w}\right)^2\right] \qquad (13)$$

We have fitted our experimental values of the magneto-PL intensities with this model and the results are shown in Fig. 10(b) as the broken curves. From these fits one obtains the value of $r_0 = 9.4$ nm at atmospheric pressure. It increases to 15 nm at the pressure of 2.6 GPa. Compared with the typical size of domains measured by TEM techniques these values are about 10 times smaller. This is to be expected since within the domains imaged by TEM the degree of CuPt ordering is not homogenous and hence the region with type II offset should be smaller than the domain size in the TEM image. The increase in quantum box size with pressure is understood in terms of an increase in the conduction band discontinuity at the GaInP/GaAs interface in favor of type II band offsets. The pressure dependence of the band offset estimated from these fits (4.6 ± 0.6 meV/GPa [24]) is quite consistent with the value of 5.1 meV/GPa measured directly from the T700 sample [34].

This fitting procedure has allowed us to extrapolate to $B_c = 0$T to obtain $r_0(0)$ ≅ 20 nm, the minimum radius r_0 required to capture the electrons without an applied magnetic field. Obviously this condition is most likely to be fulfilled in the sample T700 even at zero field where strong UPL is observed at atmospheric pressure.

The quantum box model is probably oversimplified since it fails to explain the decrease in the peak intensity for $B >> B_c$ which is observed for some of the pressures (see Fig. 10(b)). One possible way to resolve this disagreement is to invoke many particle effects such as assuming that at high fields there is a higher probability to capture more than one electron into each quantum box. It has been shown that such correlation effects produce a "hole" in the electron density at the center of the box [35,36], whose size increases with the field so that the peak in the electron density is pushed towards the edge of the quantum box where the non-radiative recombination rate is higher.

In summary, the capture behavior of electrons into type II domains in GaInP under high magnetic field suggests the presence of "quantum boxes" within partially ordered GaInP domains at the GaInP/GaAs interface. On the other hand, the domain boundaries can act as non-radiative recombination centers for photo-excited electrons. The presence of these quantum boxes which can localize carriers in GaInP is necessary for efficient light up-conversion in the GaAs/GaInP samples. The ability of hydrostatic pressure to change the band offset also makes it possible to vary the average size of the quantum boxes. It appears that in some samples these quantum boxes are sufficiently large at atmospheric pressure to trap electrons efficiently without magnetic field while in less ordered GaInP samples these boxes are less than 20 nm in size and therefore require a high magnetic field to shrink the electron orbits to allow efficient trapping into these boxes. In some samples the average radius of these optically active boxes was found to be only about 10 nm, as much as a factor of 10 smaller than the size of the CuPt domains measured by TEM techniques. Thus the degree of ordering within the CuPt domains in partially ordered GaInP is highly non-uniform.

5.3   Band Alignment and Emission from po-GaInP/GaAs/ po-GaInP Quantum Wells (V Series)

The question of band alignment between po-GaInP and GaAs appear also in the study of emission from GaAs quantum wells which use GaInP as the barrier layers. Such quantum wells have several advantages over the more popular GaAs/AlGaAs quantum wells. For example, GaInP is more stable than AlGaAs and the band gap difference between GaAs and GaInP is larger than that between GaAs and direct gap AlGaAs alloys. However, early efforts to observe quantum well emission from po-GaInP/GaAs/po-GaInP structures did not always succeed. Emission peaks well below the band gap of GaAs were observed in several reports instead of quantum confined emission [37,38,39,7,40]. Two different explanations have been proposed to account for the occurrence of this "deep emission". In one theory [40] the emission is produced by deep centers which are present in the GaAs well layer grown on po-GaInP. Another explanation [7] suggests a type II band alignment between GaAs and po-GaInP and the observed deep emission is spatially indirect. Subsequent work found that the deep emission can be suppressed when thin layers of GaP are grown as a buffer layer between GaAs and GaInP [40].

To understand the origin of the deep emission and the role of the GaP buffer layer Kwok *et al.* [41] have studied the pressure dependent PL in the V-series of po-GaInP/GaAs/ po-GaInP quantum well samples. The difference between the samples in this series lies in the presence or absence of a GaP layer at the GaAs/po-GaInP interface. The sample V4 which contains GaP on both sides of the GaAs well is the only sample found to produce quantum well emission above the band gap of GaAs. All the other three samples show only the deep emission below the band gap. The idea behind the experiment of Kwok *et al.* [41] was to determine the pressure coefficient of these emissions. From these pressure coefficients we can determine whether the deep emission is a spatially indirect transition or not. Pressure can also bring the X conduction band minima of the GaP and GaInP layers below the Γ conduction band minimum of GaAs so that their emission can be observed and identified. As pointed out in Section 2.2, the X conduction band in the zincblende-type semiconductors has a small but negative pressure coefficient. With sufficient pressure one can induce a Γ to X crossover causing a direct band gap semiconductor to convert to an indirect one. Even if the GaInP/GaAs/GaInP QW sample has a type I band alignment at atmospheric pressure, high pressure can lower the *X conduction band* of GaInP below the Γ conduction band of GaAs. As a result, the heterojunction becomes type II and one can observe a spatially indirect transition from GaInP to GaAs. By identifying these indirect transitions it will be possible to determine the band alignment of the conduction bands in all the heterojunctions present in these complex structures.

First we shall describe the results in the sample V4 which has the structure shown in Fig. 5(c). Fig. 13 shows its PL spectra at several pressures measured by Kwok *et al.* [41]. These results display a very complex dependence on pressure with peaks appearing and disappearing. To understand better their origin Kwok *et al.* [41] plotted (reproduced in Fig. 14) the pressure dependence of all the observed peaks which have been labeled T_1 to T_8 for convenience.

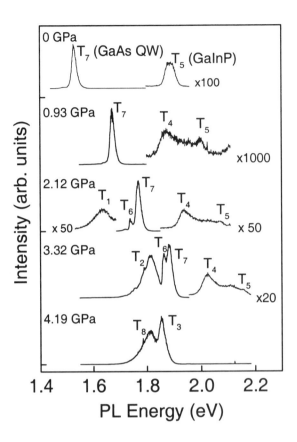

Figure 13. The PL spectra of the sample V4 at atmospheric pressure and at four different pressures. The identifications of these peaks are discussed in the text and summarized in Table 2. Reproduced from [42].

We note first that at $P = 0$ GPa the peaks T_5 and T_7 in Fig. 13 must be identified with the GaInP and GaAs QW layers, respectively, simply based on their emission energies. These identifications are further supported by their pressure coefficients (α) of 107 and 82 meV/GPa, respectively. These pressure coefficients and those of the other PL peaks in sample V4 are listed in Table II. The peak T_7 is accompanied by a weaker peak T_6 at slightly lower energy. It has the same energy and pressure coefficient as bulk GaAs and is therefore identified with transition from the GaAs buffer layer. Except for its use as an independent check on our pressure calibration, we shall not be interested in this peak.

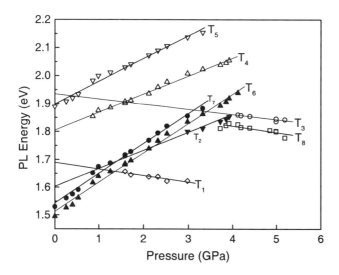

Figure 14. The pressure dependence of the observed PL peak energies in the sample V4. Reproduced from [42].

Table 2. Ambient pressure energies E(0) and linear pressure coefficients α of PL transitions in the po-GaInP/GaP/GaAs sample V4 measured at 77 K. The identification of the peaks is discussed in the text. 1h and 1e denote the lowest energy confined hole and electron states, respectively. Spatially indirect transitions are listed in italics. Reproduced from [41].

Transition	Peak	E(0) (eV)	α (meV/GPa)
GaP(X^{1e}_z) $\rightarrow$ GaAs(Γ^{1h}_v)	T_1	1.689	−24
GaInP(Γ_c) $\rightarrow$ GaAs(Γ^{1h}_v)	T_2	1.603	65[a]
GaInP(X_c) $\rightarrow$ GaAs(Γ^{1h}_v)	T_3	1.935	−18
GaInP(Γ_c) $\rightarrow$ (Γ_v) (more ordered)	T_4	1.805	65
GaInP(Γ_c) $\rightarrow$ (Γ_v) (less ordered)	T_5	1.898	82
GaAs(Γ_c) $\rightarrow$ (Γ_v) (buffer layer)	T_6	1.513	105
GaAs(Γ^{1e}_c) $\rightarrow$ (Γ^{1h}_v) (QW emission)	T_7	1.543	107
GaAs(X_c) $\rightarrow$ (Γ^{1h}_v) (impurity)	T_8	1.912	−23

a) The pressure coefficient of T_4 is used in determining the α of the T_2 transition.

Under a pressure of 1 GPa another peak T_4 appeared in V4 at about 90 meV below T_5 and having a pressure coefficient of 65 meV/GPa. Both the pressure coefficient and PL peak energy suggest that T_4 is emitted by a GaInP layer with a higher degree of CuPt ordering than the layer responsible for T_5. We therefore attribute the two peaks T_5 and T_4 to the two po-GaInP barrier layers. As to why the peak T_5 appears first T_4 appears only at

$P > 1$ GPa, Kwok *et al.* [41] theorized that the top layer is less ordered and is excited more strongly.

It is, therefore, responsible for the peak T_5. The bottom layer is grown on the GaAs buffer layer and is therefore more ordered. As pressure causes the band gap of the GaAs QW layer to shift to higher energy, more of the excitation laser light can reach this bottom GaInP layer so that its PL eventually becomes observable. At pressures around 2 GPa a new peak T_1 with a negative value of $\alpha \sim -24$ meV/GPa appears. Its negative α suggests that it involves electrons in a X conduction band minimum and holes in the GaAs valence band. Since the Γ-X cross-over in both GaAs [21] and GaInP [3,23] are known to occur at pressures exceeding 3 GPa, one is left with the only possibility that these electrons come from the X conduction band of the thin GaP layers. Furthermore, Kwok *et al.* [41] propose that this transition originates from the thicker GaP layer which lies below the GaAs QW. The reason is because the energy population in the top GaP layer is expected to be lower as it is thinner and hence would have a higher electron energy as a result of the stronger confinement effect. This implies that the transition responsible for the peak T_1 *is indirect in both real and reciprocal spaces.* One normally expects such transition to be extremely weak since the overlap between the electron and hole wave functions should be quite small. Peak T_1 is observable mainly because pressure causes the X conduction band in the GaP layer to have the lowest energy so that its electron population becomes the highest among the conduction minima of all the layers present in V4. An additional factor is that the smaller thickness of the GaP layer causes the electron wave function to spill over into the neighboring GaAs layer and increases the overlap between the electron and hole.

As P increases further one notice that another weak feature which is labeled as T_2 in Fig. 13 starts to appear and grow with pressure. Its pressure coefficient suggests that it involves an electron in GaInP. Thus we conclude that the band alignment between the GaAs QW and the more ordered GaInP bottom layer has been converted from type I to II. The sudden drop in the intensity of the peak T_7 by a factor of ~ 5 at $P = 2.12$ GPa (see Fig. 13) is also consistent with this change of the band alignment between the GaAs and the GaInP layer. By extrapolating the energy of the peak T_2 back to ambient pressure Kwok *et al.* [41] deduced the conduction band offset between the GaAs QW layer and the more ordered (lower) GaInP barrier layer to be ~ 80 meV. By applying similar arguments based on pressure coefficients they have identified all the other PL peaks in V4 many of them appear only under pressure. These identifications are reproduced in Table II. The alignment between the conduction and valence bands of the various epilayers in sample V4 at ambient pressure is shown schematically in Fig. 15. In estimating the band edge energies, Kwok *et al.* [41] have taken into consideration the effect of confinement on the conduction and valence bands. These confinement effects are significant mainly in the GaAs QW [43] and for the conduction bands of the thin GaP layers.

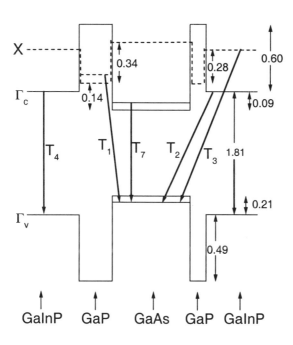

Figure 15. The ambient pressure conduction band alignment in sample V4 determined by extrapolating the observed PL peak energies to $P = 0$ GPa. The unit for all energy separations in the figure is eV. Reproduced from [42].

We note that the band offset of ~ 80 meV between the lower (more ordered) GaInP layer and the GaAs QW is larger than one would expect from the band gap of the po-GaInP layer. Without the GaP layer Kwok *et al.* [41] estimated from the GaInP PL peak energy that the band offset should be type II with a value of around –20 meV. To explain this discrepancy between their estimate and the experimental result, Kwok *et al.* [41] suggested that the band offset between GaAs and po-GaInP must be affected by the presence of the thin GaP layer.

In the remaining samples of the V-series, the GaAs QW emission was not observed. Instead a broad peak at energies below the band gap of bulk GaAs was observed [44]. Its energy varies between 1.45 to 1.48 eV depending on the excitation power. This emission has been referred to as the "deep emission" in the literature [40]. Liu *et al.* [7] have suggested that this PL peak is due to a spatially indirect type II transition. One reason in support of this identification is the power-dependent blue shift of the peak energy. To resolve this controversy Kwok *et al.* [44] measured the pressure dependence of this "deep emission." The resultant pressure dependence of its peak energy and intensity are reproduced in Fig. 16.

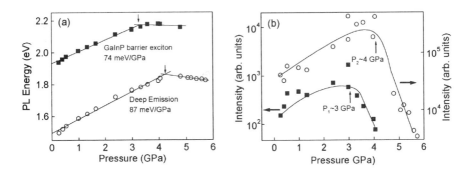

Figure 16. The pressure dependence of a) the PL peak energy and b) intensity of the deep emission from the V series sample compared with the corresponding quantities for the emission from the GaInP barrier layers. The arrows highlight the different pressures at which the Γ-X crossover in the conduction band occurs for the two PL peaks. Reproduced from [44].

While the "deep emission" has a positive pressure coefficient at low pressure, it becomes negative beyond ~ 4 GPa. At about the same pressure the peak intensity also drops abruptly. These results are consistent with a pressure-induced Γ - X crossover in the conduction band valleys. We note that a similar crossover occurring in GaInP is also observed in Fig. 16. The fact that the crossover for the deep emission occurs at a higher pressure than for GaInP is a strong indication that *the deep emission is not a spatially indirect transition* involving electrons in GaInP and holes in GaAs.

As an alternative explanation, Kwok *et al.* [44] suggested that the deep emission is a donor-acceptor-pair transition involving a shallow donor and a *deep* acceptor. This explanation accounts for the blue shift of the PL peak energy with excitation power, the time-dependent decay and also the different pressure coefficient than that of the GaAs band gap. Kwok *et al.* [42] found that the deep emission is greatly suppressed but still observable at low excitation power even in the sample V4. Based on this fact these authors [44] suggested that the deep acceptor is probably associated with the GaAs/GaInP interface and its concentration is decreased by the presence of the GaP layer. Since most of the photo-excited electrons and holes recombine at these deep acceptors not much can be learned about the band alignment from the PL of the in the V-series samples other than V4.

6.   DISCUSSION

The existence of CuPt ordered domains in GaInP has been found to lead to fundamental changes in the optical and electronic properties of these alloys. The discovery of photoluminescence upconversion in some GaInP samples has led to much interest in the band alignment between the GaAs substrate and the GaInP layer and the role it plays in determining the upconversion efficiency. The size of the band offset not only affects the upconversion efficiency but also has important implication for devices relying on properties of electrons at heterojunctions, for example, the formation of quantum well lasers as shown in the V-series of samples. Theoretically the band structure and band alignment in both the perfectly ordered or disordered alloys have been calculated and appear to be in good agreement with experiment. However, many of the samples studied in the literature are only partially ordered. To understand the results in these partially ordered samples, the idea of an "order parameter" η was proposed [11]. In this model all the physical properties of the partially ordered GaInP samples are assumed to vary quadratically with η. Once η is determined, such as from the band gap, the other properties, like the band alignment, of partially ordered samples would presumably be predictable. So far in the course of studying band alignment in various partially ordered GaInP samples grown under different conditions, this has not been found to be the case as can be seen from Table I. One possible explanation is that the degree of ordering in most GaInP samples is not spatially uniform, being lower near the interface and becoming progressively higher away from the interface. This variation means the η usually determined from the PL excited by above band gap excitation energies is valid only for top surface region of the GaInP sample. This order parameter may be quite different from that near the interface between GaInP epilayer and the GaAs substrate. As a result it is not possible to use the order parameter determined from one region of the sample to predict the properties of another region. To avoid this difficulty, some authors [45] have suggested the idea of an "average order parameter" and the use of techniques like NMR as one of the ways to determine this average order parameter. However, there is now ample experimental evidence indicating that local structural order other than CuPt may exist in many po-GaInP samples [46,47,48,49]. The effect of such alternate ordering on band alignment is not understood. In addition, they call into question the use of a single ordering parameter to characterize po-GaInP samples. Even techniques like NMR which probes the entire sample may not be able to distinguish between domains of different kinds of local ordering. While optical measurements presented in this article alone also cannot determine completely the complex ordering in GaInP samples, such techniques in combination with other techniques like electron and X-ray diffraction [48] and resonant Raman scattering [49] may hold the key to a complete understanding of this alloy system.

7.   CONCLUSIONS

The band alignment between partially ordered GaInP layers grown on GaAs substrate became a controversial issue after the observation of efficient photoluminescence upconversion in some samples while not in others and the absence of quantum well emission in samples formed with GaAs as the well surrounded by GaInP barriers. To resolve this issue it was found necessary to measure the conduction band offset at the GaAs/GaInP interface with precision of a few meV. We demonstrated in this article that low temperature photoluminescence measurements performed at high magnetic field and/or under high pressure has the sensitivity and accuracy required to resolve this controversy in GaInP. It was found that the band alignment between partially ordered GaInP and GaAs can be either type I or II with small conduction band offsets in both cases. We show convincingly that a type II band offset is one of the reasons responsible for the occurrence of highly efficient photoluminescence upconversion in this system. In order to form good quantum wells with GaAs and GaInP we found that a thin layer of GaP at the interface helps to suppress surface defects and promote a type I band alignment with sizable band offset. However, a detailed understanding of how ordering determines the type of band alignment and the size of the band offset is still lacking.

ACKNOWLEDGMENTS

The Grenoble High Magnetic Field Laboratory is "Laboratoire conventionné à l'UJF et l'INPG de Grenoble." GM gratefully acknowledges the Miller Institute for Research in Basic Sciences of UC-Berkeley for a Visiting Miller Professorship during which part of this work was carried out. JZ acknowledges partial support from the grant ERBCHGECT 930034 of the European Commission, Grant 202/99/0410 from the Grant Agency of the Czech Republic and Grant A1010806 of the Academy of Sciences of the Czech Republic. The work at Berkeley was supported by the Director,

Office of Energy Research, Office of Basic Energy Sciences, Materials Sciences Division, of the U.S. Department of Energy under Contract No. DE-AC03-76SF00098. This work was initiated when PYY was a J.S. Guggenheim Memorial Foundation Fellow and visiting professor at the Université de Joseph Fourier of Grenoble, France.

REFERENCES

1. J. I. Hashimoto, T. Katsuyama, J. Shinkai, I. Yoshida and H. Hayashi, J. Appl. Phys. **61**, 1713 (1991).
2. W. Liu and S. K. Fan, IEEE Electron Device Lett. **13**, 510 (1992).
3. J. Chen, J. R. Sites, I. L. Spain, M. J. Hafich and G. Y. Robinson, Appl. Phys. Lett. **58**, 744 (1991).
4. A. Gomyo, T. Suzuki, K. Kobayashi, S. Kawata and I. Hino, Appl. Phys. Lett. **50**, 673 (1987).
5. M. Kondow, H. Kakibayashi, S. Minagawa, Y. Inoue, T. Nishino and Y. Hamakawa, Appl. Phys. Lett. **53**, 2053 (1988).
6. P. Y. Yu and M. Cardona, Fundamentals of Semiconductors: Physics and Materials Properties, 2nd Edition (Springer-Verlag, Berlin, 1999).
7. Q. Liu, S. Derksen, A. Lindner, F. Scheffer, W. Prost, and F.-J. Tegude, J. Appl. Phys. **77**, 1154 (1995).
8. F.A.J.M. Driessen, Appl. Phys. Lett. 67, 2813 (1995).
9. Z. P. Su, K. L. Teo, P. Y. Yu and K. Uchida, Solid State Commun. **99**, 933 (1996).
10. S. Froyen, A. Zunger and A. Mascarenhas, Appl. Phys. Lett. 68, 2852 (1991).
11. D. Laks, S-H. Wei, and A. Zunger, Appl. Phys. Lett. **69**, 3766 (1992).
12. J. R.Waldrop, and R.W. Grant, Appl. Phys. Lett. **68**, 2879 (1996).
13. J. Almeida, L. Sirigu, G. Margaritondo, P. Da Padova, C. Quaresima and P. Perfetti, J. Phys. D **32**, 191 (1999).
14. J. J. O'shea, C. M. Reaves, S. P. DenBaars, M. A. Chin, and V. Narayanamurti, Appl. Phys. Lett. **69**, 3022 (1996).
15. J. J. O'shea, E. G. Brazel, M. E. Rubin, S. Bhargava, M. A. Chin, and V. Narayanamurti, Phys. Rev. **56**, 2026 (1997).
16. Y. C. Yeo, M. F. Li, T. C. Cheong and P. Y. Yu, Phys. Rev. B **55**, 16414 (1997).
17. S-H. Wei and A. Zunger, Appl. Phys. Lett. **56**, 662 (1990).
18. S-H. Wei, A. Zunger, I-H Choi and P. Y. Yu, Phys. Rev. B **58**, R1710, (1998).
19. See for example, the review by W. Paul, in Proceedings of the International Symposium on Physical Properties of Solids under High Pressure, Grenoble, 1969 (CERN, Paris, 1970) p. 199 and references therein.
20. G. Martinez, in *Optical Properties of Semiconductors under Pressure, Handbook of Semiconductors,* Vol. 2, ed. M. Balkanski, (North Holland Publishing Company, 1980). p. 181.
21. P. Y. Yu and B. Welber, Solid State Commun. **25**, 209 (1978).
22. D. Patel, J. Chen, S. R. Kurtz, J. M. Olson, J. H. Quigley, M. J. Hafich and G. Y. Robinson, Phys. Rev. B **39**, 10978 (1989).
23. K. Uchida, P. Y. Yu, N. Noto, and E. R. Weber, Appl. Phys. Lett. **64**, 858 (1994).
24. J. Zeman, S. Jullian, G. Martinez, P. Y. Yu, K. Uchida, Europhys. Lett. **47**, 260, (1999).

25. P. Emanuelsson, M. Drechsler, D. M. Hofmann, B. K. Meyer, M. Moser and F. Scholz, Appl. Phys. Lett. **64**, 2849 (1994).
26. C. G. Darwin, Proceedings Cambridge Philos. Soc., **27**, 86 (1930).
27. J. Zeman, G. Martinez, P.Y. Yu and K. Uchida, Phys. Rev. B **55**, R13428(1997).
28. J. Zeman, G. Martinez, P.Y. Yu, and K. Uchida, Proc. ICPS-23, Berlin 1996, eds. M. Scheffler, R. Zimmermann, (World Scientific, Singapore 1996), p. 493.
29. K. Uchida, P.Y. Yu, J. Zeman, S.H. Kwok, K.L. Teo, Z.P. Su, G. Martinez, T. Arai, K. Matsumoto, High Pressure Mateirals Research Symposium, 1-4 Dec. 1997, Boston, MA.
30. S-H. Wei, D. B. Laks and A. Zunger, Appl. Phys. Lett. **62**, 1937 (1993).
31. J. Zeman, G. Martinez, P. Y. Yu, K. Uchida, Proc. ICPS-23, Berlin 1996, eds. M. Scheffler, R. Zimmermann, (World Scientific, Singapore 1996), p. 493.
32. S. Zemon *et al.*, J. Appl. Phys. **59**, 2828 (1986).
33. H. M. Cheong, A. Mascarenhas, S. P. Ahrenkiel, K. M. Jones, J. F. Geisz, and J. M. Olson, J. Appl. Phys. **83**, 5418 (1998).
34. J. Zeman, G. Martinez, P. Y. Yu, S. H. Kwok, K. Uchida, Phys. Stat. Sol. (b) **211**, 239 (1999).
35. D. Pfannkuche and R. R. Gerhards, Phys. Rev. B **44**, 13132 (1991).
36. D. Pfannkuche, V. Gudmundsson and P. A. Maksym, Phys. Rev. B **47**, 2244 (1993).
37. F. E. G. Guimaraes, B. Elsner, R. Westphalen, B. Spangenberg, H. J. Geelen, P. Balk and K. Heime, J. Cryst. Growth **124**, 199 (1992).
38. R. Bhat, M. A. Koza, M. J. S. P. Brasil, R. E. Nahory, C. J. Palmstrom and B. J. Wilkens, J. Cryst. Growth **124**, 576 (1992).
39. C. Y. Tsai, M. Moser, C. Geng, V. Harle, T. Forner, P. Michler, A. Hangleiter and F. Scholz, J. Cryst. Growth **145**, 786 (1994).
40. K. Uchida, T. Arai and K. Matsumoto, J. Appl. Phys. **81**, 771 (1997).
41. S. H. Kwok, P. Y. Yu, K. Uchida and T. Arai, Appl. Phys. Lett. **71**, 1110 (1997).
42. S. H. Kwok, P. Y. Yu, J. Zeman, S. Jullian, G. Martinez, and K. Uchida. J. Appl. Phys. **84**, 2846 (1998).
43. P. Perlin, W. Trzeciakowski, E. Litwin-Staszewska, J. Muszalski and M. Micovic, Semicond. Sci. Technol. **9**, 2239 (1994).
44. S. H. Kwok, P. Y. Yu, K. Uchida and T. Arai, J. Appl. Phys. **82**, 3630 (1997).
45. D. Mao, P. C.Taylor, S. R. Kurtz, M. C. Wu and W. A. Harrison. Phys. Rev. Lett., **76**, 4769 (1996).
46. T. Suzuki et al, Jpn. J. Appl. Phys. **27**, 2098 (1988).
47. S. R. Kurtz, J. Appl. Phys. **74**, 4130 (1993).
48. D. Munzar, E. Dobrocka, I. Vavra, R. Kudela, M. Harvanka and N. E. Christensen, Phys. Rev. B **57**, 4642 (1998).
49. S.H. Kwok, P. Y. Yu and K. Uchida. Phys. Rev. B **58**, R13395 (1998).

Chapter 13

Polarization Effects in the (Electro)absorption of Ordered GaInP and their Device Applications

Peter Kiesel, Thomas Kippenberg, and Gottfried H. Döhler
Institut für Technische Physik I, Universität Erlangen-Nürnberg, Erwin-Rommel-Str.1
91058 Erlangen, Germany

Key words: Polarization anisotropy, polarization detectors and -switches

Abstract: Electro-absorption measurements provide a powerful tool to obtain information on the electronic structure of ordered GaInP. Due to their high resolution of approximately 1 meV, they allow the exact determination of transition energies at the Γ point. We are able to give accurate values for ordering-relevant parameters like the band gap reduction, valence band splitting and the position of the split-off valence band. Moreover, the exact energetic position of zone edge folded states could be determined. We deduce that only about 80% of the band gap reduction is due to the ordering-induced lowering of the conduction band. The polarization anisotropy of ordered GaInP is used to demonstrate novel device concepts such as detectors and (bistable) switching devices, which are sensitive to the polarization direction of the optical input signal. The electrical output signal of the switches is independent of the light intensity. It changes by many orders of magnitude as a function of the polarization angle of the incident light. A switching contrast of 50 dB and a maximum sensitivity of about 10 dB/degree have been demonstrated.

1.   INTRODUCTION

The ordering-induced modifications of the electronic structure of GaInP lead to a strong polarization dependence of its optical properties. In this chapter, two important aspects of the optical anisotropy will be discussed. In sections 2 and 3 it will be shown that significant information on the electronic structure of ordered GaInP can be obtained from studies of the polarization dependence of optical interband transitions. In section 4 it will

365

be shown that the anisotropy of the optical properties also implies interesting applications for novel opto-electronic devices.

In Section 2 the facts about the modifications of the crystal structure and the resulting modifications of the electronic band structure will be briefly reviewed, followed by a more detailed discussion of the resulting anisotropy with regard to optical interband transitions. In Section 3 it will be shown that polarization dependent electro-absorption measurements represent a very powerful tool for obtaining precise information on the electronic structure. Recent experimental results, which strongly support previous predictions of theoretical studies, will be reviewed. These experimental studies have not only confirmed the expected relation between ordering-induced band gap reduction and valence band splitting. Moreover, it has become possible for the first time to observe systematically the relation between the order parameter and the energy shift of the back-folded L-point of the zinc-blende structure. Section 4 deals with device applications of the polarization anisotropy in ordered material. This section will focus on recently demonstrated polarization sensitive detectors and switching devices operating at normal light incidence.

2.   ORDERING INDUCED EFFECTS ON THE OPTICAL TRANSITIONS IN GaInP

2.1   Crystallographic and Electronic Structure

As discussed in previous chapters, the ordering in GaInP results in a transition from zinc-blende to CuPt$_B$ structure. The corresponding unit cells are shown again in the left part. On the right hand side of this figure the band structure near the Γ-point is shown for disordered and ordered GaInP, respectively, labeled in Slater-Koster notation as used also in the discussions below.

For the later consideration of optical transitions, it is important to point out the following features:

(1) The lowest $\Gamma_{4C}(\Gamma_{6C})$ conduction band of the CuPt$_B$ structure corresponds to the former Γ_{6C} conduction band of the zinc-blende structure, whereas the upper $\Gamma_{4C}(L_{4C})$ conduction band derives from the back-folded L-point conduction band minimum. Because of the same symmetry of these states their energies are shifted compared to those of the ordered GaInP due to the level repulsion by an amount the value of which increases with

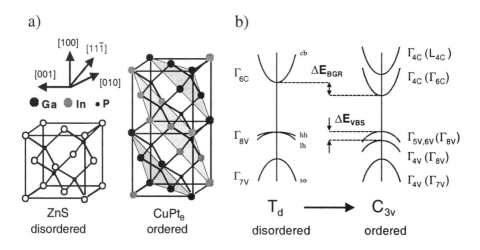

Figure 1. a) Unit cell and b) schematic band diagram of the zinc-blende and the CuPt$_B$-structure.

increasing order parameter η. According to [1,2], this leads to a reduction of the gap between the upper valence band $\Gamma_{5V,6V}(\Gamma_{8V})$ and the $\Gamma_{4C}(\Gamma_{6C})$ conduction band by an amount ΔE_{BGR}, whereas the gap to the $\Gamma_{4C}(L_{4C})$ conduction band increases.

(2) Due to reduction of symmetry with regard to the [11-1] direction, the degeneracy of the $j = 3/2$-states at the Γ-point is lifted. The $\Gamma_{5V,6V}(\Gamma_{8V})$ correspond to the $(3/2, \pm 3/2)$ (heavy) hole and the $\Gamma_{4V}(\Gamma_{8V})$ to the $(3/2, \pm 1/2)$ (light) hole band. The valence band splitting energy ΔE_{VBS}, again, increases with increasing ordering parameter η, according to [3]. For the $\Gamma_{4V}(\Gamma_{7V})$ band, which corresponds to the spin split-off valence band, the theory predicts a rather weak dependence on the ordering parameter for usual values of $\eta < 0.5$.

2.2   Polarization Anisotropy of the Optical Properties

The most obvious effect of the ordering-induced changes of the electronic structure in GaInP is a red shift of the absorption edge corresponding to the band gap reduction energy ΔE_{BGR}. From this red shift, information on the ordering parameter can be directly obtained, according to [3]. More detailed information on the symmetry changes, however, follows from the valence band splitting as it implies a polarization dependence of the interband transitions.

Due to the heavy hole character of the $\Gamma_{5V,6V}(\Gamma_{8V})$ valence band, the lattice periodic part of the Bloch states has $|m_z| = 1$ p-like character with the corresponding orbits oriented normal to the [11-1] direction. Therefore, only the electric field components normal to the ordering direction contribute to optical transitions into the s-like conduction bands. For the usual case of a [100] oriented surface and normal incidence of the electromagnetic radiation, this means that maximum absorption is possible for light polarized along the [011] direction (see *Figure 2*). If the light is polarized along the [01-1] direction, only the projection of the electric field vector onto the plane normal to the [11-1] direction contributes to the absorption. It can easily be seen from *Figure 2* that this yields only 1/3 of the [011] - contribution to the absorption.

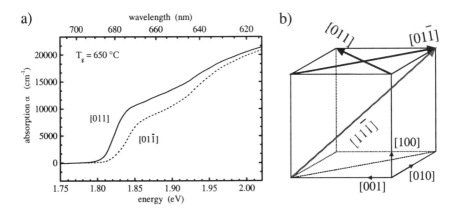

Figure 2. a) Room temperature absorption spectra of an ordered GaInP sample, grown at 50°C on 6° toward [111]$_B$ tilted substrate for [011] and [01-1] polarized light. b) Illustration of the important crystal directions in ordered GaInP.

The polarization dependence is more complicated for the contribution of the $\Gamma_{4V}(\Gamma_{8V})$ band as the light-hole states are composed of a small $|m_z| = 1$ component and a larger $m_z = 0$ component. A simple calculation [4] yields a polarization dependence which now favors the [01-1] direction compared to the [011] direction by a factor of 3. Neglecting the anisotropy of the effective masses parallel and normal to the [11-1] direction, one can express the absorption coefficient for light polarized along the [01-1] and [011] direction by a weighted sum of the heavy and light hole contributions to the optical transitions

$$\alpha_{[011]}(\omega) = 3/4\ \alpha_{low}\ (\omega) + 1/4\ \alpha_{high}(\omega) \tag{1}$$

and

$$\alpha_{[01\text{-}1]}(\omega) = 1/4\ \alpha_{low}\ (\omega) + 3/4\ \alpha_{high}(\omega) \tag{2}$$

respectively, where $\alpha_{low}(\hbar\omega)$ and $\alpha_{high}(\hbar\omega)$ correspond to the low and high energetic transitions from the $\Gamma_{5V,6V}(\Gamma_{8V})$ and $\Gamma_{4V}(\Gamma_{8V})$ valence band into the $\Gamma_{4C}(\Gamma_{6C})$ conduction band, respectively.

In a single particle effective mass approximation, both $\alpha_{low}(\hbar\omega)$ and $\alpha_{high}(\hbar\omega)$ would increase as the square root of excess photon energy, with $\alpha_{high}(\hbar\omega)$ being shifted to higher energies by ΔE_{VBS} compared to $\alpha_{low}(\hbar\omega)$

$$\alpha_{low}(\hbar\omega) = C_{low}\left[\hbar\omega - E_{gap}^{\,ord}\right]^{\frac{1}{2}} \tag{3}$$

$$\alpha_{high}(\hbar\omega) = C_{high}\left[\hbar\omega - (E_{gap}^{\,ord} + \Delta E_{VBS})\right]^{\frac{1}{2}} \tag{4}$$

where $E_{gap}^{\,ord} = E_{gap}^{\,disord} - \Delta E_{BGR}$ is the band gap of ordered GaInP.

In experiment, deviations from this behavior are observed because of excitonic effects, deviations from the effective mass approximation and due to inhomogeneous broadening resulting from spatial variations of the ordering parameter and the domain size. The broadening, in particular, is so strong that the important quantities $E_{gap}^{\,ord}$ and ΔE_{VBS} cannot be extracted from experimentally measured data for $\alpha_{[011]}(\hbar\omega)$ and $\alpha_{[01\text{-}1]}(\hbar\omega)$ by inverting Eq.s (1) and (2) for a determination of $\alpha_{low}(\hbar\omega)$ and $\alpha_{high}(\hbar\omega)$ according to

$$\alpha_{low}(\hbar\omega) = \quad 3/2\ \alpha_{[011]}(\hbar\omega) - 1/2\ \alpha_{[01\text{-}1]}(\hbar\omega) \tag{5}$$

$$\alpha_{high}(\hbar\omega) = -1/2\ \alpha_{[011]}(\hbar\omega) + 3/2\ \alpha_{[01\text{-}1]}(\hbar\omega) \tag{6}$$

The data for the polarization resolved absorption spectra depicted in *Figure 2* clearly confirm the strong polarization dependence. At the same time, it becomes also obvious that an expected valence band splitting of the order of 20 meV can hardly be resolved in this data with the accuracy required using Eq.s (5) and (6). In the next section it will be demonstrated that the problems resulting from the inhomogeneous broadening can be overcome in an elegant way by using the electro-modulation technique for absorption studies.

3.   POLARIZATION DEPENDENT ELECTRO-ABSORPTION

As shown in the previous section, ordering in GaInP induces significant changes of the band structure. As a result, the absorption spectra reveal a strong anisotropy for light polarized parallel to the [011] and [01-1] direction (*Figure 2*). Details of the ordering-induced band structure changes, however, can hardly be extracted from measurements of the absorption spectra, in particular because of both inhomogeneous and homogeneous broadening of the spectra.

Modulation techniques, however, are well known to provide much more and more precise information on the electronic band structure. The electro-absorption measurements we have carried out allow us to measure very small absorption changes superposed to a strong background absorption and to determine transition energies with a resolution of approximately one meV.

This section starts with a brief introduction about the theoretical background of electro-absorption (Franz-Keldysh effect) measurements, followed by experimental details. After a comparison between simple absorption and electro-absorption spectra, we subsequently present in detail measurements at the fundamental band gap, including transitions from the spin orbit split-off band. At the end of this section, we will demonstrate the power of electro-absorption measurements by investigating high energy transitions, which occur due to the zone-folding in ordered material. By this method, we gain very clear fingerprints of ordering in GaInP.

3.1   Franz-Keldysh Effect

The Franz-Keldysh effect [5,6] describes the characteristic changes of the absorption α, which occur in bulk semiconductors in the presence of an electric field. Obviously, the translation symmetry is broken along the direction of the electric field. The eigenstates of electrons and holes can no longer be described by Bloch functions. In the case of a homogenous electric field, the envelope functions with regard to the field direction become Airy functions. *Figure 3a* illustrates this situation for the fundamental band gap. Due to a non-vanishing overlap between valence- and conduction-band wave functions, absorption results occurs even for photon energies less than the band gap. With increasing photon energies, the absorption coefficient α also increases. At photon energies above band gap, it oscillates around the zero-field absorption as the Airy-functions of hole and electron interfere constructively or destructively. A very important observation is that all absorption curves for different electric fields happen to coincide at an energy corresponding to the band gap E_{gap} (see *Figure 3b*). In this way, E_{gap} can be determined with high precision even under experimental situations where the electric field in the sample is spatially non-uniform.

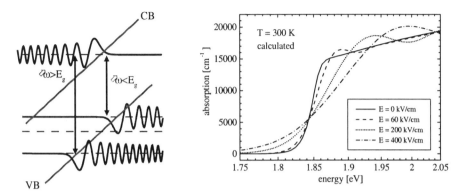

Figure 3. (a) In presence of an electric field the band edges (CB and VB) tilt parallel to the field direction and the wavefunctions of electrons and holes are changed significantly. (b) As a consequence, the absorption is increased below the band gap and shows an oscillating behaviour above E_{gap}.

Up to now, we have only discussed the Franz-Keldysh effect for transitions at the fundamental band gap. However, the situation is quite similar for transitions between states, which are separated by higher energies E_t, and for other critical points, for example E_1 or $E_1+\Delta$. In each case, the electric field affects the absorption qualitatively in the same way as described above and especially leads to a common intersection point of all α curves at exactly the transition energy E_t. By this means, E_t can be determined with high accuracy even at room temperature, provided the results are not obscured by the Franz-Keldysh oscillations (FKO) from lower lying transitions. In this case, a non-uniform field even turns out to be of advantage. With increasing energy above the band gap, the absorption changes that result from a superposition of the envelope functions belonging to different electric fields tend to be strongly damped by the FKOs by destructive interference.

3.2   Experimental Setup

To measure the Franz-Keldysh effect, we have performed transmission measurements. When designing the samples, we had to take into account that some of the transitions in ordered GaInP only differ very slightly in energy. This means that the FKOs of the lower transition disturb the signal at the slightly higher transitions. This problem can be overcome by using bulk metal-semiconductor-metal (MSM) structures with lateral finger contacts, which possess a very inhomogenous field distribution leading to a strong damping of the FKOs (see section 3.1). Moreover, these samples are simple bulk structures without any disturbing cladding layers.

The investigated samples consist of an approximately 2-μm-thick GaInP layer grown lattice-matched ($\Delta a/a < 3\times10^{-4}$) on GaAs substrates tilted 6° towards the $[111]_B$ direction. The growth temperatures in the vertical MOCVD reactor were $T_g = 650°C$, $690°C$, and $710°C$ for sample A, B, and C, respectively. As discussed in previous chapters, this substrate tilt leads to single variant ordered crystals, in which only one $[111]_B$ ordering direction is observed. The highest T_g corresponds to the lowest degree of ordering η.

Ti/Au interdigital contacts were defined by photolithography and evaporated using standard lift-off technology. The device area was 240X240 μm^2, the spacing and width of the fingers are 4 μm and 1.5 μm, respectively. This wide spacing allows one to neglect finger-induced polarization effects of the incident light with wavelengths $\lambda<1\mu m$. The GaAs substrate below the devices was removed by selective wet chemical etching. Front and back side of the devices were anti-reflection coated with SiO_x.

The transmission measurements were performed by focusing the light of a tungsten lamp that was spectrally dispersed by a monochromator on the sample. The polarization dependence of the transmitted intensity was detected using a Si detector and a Glan-Thompson polarizer. To avoid the detection of scattered light, several high and low pass filters were used. The changes of the absorption were determined in the transmitted light using a lock-in technique. We used voltages of 4, 8, 12, and 16 V, which correspond to average electric fields of approximately 10, 20, 30, and 40 kV/cm, respectively. The measured signal is proportional to $\Delta T = T_0 - T_1$, where T_1 and T_0 are the transmitted intensities with and without an applied electric field, respectively. In our case of weak absorption changes, $\Delta T/T_0$ is approximately proportional to the absorption change $\Delta\alpha$. Additionally, measurements of zero field absorption spectra were carried out by transmission measurements through devices without finger contacts.

3.3   Comparison between Absorption and Electro-Absorption

Figure 4(a) shows the absorption spectrum of sample A grown at 650°C for unpolarized light taken at photon energies between 1.7 and 2.4 eV. Basically, one can observe three kinks in this spectrum, each one corresponding to the onset of absorption from additional valence or conduction band states, respectively. At the fundamental band gap ($E_{gap} \approx$

1.8 eV) transitions from $\Gamma_{4V}(\Gamma_{8V})$ and $\Gamma_{5V,6V}(\Gamma_{8V})$ to the lowest conduction band $\Gamma_{4C}(\Gamma_{6C})$ contribute to the absorption. The small ordering-induced valence band splitting between the different hole states cannot be resolved in this simple absorption experiment (see section 2). At $E_{so} \approx 1.9$ eV, i.e., approximately 0.1 eV above the band gap, one can see a second kink due to the onset of absorption from the split-off valence band $\Gamma_{4V}(\Gamma_{7V})$ to the lowest conduction band $\Gamma_{4C}(\Gamma_{6C})$. At even higher photon energies, around 2.3 eV, one can recognize an additional kink of the absorption coefficient, which originates from transitions from $\Gamma_{4V}(\Gamma_{8V})$ and $\Gamma_{5V,6V}(\Gamma_{8V})$ to the backfolded conduction band $\Gamma_{4C}(L_{4C})$. The spectra of the samples *B* and *C* (grown at 690°C and 710°C, respectively) basically show the same features. However, the position of the fundamental band gap and the split-off band shift to higher energies, while the kink around 2.3 eV gets weaker and shifts to lower energies.

The field induced transmission changes $\Delta T/T_0$ of sample *A* are shown in *Figure 4b*. We find a typical Franz-Keldysh band-edge-transmission change structure at each one of the kink energies of the absorption spectrum. Note the common intersection points at approximately 1.8 eV, 1.9 eV, and 2.3 eV, respectively. However, as shown below, for an exact analysis, one has to keep in mind that the FK-spectra around 1.8 eV and 2.3 eV consist of a superposition of two energetically shifted transitions, originating from the $\Gamma_{4V}(\Gamma_{8V})$ and the $\Gamma_{5V,6V}(\Gamma_{8V})$ valence band, respectively, as discussed in section 2.2.

3.4   Fundamental Band Gap

To investigate the situation at the fundamental band gap, we have performed polarization-dependent electro-absorption measurements for [011] and [01-1] polarized light. This is shown for sample *A* in *Figure 5a*. Due to the polarization anisotropy of optical transitions from $\Gamma_{4V}(\Gamma_{8V})$ and $\Gamma_{5V,6V}(\Gamma_{8V})$, a small polarization shift ΔE_{pol} of approximately 6 meV occurs between the spectra taken for [01-1] and [011] polarized light, respectively. When interpreting this shift, however, one has to be very careful to keep in mind that both directions of polarization result in a superposition of both intrinsic transitions. Additionally, the valence band splitting ΔE_{VBS} is usually much smaller than the FK-oscillation period – making the observation of undisturbed Franz-Keldysh signals impossible. Thus, in general, the polarization splitting ΔE_{pol} differs from the valence band splitting ΔE_{VBS}. Because, however, the valence band splitting is a key parameter of ordered material, we have to isolate both intrinsic transitions. Depending on the degree of ordering in the sample, two methods can be applied.

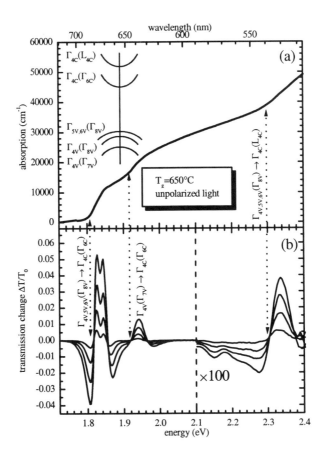

Figure 4. (a) Room temperature absorption spectra and (b) field induced transmission changes $\Delta T/T_0$ of sample A (T_g=650°C). The contributions due to the fundamental band gap, the split-off band and the back-folded L point are visible. The inset schematically shows the band structure of ordered material in the notation of Koster et al. [7].

Very small electric fields F can be applied to diminish the FK oscillations width, which is proportional to $F^{-2/3}$. Typically, fields of F<1 kV/cm are sufficient to resolve a splitting $\Delta E_{VBS} > 20$ meV, which occur in strongly ordered samples, like sample A. However, for less ordered samples (*B* and *C*), which exhibit a smaller ΔE_{VBS}, this is impossible. In this case, we take advantage of the simple model described in section 2. Within this simple quasi cubic model, the transition strength ratios for transitions from $\Gamma_{5V,6V}(\Gamma_{8V})$ and $\Gamma_{4V}(\Gamma_{8V})$ are 1:3 or 3:1 for [011]- and [01-1]-polarized light, respectively. By using these ratios, the contributions of both transitions can be separated analytically using Eq.s (5) and (6) [4] (*Figure 5b*). In fact, a comparison with the directly measured values (which is possible for sample *A*) shows an excellent agreement within 2 meV, although a detailed theory (see e.g. [8,9]) yields slightly different transition ratios.

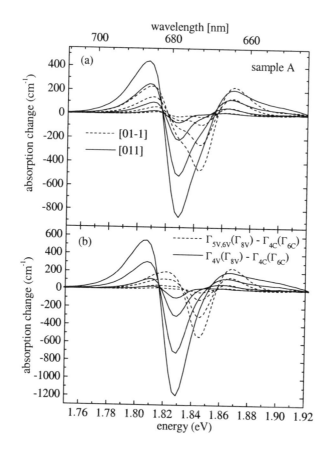

Figure 5. (a) Polarization dependent electro-absorption spectra of ordered GaInP (sample A). The measured spectra for [011] and [01-1] polarized light represent a superposition of the intrinsic transitions, $\Gamma_{5v,6v}(\Gamma_{8v}) \rightarrow \Gamma_{4C}(\Gamma_{6C})$ and $\Gamma_{4v}(\Gamma_{8v}) \rightarrow \Gamma_{4C}(\Gamma_{6C})$, which have been deconvoluted, using a quasi-cubic model, in part (b).

After either directly measuring or analytically decomposing the measured spectra into the transitions from the valence bands $\Gamma_{5v,6v}(\Gamma_{8v})$ and $\Gamma_{4v}(\Gamma_{8v})$ to conduction band $\Gamma_{4C}(\Gamma_{6C})$, we finally obtain the band gap energy E_{gap} and valence band splitting ΔE_{VBS}. The value of the band gap reduction ΔE_{BGR} can then be calculated by subtracting E_{gap} from the value of a totally disordered sample $(E_{gap}(0) = 1.915$ eV, at room temperature) [10]. The ordering parameter η can finally be determined from the band gap reduction, using the relation

$$\Delta E_{BGR}(\eta) = E_{gap}(0) - E_{gap}(\eta) = \eta^2 [E_{gap}(0) - E_{gap}(1)] \qquad (7)$$

Concerning the fundamental band gap, the question on the influence of superlattice ordering on the split off band remains. Looking at the electro-absorption spectra of sample *A* around 1.9 eV, we see an additional common intersection point of all curves at E_{SO} (see *Figure 4b*), which does not depended on the polarization state of the incident light. The spin orbit splitting $\Delta E_{SO} = E_{SO} - E_{gap}$ of approximately 100meV is nearly independent of the degree of ordering (see table 1 in section 3.6). This is in good agreement with the theory of Wei and Zunger [8].

3.5   Backfolded States - Fingerprints of Ordering

Until now, it has been very difficult to perform exact measurements of the back-folded conduction band L point minima. Several authors have reported peaks in electro-reflectance measurements [11,12]. Very recently, folded zone-edge modes have also been observed using Raman scattering [13]. Neither of these results, however, allowed an exact determination of the transition energy, nor did they provide information on polarization-dependent effects.

As mentioned in section 3.3 in connection with *Figure 4*, an additional FK-structure can be observed around 2.3eV. It is attributed to transitions from the $\Gamma_{5v,6v}(\Gamma_{8v})$ and $\Gamma_{4v}(\Gamma_{8v})$ valence band to the backfolded conduction band $\Gamma_{4C}(L_{4C})$. Its energetic position and strength significantly depends on the degree of ordering (see *Figure 6*). The situation is analogous to that observed at the fundamental band gap, as the symmetries of valence and conduction band states are identical. Thus, we again observe a polarization dependent superposition of two transitions, which are energetically separated by ΔE_{VBS}. In fact, polarization dependent measurements show a similar shift between the spectra for [011] and [01-1] polarized light like that observed at the fundamental band gap. This is shown in *Figure 7* for sample A.

Once more, this shift differs from the valence band splitting ΔE_{VBS}. In order to separate both transitions, we cannot reduce the electric field, since in this case the $\Delta T/T_0$ signal would not be detectable. The higher effective mass (in comparison to that at the fundamental band gap) also leads to much larger FK oscillation periods. Therefore, we have to perform the analytical separation described in section 2.2. For all three samples, we get an excellent agreement of the valence band splitting determined at the fundamental band gap and the backfolded state, respectively (see Table 1 in section 3.6).

A closer look at the spectra in *Figure 6* suggests an additional transition for lower energies around E=2.15eV. This transition might be due to the occurrence of back-folded states along the Σ direction, which is to be expected in the presence of non-single variant CuPt$_B$ ordering.

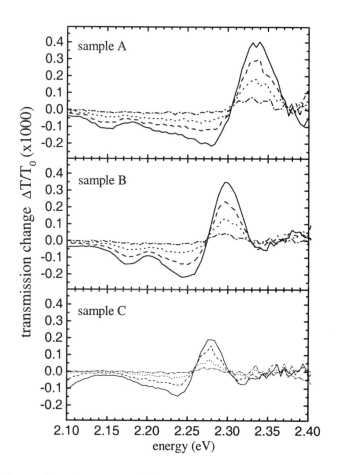

Figure 6. Electro-absorption spectra of the zone-edge folded L-state. With increasing ordering, the intensity and energy of the transition increases.

3.6   Discussion of the Results

Table 1 summarizes the results of our measurements. With increasing temperature, the ordering effects diminish. In particular, the band gap reduction ΔE_{BGR} and the valence band splitting ΔE_{VBS} reach their maximum values for the lowest growth temperature of $T_g = 650°C$. In contrast, the split-off separation ΔE_{SO} changes only slightly with varying T_g.

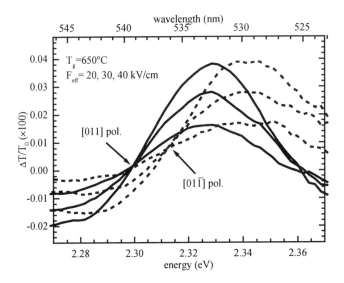

Figure 7. Polarization dependent electro-absorption spectra of sample A in the spectral range of zone-edge folded L-state.

Table 1. Analysis of the electro-absorption measurements. With increasing growth temperature T_g the ordering induced band gap reduction ΔE_{BGR} decreases while the transition energy ΔE_L between $\Gamma_{4V,5V,6V}(\Gamma_{8V})$ and $\Gamma_{4C}(L_{6C})$ decreases. The energies are given in meV.

Sample	T_g	η	E_{gap}	ΔE_{BGR}	ΔE_{VBS}	ΔE_{SO}	E_L
A	650°C	0.45	1819	96	21	105	2296
B	690°C	0.33	1865	50	11	97	2270
C	710°C	0.23	1890	25	8	95	2256

The analysis of these results provides for the first time reliable data about the band gap E_L corresponding to the difference between the uppermost valence band to the back-folded conduction band L point minimum and its dependence on the ordering parameter η. These data are depicted in *Figure 8*. By extrapolating to $\eta=0$, we find a value of $E_L(0) = 2.243$ (± 0.001) eV for the energetic distance between the upper valence band and the conduction band L point minimum for disordered GaInP. This value is in excellent agreement with a recently obtained value by BEEM [14] The observed linear increase of E_L vs. ΔE_{BGR} can be explained by noting that the lowest conduction band state $\Gamma_{4C}(\Gamma_{6C})$ and the new state $\Gamma_{4C}(L_{4C})$ have the same symmetry, which leads to a level repulsion. This repulsion increases with increasing degree of order and is expected to be symmetric, as both of these

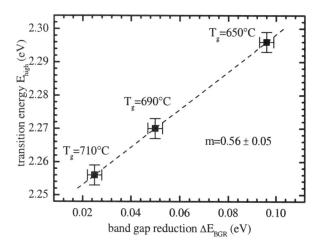

Figure 8. Energetic position of the high energetic transition ΔE_{high} versus the band gap reduction ΔE_{BGR}. The dashed line represents a linear fit, yielding a slope of m ≈ 0.56.

states have the same degeneracy. Using the uppermost $\Gamma_{5V,6V}$ (Γ_{8V}) valence band as reference, one would thus expect a slope of m=1 in *Figure 8*. However, we observe a slope of only m=0.56 ± 0.05. This indicates that only (1+0.56) / 2 = 0.78 of the observed band gap reduction is due to the (symmetric) level repulsion of the conduction bands. Therefore, the remaining 22% have to be attributed to the uppermost valence band edge being shifted up. In *Figure 9*, the evolution of the three valence and two conduction band energies is shown vs. ΔE_{BGR} as a measure of the ordering. For this plot, a symmetric level repulsion of the two conduction bands is assumed. Whereas the $\Gamma_{5V,6V}$ (Γ_{8V}) valence band increases by an amount which nearly equals the value found for the valence band splitting, it turns out that the Γ_{4V} (Γ_{8V}) valence band edge only slightly changes.

4. POLARIZATION-SENSITIVE DEVICES

In this section, we will report on some device oriented work. Control and detection of the polarization state of the incident light is of fundamental importance in many applications. Currently, this is achieved by the hybrid combination of polarization-insensitive active devices (photodiodes, light emitting diodes) with polarization-sensitive passive elements (sheet polarizer, polarization beam splitter, micro-optical gratings). Potential disadvantages of such hybrid systems are the relatively large size that limits the packaging density, as well as the necessity of accurate alignment and

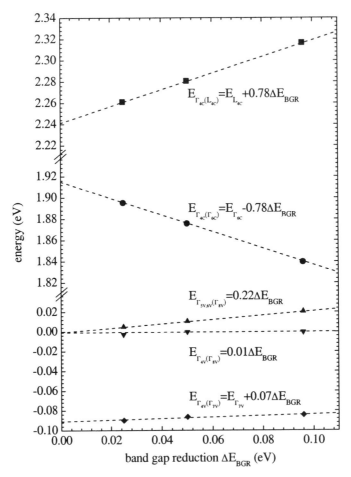

Figure 9. Energetic position of the three valence band maxima and two conduction band minima in terms of the ΔE_{BGR} as a measure of the ordering. About 78% of the band gap reduction is due to a level repulsion between the backfolded L point states and the original conduction band Γ point states. Around 22% is due to the energetic increase of the valence band maximum being shifted upwards.

usage of mechanical elements. In this section, we report on intrinsically polarization-sensitive semiconductor devices that take advantage of the polarization anisotropy of ordered GaInP.

The strong polarization anisotropy of the absorption coefficient, shown in *Figure 2*, can be exploited to realize polarization-sensitive detectors and switches [15]. The photocurrent I_{ph} of a p-i-n-detector is defined as

$$I_{ph} = R_{pin} P_{opt} = (\eta^{ext} e / \hbar\omega) \ P_{opt} \qquad (8)$$

where R_{pin} is the responsivity, P_{opt} the incident optical power, and $\hbar\omega$ the photon energy. The external quantum efficiency η^{ext} of a p-i-n photodetector can be expressed as

$$\eta^{ext} = (1-R)\,\eta^{int}\,[1 - exp(-\alpha d)] \qquad (9)$$

Herein $(1-R)$ and η^{int} account for the reflection and internal losses, while α and d denote the absorption coefficient and the thickness of the absorption layer, respectively. Since the absorption coefficient of ordered material depends on the polarization of the incident light, it is obvious that the photocurrent of p-i-n-detectors based on ordered material also depends on the polarization state of the optical input. In order to realize polarization-sensitive devices with high opto-electronic gain enabling an intensity independent operation, we have designed a special photoconductive device.

Our detector basically represents a p-i-n photodiode and a field effect transistor integrated into a single device. Its principle is based on the light-induced increase of the n-layer conductance. A schematic picture of the device is shown in *Figure 10*. The absorption layer consists of ordered GaInP. Two selective n-type contacts, acting as source and drain of a junction field effect transistor, and a back gate contact to the p-layer are provided. Thickness and doping concentration of the n-layer are adjusted such that this layer becomes depleted under a suitable reverse bias ("pinch-off" voltage).

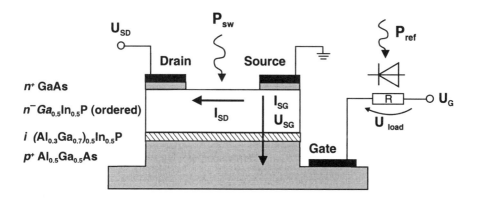

Figure 10. Schematic picture of our polarization-sensitive device which basically represents a p-i-n photodiode and a field effect transistor integrated into a single device.

Figure 11a shows the conductance of the n-layer as a function of the source-gate voltage in a mixed linear and logarithmic plot for one of our devices. The conductance decreases nearly linearly with the applied reverse bias, followed by an exponential tail beyond the pinch-off voltage (for this device $\approx$ -2V). *Figure 11b* shows the pn-characteristics (I_{SG} vs. U_{SG}) of the device for illumination with light (P_{opt}=10nW) linear polarized along the [011] and [01-1] direction, respectively. The photocurrent for [011] polarized light is almost twice as high as for [01-1] polarization, which is the direct consequence of the absorption anisotropy in the ordered material (see *Figure 2*). For operation, the source-gate circuit is connected in series to a load (resistor or diode) as indicated in *Figure 10*. According to Kirchhoff's law, the voltage drop across the source-gate contacts and thus the corresponding n-channel conductance is given by the intersection point of the source-gate I-V curve with the loadline. This is shown in *Figure 11b* for a load resistor and a load diode, respectively.

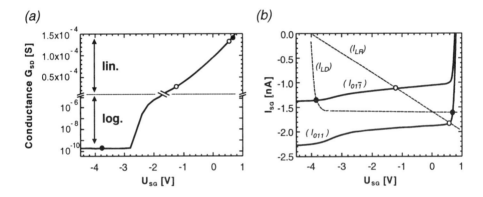

Figure 11. (a) n-channel conductance vs. source-gate voltage U_{SG}. Note that the upper and lower part of the y-axis show a linear and logarithmic plot, respectively. (b) Source-gate current under illumination with [011] and [01-1] polarized light, respectively. The dashed lines indicate the loadline for case of a load resistor and a load diode.

4.1   Polarization Detector

For the operation as polarization detector, an ohmic resistor is used as a load. In this case, the device operates in the "linear regime." That means that the n-channel (source-drain) conductance changes within the linear part of *Figure 11a*, if the polarization of the incident light is changed from the [01-1] to the [011] direction. The resulting voltage drop and n-channel conductance for these two polarizations are marked with open circles in

Figure 11. As shown in Ref. [16], the n-channel current I_{SD} is expected to exhibit a $\sin^2$-dependence on the polarization angle Θ of the incident light. The measurement in *Figure 12* confirms that this is fulfilled with high accuracy and shows that the source-drain current is a sensitive measure for the polarization state of the optical input even at low optical powers (in this case $P_{sw} \approx 25$ nW). The photoconductive gain defined by g = I_{SD} / I_{SG} was approximately 7000 in this case. By using an improved contact design (e.g. interdigited finger contacts), the opto-electronic gain of this device can be significantly increased. As mentioned above, our device represents an integrated combination of a p-i-n photodiode and a junction FET. Therefore, it behaves like a polarization-sensitive photo transistor. The set of characteristic curves of our non-optimized test device is shown in *Figure 12b*. Current saturation does not occur in this measurement due to the relatively small applied voltage and the large separation (100 µm) of the source and drain contacts. Note whereas in conventional junction FETs the slope of the I-V curves can be tuned by the gate voltage, in our circuit, the same effect is achieved by rotating the polarization direction of the incident light.

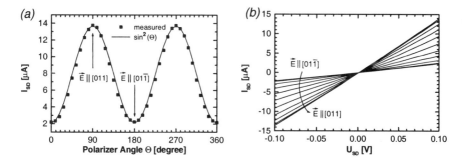

Figure 12. (a) Variation of the source drain current proportional to $\sin^2(\Theta)$ by rotating the linear polarization of the optical input power. (b) The slope of the source-drain characteristics can be tuned by changing the polarization of the incident light.

4.2 Polarization Switch

If a second p-i-n photodiode is used as load (see *Figure 10*), threshold switching behavior can be observed. In this case, the electrical output signal can be switched "on" and "off" by a slight variation of the linear polarization of the optical input P_{sw}. The principle of operation can again be understood from *Figure 11b*, where the situation for the load diode is shown, as well. The voltage drop across the source- and gate contact and the corresponding n-channel conductance are marked with full circles. If the incident light P_{sw} is polarized along the [01-1] direction, the voltage drop is larger than the pinch-off voltage ($\approx$ -2V). Thus, the n-layer is depleted and the electrical output signal (n-channel current) is very small. However, if the polarization is rotated by 90° into the [011] direction, the working point flips to +0.5V causing a drastic increase of the n-channel current. As shown in *Figure 13*, the n-channel current I_{SD} in our device can be changed by five orders of magnitude (switching contrast of 50dB) with a maximum sensitivity of 3dB/degree by turning the polarization angle of the optical input. The angle at which "on/off" switching occurs is determined by the position of the loadline (see *Figure 11*b). The switching angle can be tuned by changing the optical power P_{ref} incident on the load diode.

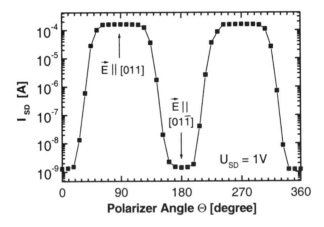

Figure 13. Polarization threshold switching of the n-channel current. By turning the linear polarization of the optical input power, the n-channel current can be switched "on" and "off" over five orders of magnitude.

4.2.1   Intensity Independent Switching Behavior

In this section we will demonstrate that the switching characteristic presented in *Figure 13* is independent of the intensity of the optical input. The I-V curves of both, load and switch, scale linearly with the optical power if the dark current in the diodes can be neglected. Thus, the voltage drop across the switch, determined by Kirchhoff's law, is independent of the optical power. As shown in *Figure 14*, the switching angle remains the same by changing the optical input from 3 nW to 100 nW. In principle, intensity independent switching should be possible over many orders of magnitude. The limitation in this measurement to an optical power less than 100nW is only due to our experimental setup.

The measurements presented are taken from a hybrid combination of our polarization-sensitive switch with a polarization-insensitive load diode. However, a monolithic integration into a single device can easily be realized by growing both devices on top of each other. Work on such a version is in progress. In combination with the intensity independent switching behavior, this device allows for a very uncritical alignment of the incident light beam. For possible applications of our polarization switch, a fast response and a large electrical output signal in the on-state is desirable. As demonstrated for photo conductive switches in the AlGaAs system [17], such an optimization should be straightforward. By using interdigitated contacts, the n-layer conductance in the on-state can be increased by several orders of magnitude. The principle of switching behavior is independent of the material and can be observed in all material systems where spontaneous superlattice ordering occurs (e.g InGaAs, InGaN, InGaAsP). Therefore, the operation wavelength can be chosen within a wide range.

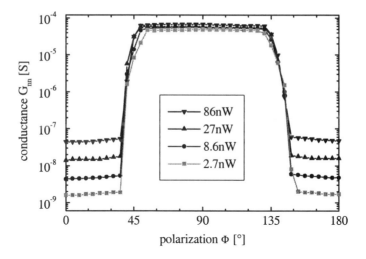

Figure 14. Intensity independent switching behavior.

4.2.2   Bistable Polarization Switch

At appropriate wavelengths, the absorption decreases with increasing field, leading to a "negative differential resistance" of the I-V curve of the switch. In such a case, the photo current at zero bias is larger than under reverse bias. This behavior is due to the wavelength dependence of the FKE in the spectral range slightly above the band gap energy (see Section 3.1). This behavior allows the design of bistable polarization switches[18].

The principle of operation is nearly the same as for the threshold switch. In order to get bistable switching with a broad hysteresis, we have used a load diode consisting of ordered material in this case. The load diode was mounted skewed by 90° with respect to the photoconductive switch in order to show the opposite polarization behavior. For illumination with polarized light with a polarization angle of $\Phi = 0°$ and 90° (corresponding to [011] and [01-1] polarized light with respect to the switch), we get one defined intersection point between switch and load line, as shown in *Figure 15*. For a polarization angle in between, there is a range with three intersection points, two of them being stable. The actual solution now depends on whether one is moving from $\Phi = 0°$ or 90°. Rotating the polarization clock- or counter-clockwise leads to a hysteresis in the n-channel conductance. *Figure 16* shows the hysteresis in the switching characteristics for three different wavelengths λ of the optical input. The hysteresis loop is as large as 60° for $\lambda=662$ nm. By choosing the wavelength of the optical input, one can tune the width of the hysteresis between 0° and 60°, as shown in the inset of *Figure 16*. Like in the case of the threshold version, the switching points in this version are again independent of the intensity of the optical input.

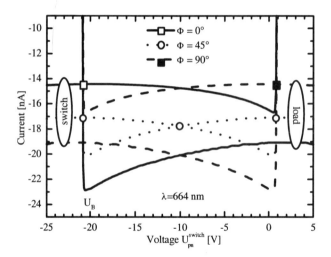

Figure 15. Graphical solution of Kirchhoff's law for series connection of switch and ordered load diode mounted skewed by 90° to each other. Both devices are illuminated with linear polarized light with a polarization angle of 0°, 45° and 90°, respectively. A polarization angle of 0° denotes light polarized parallel to the [01-1] ([011]) crystal direction of the switch (load).

Recently, we have shown that these bistable switching devices can be used for polarization-coded logic elements, like a polarization-controlled reset-set (RS-) flip flop [19]. The Boolean input states are coded by the polarization direction of the incoming light.

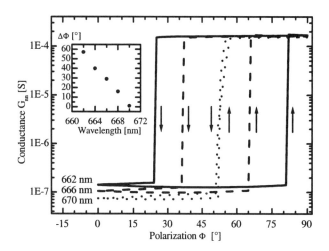

Figure 16. Bistable switching behavior of the n-channel conductance as a function of the polarization angle of the optical input. By choosing the wavelength of the light between 662 and 670 nm, it is possible to tune the width of the hysteresis between 0° and 60° (see inset).

5.   CONCLUSIONS AND OUTLOOK

In this chapter, we have shown that the polarization effects in the absorption yield, first of all, important information on the electronic structure of ordered GaInP. But we have seen as well that they also allow for interesting novel device applications.

With regard to the first aspect, we have been able to demonstrate that electro-absorption measurements are particularly suitable to verify quantitatively the theoretically expected effects of ordering. By analyzing the observed polarization dependent shift of the electro-absorption spectra, we have been able to deduce the ordering-induced valence band splitting. This valence band splitting has been directly correlated to the ordering-induced (direct) band gap reduction. Moreover, it has also been possible to observe the systematic shift of the back-folded conduction band L-point minimum as a function of increasing ordering. As expected, the corresponding transitions from the two uppermost valence bands also exhibit a strong polarization anisotropy. Interestingly, these investigations have shown that the changes of the direct and the back-folded conduction band edge energies, which are attributed to level repulsion between states of the same symmetry, are not symmetric with reference to the uppermost valence bands. Finally, we have found additional structures in the electro-absorption spectra, which are hints for the occurrence of back-folded states with regard to the Σ direction. A systematic investigation of non-single variant samples (with suitable tilt angles of the crystal axis in the substrate) could further improve our understanding of the ordering induced effects on the electronic structure of ordered GaInP in the future.

In our discussion of applications of the polarization-dependent absorption in ordered GaInP, we have presented results on a few simple polarization-sensitive devices. The advantages of these devices include the option between more or less steep threshold switching or bistability, small lateral dimensions, high speed size and the potential for monolithic integration into two-dimensional arrays. A particularly interesting aspect of the polarization-dependent switch is the fact that the switching point depends only on the polarization angle, but not on the intensity of the incident optical signal. During the conclusion of this report, the operation of a polarization switch with its two components packaged monolithically on top of each other (instead of side by side) has been successfully demonstrated as well. This achievement makes the realization of two-dimensional polarization switch arrays quite easily accessible. We have not discussed the interesting device applications of wave-guide structures based on GaInP as active material, for which voltage dependent changes of the polarization direction (polarization converter) have been demonstrated by Wirth et al. [20]. As this chapter is devoted to absorption effects, we have excluded the (strongly related) polarization-dependent luminescence phenomena [21]. We only mention that in the case of the polarization anisotropy of the luminescence a very attractive application may derive from the possibility of fabricating polarization-stable vertical cavity surface emitting lasers (VCSELs) [22] and micro-cavity LEDs.

REFERENCES

1. S.-H.Wei and A. Zunger, Phys. Rev. B 39, 3279 (1989).
2. A. Gomyo, T. Suzuki, K. Kobayashi, S. Kawata, I. Hino, and T. Yuasa, Appl. Phys. Lett. 50, 673 (1987).
3. P. Ernst, C. Geng, F. Scholz, H. Schweizer, Y. Zhang, and A. Mascarenhas, Appl. Phys. Lett. 67, 2347 (1995).
4. P. Kiesel, T. Kippenberg, E. Greger, M. Moser, U. Hilburger, J. Krauss, G. Schmiedel, and G. H. Döhler, Physica E 2, 599 (1998).
5. W. Franz and Z. Naturforsch. 13a, 484 (1958).

6. L.V. Keldysh, Soviet Phys. JETP 34, 788 (1958).
7. G. F. Koster, J. O. Dimmock, R. G. Wheeler, and H. Statz, "Properties of the Thirty-Two Point Groups", MIT Cambridge MA,(1963).
8. S.-H. Wei and A. Zunger, Appl. Phys. Lett. 64, 1676 (1994).
9. S.-H. Wei and A. Zunger, Phys. Rev. B 57, 8983 (1998).
10. E. Greger, K.H. Gulden, M. Moser, G. Schmiedel, P. Kiesel, and G.H.Döhler, Appl. Phys. Lett. 70, 1459 (1997).
11. S. R. Kurtz, J. Appl. Phys. 74, 4130 (1993).
12. T. Nishino, Y. Inoue, Y. Hamakawa, M. Kondow, and S. Minagawa, Appl. Phys. Lett. 53, 583 (1988).
13. S. H. Kwok, P. Y. Yu, and K. Uchida, Phys. Rev. B 58, R13395 (1998).
14. M. Kozhevnikov, N. Narayanamurti, A. Mascarenhas, Y. Zhang, J.M. Olsen, and D.L. Smith, Appl. Phys. Lett. 75, 1128 (1999).
15. E. Greger, P. Riel, M. Moser, T. Kippenberg, P. Kiesel, and G. H. Döhler, Appl. Phys. Lett. 71, 3245 (1997).
16. E. Greger, T. Kippenberg, P. Kiesel, M. Moser, K.H. Gulden, and G. H. Döhler, SPIE Proc. 3283, 666 (1998).
17. M. Kneissl, P. Kiesel, P. Riel, K. Reingruber, K.H. Gulden, S.U. Dankowski, E. Greger, A. Höfler, B. Knüpfer, X.X. Wu, J.S. Smith, and G.H. Döhler, SPIE Proc. 2139, 115 (1994).
18. T. Kippenberg, J. Krauss, J. Spieler, P. Kiesel, E. Greger, M. Moser, and G.H. Döhler, IEEE Photon. Techn. Lett. 11, 427 (1999).
19. J. Krauss, T. Kippenberg, J. Spieler, P. Kiesel, G.H. Döhler, and M. Moser, Electron. Lett. 35, 1878 (1999).
20. R. Wirth, A. Moritz, C. Geng, F. Scholz, and A. Hangleiter, Appl. Phy. Lett. 69, 2225 (1996).
21. E. Greger, K.H. Gulden, P. Riel, H.P. Schweizer, M. Moser, G. Schmiedel, P. Kiesel, and G.H. Döhler, Appl. Phys. Lett. 68, 2383 (1996).
22. E. Greger, Development of VCSELs and polarization switches based on superlattice ordered GaInP, in Physik Mikrostrukturierter Halbleiter Vol. 5 (ISBN: 3-932392-11-6).

Chapter 14

Phonons in Ordered Semiconductor Alloys
Raman and Infrared Spectroscopy

Angelo Mascarenhas†, Hyeonsik M. Cheong†*, M. J. Seong†, and Francesc
Alsina†**
†*National Renewable Energy Laboratory, Golden, Colorado, U.S.A.*
*Department of Physics, Sogang University, Seoul, Korea
**University of Barcelona, Bella Terra, Spain

Key words: ordering, phonons, Raman, infrared, selection rules

Abstract: The lattice dynamics of ordered semiconductor alloys is described with respect
to studies based on Raman and infrared spectroscopy. The Raman and infrared
spectra reflect the structural change and symmetry of these alloys that are
induced by ordering. Micro-Raman techniques using several different
scattering geometry enable symmetry assignments of all the phonon modes
observed in ordered $GaInP_2$. The dominance of the long-range electrostatic
forces with respect to the short-range interatomic forces is discussed with
respect to the lattice-dynamics of this alloy.

1.   INTRODUCTION

Light scattering studies have made significant contributions to the
understanding of the dynamics of structural phase transformations in
crystalline materials. When a crystal changes structure it always has either
one symmetry or another. Such a phase change may occur discontinuously
through a sudden rearrangement of the atoms in the crystal (first-order phase
change), or the symmetry may be changed by an arbitrarily small
displacement of the atoms from their lattice points (second-order phase
change). At a first-order phase transition, two different states are in
equilibrium, but there is no predictable symmetry relationship between them.
On the other hand, in second-order phase transition the states of the two
phases are the same at the transition point, and therefore the symmetry of the
crystal at the transition point must contain all the symmetry elements of both
phases. It follows that the symmetry of one phase must be higher than that of

the other, and the symmetry group of the lower-symmetry phase must be a subgroup of the symmetry group of the higher-symmetry phase [1]. There are generally two kinds of structural second-order phase transitions, the displacement type and the order-disorder type, depending on how the atoms in the lattice reconstruct. The phase transition in $BaTiO_3$ and the martensite transformation are well-known examples of displacement type transitions. Spontaneous ordering in semiconductor alloys, characterized by an order parameter η that can have any value between 0 and 1, belongs to a second-order phase transition of the order-disorder type.

In the study of phonons in semiconductor alloys, the issue of alloy disorder plays an important role. Theoretical treatment of phonons in alloys is extremely complicated due to the lack of perfect translational symmetry. Experimentally, alloys are classified into two classes, as shown in Figures 1(a) and 1(c), according to the dependence of the optical phonon modes on alloy concentrations [2]. One class exhibits the so-called 'one-mode behavior' in which the $\mathbf{k}=0$ phonon frequencies of each of the phonon modes vary continuously with concentration from the phonon mode frequencies of one end component to that of the other. This class includes such alloys as $Na_xK_{1-x}Cl$, $Ni_xCo_{1-x}O$, $Cd_xZn_{1-x}S$, $ZnSe_xTe_{1-x}$, Se_xTe_{1-x}, and $GaAs_{1-x}Sb_x$ ($x<0.89$). The other class exhibits the 'two-mode behavior' in which for an intermediate mixing ratio two sets of phonon modes occur at frequencies close to those of the end components, the strength of each mode being approximately proportional to the mole fraction of each component. For example, in $Si_{1-x}Ge_x$ alloys two $\mathbf{k}=0$ optical phonon modes, one characteristic of Si and one of Ge, are observed. This class includes such alloys as $Si_{1-x}Ge_x$, CdS_xSe_{1-x}, ZnS_xSe_{1-x}, GaP_xAs_{1-x}, $In_xGa_{1-x}As$, and $Al_xGa_{1-x}As$. The study of the lattice dynamics of spontaneously ordered semiconductor alloys presents a unique opportunity for enhancing the understanding of the effects of statistical alloy fluctuations on the phonon spectra. By varying the order parameter, one can in principle continuously vary the degree of alloy disorder, and thus can map out the effect of alloy fluctuations on the lattice dynamics. In this chapter, the current status of the study of lattice dynamics of spontaneously ordered semiconductor alloys is surveyed, with the emphasis on the most widely studied alloy, $Ga_{0.52}In_{0.48}P$ (abbreviated $GaInP_2$).

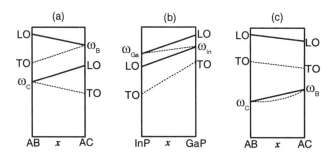

Figure 1. Schematic diagram illustrating different vibrational mode behaviors in ternary alloys $AB_{1-x}C_x$; (a) Two-mode behavior, (b) $GaIn_{1-x}P_x$ as an example of modified two-mode behavior, and (c) one-mode behavior.

The random alloy $Ga_xIn_{1-x}P$ is a mixed III-V semiconductor alloy for which the classification of phonon mode behavior has been the subject of controversy. Experimentally it displays, over the entire concentration range, a pair of longitudinal optical (LO) phonons and a transverse optical (TO) phonon. The presence of an additional transverse mode has been somewhat controversial. Although the analysis of the reflectivity spectra[3] implied the presence of two TO-LO pairs, the second TO mode has not been clearly observed in Raman measurements [4,5]. Galtier *et al.*[6] argued that this second transverse mode is an artifact due to the fact that the models used in the analysis of reflectivity spectra explicitly assumed two oscillators. From an extrapolation of the phonon frequency vs composition curves, Jahne *et al.*[7] on the other hand concluded that the longitudinal modes will show a one-mode behavior. Conversely, Jusserand and Slempkes[8] in a more careful study established that the $Ga_xIn_{1-x}P$ alloy has to be ascribed the 'modified two-mode behavior' in which two LO and two TO modes associated with GaP and InP are present for the entire composition range. It is different from the aforementioned 'two-mode behavior', as schematically illustrated in Figure 1(b) , in that a strong low-frequency TO mode joins the TO mode frequencies of GaP and InP whereas the higher-energy one connects the frequencies of two impurity modes and is expected to be very weak. However, the existence of the second (weaker) TO phonon mode had not been experimentally established until an infrared transmission measurement clearly demonstrated its existence [9].

In CuPt$_B$-ordered GaInP$_2$, the doubling of the unit cell along the ordering axis doubles the number of atoms per unit cell and breaks the cubic symmetry of the zinc-blende structure of the random-alloy GaInP$_2$. When the ordering is single variant, the ordered structure has trigonal symmetry with point group C_{3v}, which is a subgroup of point group T_d for the random alloy of GaInP$_2$ with the cubic zinc-blende structure. As a consequence, ordering changes the phonon spectrum in the following ways: (1) the doubling of the number of atoms in a unit cell doubles the number of phonon branches; (2) the lowering of symmetry lifts some degeneracies in the phonon branches of the random alloy and introduces anisotropy depending on the phonon wave vector with respect to the ordering axis; and (3) the difference in the local arrangement of atoms changes the frequencies of phonons with respect to those of the random alloy. Experimentally, these changes affect the measured Raman or infrared spectra of ordered alloys. Firstly, additional phonon modes that are not observed in the spectra for random alloy appear in the spectra for ordered alloys. Secondly, the frequencies of some phonon modes change with ordering. Thirdly, the observed phonon modes follow the selection rules for the point group C_{3v} and exhibit angular dispersion as the phonon wave vector changes its direction with respect to the uniaxis (ordering direction), whereas the phonons of the random alloy obey the selection rules of the point group T_d and these phonon frequencies are isotropic.

2.   RAMAN SCATTERING STUDIES

In 1988, Suzuki *et al.*[10] observed that an additional phonon peak at 207 cm^{-1} appeared in the Raman spectra of ordered GaInP$_2$ samples, but this phonon mode was not systematically analyzed. In the early 1990s, broad peaks near 207 and 60 cm^{-1} were also observed in GaInP$_2$ samples that exhibited ordering [11]. A confusing aspect due to poorer sample quality at that time was that nominally random alloys of GaInP$_2$ also exhibited features in this spectral range emanating from the activation of zone-boundary phonons due to the relaxation of the **k**=0 rule. Subsequent Raman studies concentrated on the effects of ordering on the line shape of the Raman spectrum [10,12-15]. In these studies, although some changes in the Raman spectrum were observed, no definite evidence of the symmetry change was obtained as expected. With improved sample quality, new peaks in the

Raman spectrum that definitely correlated with ordering were observed almost simultaneously by several groups between 1995 and 1996.[16-20] In these studies, (001)-backscattering Raman spectra of ordered alloys showed up to three new peaks that were not present in the Raman spectrum of the random alloy.

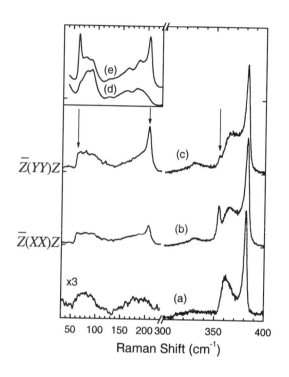

Figure 2. Comparison of the (001) backscattering Raman spectra of (a) random alloy GaInP₂; and (b) and (c) ordered GaInP₂. Three new phonon modes due to ordering are indicated by arrows. In the inset, (110) cleaved edge backscattering Raman spectra of (d) random alloy GaInP₂ and (e) ordered GaInP₂ are displayed.

Figure 2(a) is a typical Raman spectrum of a $GaInP_2$ random alloy sample taken in the (001) backscattering geometry. The strongest and sharpest peak at ~380 cm^{-1} is attributed to the GaP-like LO phonon, the broader peak at ~362 cm^{-1} to the InP-like LO phonon, and the weaker peak at ~330 cm^{-1} to a forbidden TO phonon. In the ordered alloys, the number of optical phonon modes at the zone center doubles as explained in the previous section. In the ideal case of perfect ordering ($\eta = 1$), three LO and three TO modes should exist, while in the real case of partial ordering ($0 < \eta < 1$), where each superlattice layer is a $Ga_xIn_{1-x}P$ [$x = (1\pm\eta)/2$] alloy, four LO and four TO modes are expected due to the modified two-mode behavior of these alloy layers. In addition, folded longitudinal acoustic and folded transverse acoustic phonon modes should appear at the Brillouin-zone center, due to the "folding" of the acoustic branches. Figure 2(b) and 2(c) are Raman spectra of ordered GaInP$_2$ in (001) backscattering geometries with two different polarizations. Three new phonon peaks at 62, 205, and 354 cm^{-1} are observed. The emergence of the folded acoustic phonon modes is distinctly illustrated in the inset, where (110) cleaved edge backscattering Raman spectra of random alloy GaInP$_2$, Figure 2(d), and ordered GaInP$_2$, Figure 2(e), are compared.

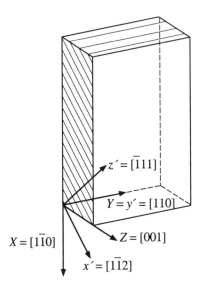

Figure 3. Sample geometry and the coordinate system. The axes *X*, *Y*, and *Z* refer to the two cleavage directions of the samples and the growth direction, respectively. The axes *x′*, *y′*, and *z′* are the principal axes of the ordered structure.

Although the ordering-induced changes are evident in these spectra, the different orientations of the ordering axis and the sample growth axis makes the interpretation of these changes in the (001) backscattering Raman spectrum quite complicated. Since the ordering direction is along the [111]$_B$ direction whereas the samples are grown on (001) substrates, conventional backscattering spectroscopic measurements on the (001) growth surface cannot be used to test the selection rules derived from group theoretical considerations. In this geometry, because the [111]$_B$ ordering axis is neither parallel nor perpendicular to the phonon wave vector ([001] in this geometry), transverse and longitudinal modes are mixed, and thus it is not possible to make assignments of the phonon mode symmetries. In order to overcome this difficulty, it is necessary to perform micro-Raman scattering experiments, where the phonon wave vector can be chosen either parallel or perpendicular to the ordering direction. The principal axes of the ordered structure and the directions used in the following sections are summarized in Figure 3.

2.1 Raman Scattering from the (001) Surface

2.1.1 Optical Phonon Range (300–400 cm^{-1})

Figures 4(a) and (b) show a series of front-surface Raman spectra for eight GaInP$_2$ samples with varying degrees of ordering in the $\bar{Z}(X, X)Z$ and the $\bar{Z}(Y, Y)Z$ configurations, respectively. Here, Porto's notation where $k_i(e_i,e_s)k_s$ refers to the scattering configuration in which k_i and k_s are the propagation directions of the incident and scattered photons, and e_i and e_s are the polarization directions of the incident and scattered photons, respectively, have been used. All samples except for the $\eta = 0$ and 0.23 samples were grown on (001) substrates misoriented by 6° toward [111]$_B$, and thus were single-variant ordered. The $\eta = 0$ and 0.23 samples were grown on (001) substrates misoriented by 6° toward [111]$_A$ For the nominally random alloy sample ($\eta = 0$), the Raman spectrum consists of a GaP-like LO phonon peak at 381 cm^{-1}, an InP-like LO phonon peak at ~362 cm^{-1}, and a broad feature at ~330 cm^{-1} due to TO phonon bands. The TO phonon modes are forbidden in the (001) backscattering for crystals with the T_d symmetry, but are observed here due to the small substrate tilt and fluctuations in the distribution of the cations in the alloy. With the increase in the order parameter η, three major effects are observed: (1) the GaP-like LO phonon peak shifts to higher energy with increasing η. For the most ordered sample ($\eta = 0.54$), this peak is blue-shifted by about 1 cm^{-1} with respect to that in the $\eta = 0$ sample. Although this shift is small compared with the overall line width of the GaP-like LO peak, a definite correlation between this shift and η exists. Note that the line width of this peak does not change appreciably with η. (2) The so-called peak-to-valley ratio in the range of 360 cm^{-1} to 380 cm^{-1} decreases with increasing η, in agreement with earlier measurements [10,12]. (3) For highly ordered samples ($\eta \geq 0.32$) an additional peak appears at 354 cm^{-1}. This peak is barely resolved in the $\bar{Z}(Y, Y)Z$ configuration, but is strongly enhanced in the $\bar{Z}(X, X)Z$ configuration.

The most evident effect of ordering on the phonon spectrum is the emergence of the additional peak at 354 cm^{-1}. This peak is observed for both $\bar{Z}(X, X)Z$ and $\bar{Z}(Y, Y)Z$ configurations, but is strongly enhanced in the first case. It is not observed in the $\bar{Z}(X, Y)Z$ configuration. This peak appears for samples with a high degree of ordering ($\eta \geq 0.32$), but its frequency is independent of η.

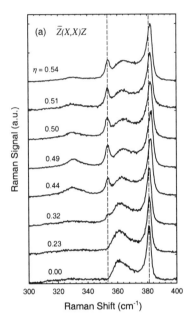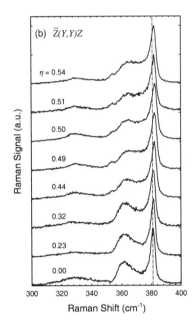

Figure 4. Raman spectra in the optical phonon range for eight GaInP$_2$ samples with varying degrees of ordering, taken in the backscattering geometries (a) $\bar{Z}$ (X, X)Z and (b) $\bar{Z}$ (Y, Y)Z. The order parameter η for each sample is indicated next to each spectrum.

2.1.2   Acoustic Phonon Range (30–250 cm^{-1})

A series of Raman spectra measured in the (001) backscattering geometry in the range of 30–250 cm^{-1} are shown in Figures 5(a) and (b). For the $\eta = 0$ sample, broad features due to disorder-activated zone-boundary longitudinal-acoustic phonons (DALA) are observed. With increasing order parameter, two sharp features appear in the spectrum: a sharp peak appears at 205 cm^{-1}, and another feature at 60 cm^{-1} appears as a sharp edge in the spectrum. The frequency of the peak at 205 cm^{-1} is close to the simple mean of the L-point longitudinal acoustic (LA) phonon frequencies of GaP and InP. This peak is attributed to an optical mode arising from the folded LA phonon branch. The frequency of the feature at 60 cm^{-1} is close to the mean of the L-point transverse acoustic (TA) phonon frequencies of GaP and InP. This peak is attributed to the folded TA phonon branch.

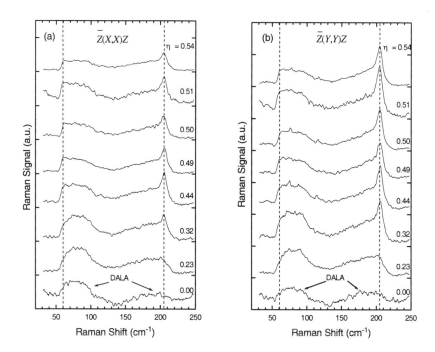

Figure 5. Raman spectra in the acoustic-phonon range for the eight samples, taken in the backscattering geometries, (a) $\bar{Z}(X, X)Z$ and (b) $\bar{Z}(Y, Y)Z$.

2.1.3 $Ga_{1-x}In_xAs$

(001) backscattering data clearly demonstrate the effect of spontaneous ordering on the phonon spectrum of $GaInP_2$ alloys. Similar effects of $CuPt_B$ ordering on phonon spectra have been observed in $Ga_{0.47}In_{0.53}As$ alloys grown on (001) InP substrates by OMVPE [21]. Figure 6 shows Raman spectra of three $Ga_{0.47}In_{0.53}As$ samples, measured in the $\bar{Z}(X, X)Z$ and the $\bar{Z}(Y, Y)Z$ configurations. By comparing transmission electron microscopy data and polarized low-temperature photoluminescence, it was determined that sample 1 was a nominally random alloy, sample 3 exhibited the strongest ordering, and sample 2 an intermediate ordering. The peaks labeled A and B arise from GaAs-like and InAs-like LO phonons, respectively. The origin of the peak labeled C has not yet been identified. With ordering, the peak labeled D, which emerges at 181 cm^{-1}, has been attributed to ordering-induced folding of the longitudinal acoustic phonon branches.

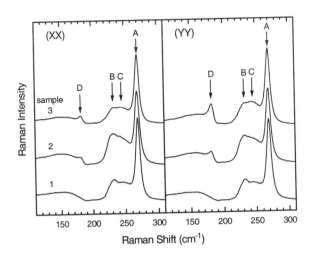

Figure 6. Polarized Raman spectra of $Ga_{0.47}In_{0.53}As$ alloys with varying degrees of ordering taken in the $\overline{Z}\,(X, X)Z$ and $\overline{Z}\,(Y, Y)Z$ polarizations.

2.1.4   (001) Raman Scattering of Ultra-Thin Ordered Layers

Since Raman scattering directly reflects the lattice symmetry of the solid being studied, it has been found to be a very sensitive probe of ordering, especially for very thin films. Raman scattering has been used to study the initial stages of growth in $CuPt_B$ ordered $Ga_{0.47}In_{0.53}As$ and $GaInP_2$ alloys [22]. It was demonstrated that spontaneous ordering in extremely thin films of $Ga_{0.47}In_{0.53}As$ and $GaInP_2$ could be detected by the emergence of ordering-induced phonon peaks in the Raman spectra for films as thin as 5 nm. This sensitivity far exceeds that of other means of detecting ordering, such as transmission electron microscopy or polarized photoluminescence.

2.2   Micro-Raman Scattering

In a polar crystal such as GaAs or $GaInP_2$, the frequency of an optical phonon is split into longitudinal and transverse components by the macroscopic electric field associated with the longitudinal phonon. This electric field serves to stiffen the force constant of the phonon and thereby raise the frequency of the LO over that of the TO. In a cubic zinc-blende crystal with large LO-TO splitting, the phonon spectrum consists of an LO mode and a twofold-degenerate TO mode. The situation is far more complicated in uniaxial crystals, as it becomes necessary to consider simultaneously two independent forces: the long-range electrostatic forces responsible for the LO–TO splitting, and the short-range inter-atomic forces, which exhibit the anisotropy of the force constants [1]. In a trigonal crystal, if the phonon propagation direction $\mathbf{q}$ is along one of the crystal principal axes x', y', or z', where the z'-axis coincides with the trigonal axis, only pure longitudinal and pure transverse phonons of well defined symmetry character are observed in Raman scattering. For $\mathbf{q}$ parallel to the z'-axis, the E modes are transverse with common frequency $\omega_{TO}(E)$ whereas the A_1 mode is longitudinal with frequency $\omega_{LO}(A_1)$. Similarly, for $\mathbf{q}$ perpendicular to the z'-axis, the A_1 mode is transverse with frequency $\omega_{TO}(A_1)$, one of the E modes is transverse with frequency $\omega_{TO}(E)$, and the other E mode is longitudinal with frequency $\omega_{LO}(E)$. When the phonon propagation is not along one of the principal axes, the situation becomes more complicated and careful consideration must be given to the competition between the effects due to the long-range (LO–TO splitting) and short-range (A_1–E splitting) forces. One can consider two extreme cases. If the electrostatic forces dominate over the anisotropy in the short-range forces, the LO–TO splitting is larger than the A_1–E splitting. This implies that the phonons under consideration will always be almost purely transverse or pure longitudinal independently of their propagation direction. For a general direction of propagation, there are three nondegenerate modes, and at least two of them have neither purely A_1 nor purely E symmetry. When the propagation is, for example, in the $x'z'$ plane, a mixing between the A_1 and $E_{x'}$ modes occurs. The phonons are pure transverse and pure longitudinal, but they have mixed A_1 and E symmetry character, since the polarization direction of the phonons are intermediate between the A_1 polarization direction z' and the E polarization direction x'. Their frequencies are located between the pure $A1$ and E frequencies. The remaining mode with y' polarization, $E_{y'}$, is a pure transverse one, and does not mix with the other two polarizations. Its frequency also remains unchanged. On the other hand, if the short-range forces dominate over the long-range forces, the splitting between the A_1 and E symmetry modes will be much larger than the LO–TO frequency splitting. This implies that the phonons under consideration will always almost maintain their A_1 or E symmetry character regardless of their propagation directions. Consider for example, the scattering from phonons propagating in the $x'z'$ plane. When an A_1 phonon, which is polarized in the z' direction, propagates along the z' direction, only the A_1-LO mode can be observed. In an intermediate direction between x' and z', since the strong interatomic forces require the polarization to be maintained in the z' direction, a pure longitudinal or transverse mode no longer exists, and its frequency is located between the pure longitudinal and pure transverse A_1 frequencies. When the observed phonon is propagating along the x' direction, it will be purely transverse A_1-TO. E phonons, polarized along the x' direction, will undergo a similar shift from transverse E-TO, when the phonon propagation is along z', to longitudinal E-LO, for propagation along x', maintaining their E symmetry character. Finally, E phonons polarized along the y' direction, perpendicular to the $x'z'$ plane, will always be purely transverse with frequency $\omega_{TO}(E)$.

The CuPt$_B$-ordered GaInP$_2$ crystal structure is uniaxial with the [111]$_B$ ordering axis as the uniaxis z'. In the (001) backscattering geometry, because this axis is neither parallel nor perpendicular to the phonon wave vector q, which is along the [001] direction in this geometry, symmetry and polar characters of the phonon modes are mixed, and therefore it is not possible to make definite assignments of the phonon modes without *a priori* knowledge of the relative strengths of the LO–TO splitting and the A_1–E splitting. Angular dispersion of phonon modes is a direct consequence of force constant anisotropies in a uniaxial crystal, providing important clues regarding whether the long-range electrostatic force or the short-range interatomic force dominates the lattice dynamics of ordered GaInP$_2$. The effect of anisotropy of phonon modes becomes maximum as q changes direction from $\phi=0°$ (parallel to z') to $\phi=90°$ (perpendicular to z'). Therefore, in order to make phonon mode assignments based on the C_{3v} symmetry of the ordered crystal and probe its anisotropy, it is necessary to perform Raman scattering measurements in geometries with q either parallel or perpendicular to the ordering axis [23].

In the C_{3v} symmetry, the three Raman scattering tensors A_1, $E_{x'}$, and $E_{y'}$ that represent vibrations along z', x', and y', respectively, are [1]:

$$\begin{pmatrix} a & 0 & 0 \\ 0 & a & 0 \\ 0 & 0 & b \end{pmatrix}, \begin{pmatrix} c & 0 & d \\ 0 & -c & 0 \\ d & 0 & 0 \end{pmatrix}, \text{and} \begin{pmatrix} 0 & -c & 0 \\ -c & 0 & d \\ 0 & d & 0 \end{pmatrix}. \tag{1}$$

The Raman scattering efficiency can be written as

$$S \propto \left| \mathbf{e}_i R \mathbf{e}_s \right|^2, \tag{2}$$

where $\mathbf{e}_i$ and $\mathbf{e}_s$ are the unit vectors along the polarizations of excitation and scattered light, respectively, and R is the tensor for a particular mode. Table 1 summarizes the values of $\left| \mathbf{e}_i R \mathbf{e}_s \right|^2$ for the scattering geometries used in this work.

Table 1. The C_{3v} Raman-scattering efficiency calculated with Eqs. (1) and (2) for the scattering geometries used in micro-Raman studies. θ is the angle between the z' direction and the direction normal to the growth surface

	A_1	E_x	E_y
$\bar{z}'(y',y')z'$, $\bar{z}'(x',x')z'$	a^2	c^2	0
$\bar{z}'(x',y')z'$, $\bar{z}'(y',x')z'$	0	0	c^2
$\bar{Z}(X,Z)\bar{X}$	$(a-b)^2 \sin^2\theta \cos^2\theta$	$\{c\sin\theta\cos\theta + d(\cos^2\theta - \sin^2\theta)\}^2$	0
$\bar{Z}(Y,Y)\bar{X}$	a^2	c^2	0
$\bar{Z}(Y,Z)\bar{X}$	0	0	$(c\sin\theta - d\cos\theta)^2$
$\bar{Z}(X,Y)\bar{X}$	0	0	$(c\cos\theta + d\sin\theta)^2$
$y'(z',z')\bar{y}'$	b^2	0	0
$y'(x',x')\bar{y}'$	a^2	c^2	0
$y'(x',z')\bar{y}'$, $y'(z',x')\bar{y}'$	0	d^2	0
$\bar{z}'(y',y')z'$, $\bar{z}'(x',x')z'$	a^2	c^2	0

2.2.1   ($\bar{1}$11) Backscattering

($\bar{1}$11) backscattering measurements were performed on a ($\bar{1}$11) surface of a 10-μm-thick ordered GaInP$_2$ sample ($\eta = 0.46\pm0.02$) [23]. The ($\bar{1}$11) surface was prepared by mechanically polishing the sample using 50-nm grit size colloidal silica. Since q is along the z' direction in this scattering geometry, the A_1 tensor represents longitudinal modes and the $E_{x'}$, and $E_{y'}$ tensors represent transverse modes. Thus, according to Table 1, both longitudinal and transverse modes are allowed in parallel polarizations [$\bar{z}'(x',\ x')\ z'$ or $\bar{z}'(y',\ y')\ z'$], while only transverse modes are allowed in crossed polarizations [$\bar{z}'(x',\ y')\ z'$ or $\bar{z}'(y',\ x')\ z'$].

Figure 7 shows Raman spectra measured in this geometry for four polarization configurations. Only the TO phonon peak at 328 cm^{-1} appears strongly in the crossed polarizations, while the InP- and GaP-like LO phonon peaks and the new peak at ~205 cm^{-1} also appear in the parallel polarizations. This is consistent with the interpretation of the 205 cm^{-1} peak as a folded longitudinal acoustic phonon mode with A_1 symmetry. The additional peak at 354 cm^{-1} is not resolved in any of these spectra. This is also consistent with the (001) backscattering result where this peak is strong and clearly resolved only when the polarization has a z' component. In the present geometry, the polarization has no z' component. For crossed polarizations, weak features appear at the frequencies of the LO phonons.

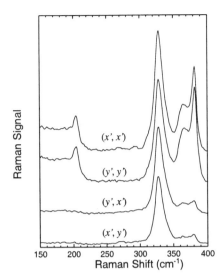

Figure 7. Raman spectra of ordered GaInP$_2$ taken in the ($\bar{1}$ 11) backscattering geometry in four different polarization configurations.

These nominally forbidden peaks appear weakly because q has some component orthogonal to z' due to the large angle of acceptance of the micro-Raman optics.

2.2.2   Right-Angle Scattering

For excitation in the $\bar{Z}$ direction and detection in the $\bar{X}$ direction, q is at 45° relative to the Z direction. For a sample grown on a 6° misoriented substrate, the angle between the z' and Z axes is 48.7°. Therefore, in this geometry it is a good approximation to assume that q is parallel to z', and as in ($\bar{1}$ 11) backscattering, the A_1 tensor represents longitudinal modes and the $E_{x'}$, and $E_{y'}$ tensors represent transverse modes. Table 1 lists the scattering efficiencies using $X = (\cos\theta, 0, -\sin\theta)$, $Y = (0, 1, 0)$, and $Z = (\sin\theta, 0, \cos\theta)$, where θ is the angle between the z' and Z axes (48.7° ≈ 45°).
Figure 8 shows four spectra taken in the four possible polarization configurations. In this geometry, the detected signal was a few orders of magnitude stronger than that for the backscattering case owing to the larger interaction volume. In the $\bar{Z}$ (X, Y) $\bar{X}$ configuration, where only TO modes are allowed, a very weak signal was detected for the TO phonon peak at 328 cm^{-1}. This implies, according to Table 1, that $c \cong -d\tan\theta \approx -d$ ($\theta \approx 45°$) for this mode. In the $\bar{Z}$ (Y, Z) $\bar{X}$ configuration, only the TO

phonon peak was detected as expected. Here, unlike the case of the $(\bar{1}11)$ backscattering, no signal from the forbidden LO phonon peaks was detected. This is due to the difference in the geometry of the optical set up; unlike the backscattering geometry, the excitation beam is almost perfectly parallel to the $\bar{Z}$ direction, and the microscope optics preferentially collect scattered photons nearly parallel to the $\bar{X}$ direction due to the high index of refraction of the material and the ~100 μm small distance between the incident beam path and the top $(\bar{1}\,10)$ surface of the sample. It should also be noted that there is no indication of a TO phonon peak between 362 and 380 cm^{-1} in this spectrum.

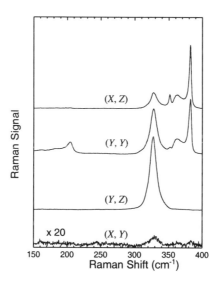

Figure 8. Raman spectra taken in the right-angle scattering geometry where the excitation is in the $\bar{Z}$ direction and the detection in the $\bar{X}$ direction.

In the $\bar{Z}\,(X,\,Z)\,\bar{X}$ and $\bar{Z}\,(Y,\,Y)\,\bar{X}$ configurations, both longitudinal and transverse modes are allowed, and the spectra show the additional peaks at ~354 cm^{-1} and at ~205 cm^{-1} in addition to the two LO phonon peaks and the TO phonon peak. This, combined with the behavior of this mode in the $\bar{Z}\,(Y,\,Z)\,\bar{X}$ and the $\bar{Z}\,(X,\,Y)\,\bar{X}$ configurations, is consistent with the result for $(\bar{1}11)$ backscattering which assigned the 205 cm^{-1} peak as an A_1 symmetry longitudinal mode. The 354 cm^{-1} peak is also an A_1 mode, because if it were an $E_{x'}$ mode, it should be about 4 times stronger in the $\bar{Z}\,(Y,\,Y)\,\bar{X}$ configuration than in the $\bar{Z}\,(X,\,Z)\,\bar{X}$ configuration ($\sin\theta \approx \cos\theta \approx 1/\sqrt{2}$). By comparing the intensities of these peaks with Table 1, one can estimate the relative magnitude of some of the tensor components of Eq. (1) for each mode. The 205 cm^{-1} mode is strong in the $\bar{Z}$ (Y, Y) $\bar{X}$, $\bar{z'}$ (x', x') z', and $\bar{z'}$ (y', y') z' configurations and weak in the $\bar{Z}$ (X, Z) $\bar{X}$ configuration. Therefore, one can deduce that $|a-b| << 2|a|$ ($\theta \approx 45°$), or equivalently, $a \approx b$ for this mode. The 354 cm^{-1} peak shows an opposite behavior, and therefore $|a-b| >> 2|a|$ for this mode. In order to check for possible effects of resonance with the band gap (~1.81eV at room temperature), the excitation energy was varied from 1.675 eV (7400 Å) to 1.785 eV (6946 Å), but the above results were consistent throughout the excitation energy range.

2.2.3   (110) Backscattering

In the geometry for backscattering from the (110) cleaved surface, q is along the y' direction; the $E_{y'}$ tensor represents longitudinal modes and the $E_{x'}$ and A_1 tensors represent transverse modes. Therefore, only transverse modes are allowed in any of the four polarization configurations according to Table 1.

Figure 9 shows Raman spectra measured in this geometry. The TO phonon peak at 328 cm^{-1} is present in all four spectra as expected. In addition, the forbidden $E_{y'}$ symmetry LO phonon peaks at 362 and 380 cm^{-1}, and the peak at 205 cm^{-1} are clearly observed in the parallel polarizations. Some signal from the forbidden modes is expected due to the geometry of the micro-Raman optics as discussed earlier for the case of the ($\bar{1}$ 11) backscattering. However, if the LO phonon signal in these spectra is solely due to the geometry of the optical setup, the intensities of the forbidden LO peaks would be the same for the two parallel polarizations; for example, very weak signal was observed from the forbidden LO phonon mode from the GaAs substrate of this sample and its intensity did not depend on the direction of the polarizations. In this case, however, there is a clear directionality; the forbidden modes are much stronger for the $y'(x', x')\bar{y'}$ configuration than for the $y'(z', z')\bar{y'}$ configuration. This also rules out a breakdown of translational symmetry due to random disorder as the cause of this selection-rule violation. Two possible mechanisms for this violation are the q-dependent Fröhlich interaction and the Fröhlich interaction due to the electric field induced by surface charges [24,25]. Both these forbidden scattering mechanisms are dependent on the excitation energy with enhancement near the resonance with an interband transition. In this experiment, the excitation energy (2.41 eV) was not far from the interband transition between the spin-orbit valence band and the folded L band state (L_{1c}'), which has energy ~2.3 eV [26]. In order to check this possibility, the (110) backscattering measurements were repeated using the 4880 Å line

(2.54 eV) as the excitation source. The forbidden scattering should be weaker for the 4880 Å excitation.

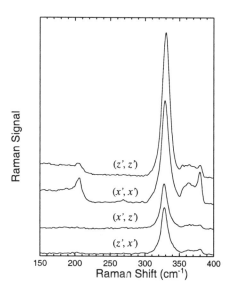

Figure 9. Raman spectra taken in the (110) backscattering geometry (Excitation energy: 2.41 eV).

Figure 10 compares two Raman spectra taken in the $y'(x', x')\bar{y}'$ configuration under nominally identical conditions except for the excitation wavelength. The overall Raman signal was weaker for the 4880 Å probably due to a combination of being off-resonance and the fact that our spectroscopy system was optimized for the 5145 Å excitation. In Figure 10, the spectra were normalized for the TO phonon peak for comparison. It is seen that the relative intensity of the LO phonon band at 360–385 cm^{-1} is significantly smaller for the 4880 Å excitation. A similar effect was observed for the $y'(z', z')\bar{y}'$ configuration. This is consistent with the above assumption that these peaks appear due to forbidden scattering. It should be pointed out that one would not observe selection-rule violations due to these forbidden scatterings in geometries with $q // z'$, because in those cases, A$_1$ longitudinal modes are already selection-rule allowed when these Fröhlich interactions are allowed (parallel polarizations).

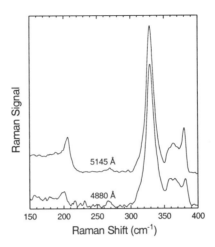

Figure 10. Comparison of Raman spectra taken in the (110) backscattering geometry with 5145- and 4880-Å excitations.

The nature of the 205 cm^{-1} phonon peak in the (110) backscattering geometry has been the subject of much controversy. Mestres *et al.* interpreted this peak as being an A$_1$ symmetry mode [27]. However, it is seen in Figure 9 that the relative intensity of the 205-cm^{-1} peak is much smaller for the 4880 Å excitation, which implies that this mode has similar characteristics as the GaP- and InP-like LO phonons at 360–385 cm^{-1}, which are E$_{y'}$ longitudinal modes in this scattering geometry. The experimental result was repeated in resonance Raman scattering studies on (110) cleaved surfaces [28]. Figure 11 shows a series of Raman spectra for excitation energies near the fundamental band gap (~1.80 eV) of an ordered GaInP$_2$ sample. It shows that the 205 cm^{-1} mode, marked FLA, is strongly enhanced relative to the intensity of the allowed TO phonon peak as the excitation energy approaches the band gap. This result is consistent with the result of the comparison of the Raman spectra for 5145- and 4880-Å excitations, and indicates that in the (110) backscattering geometry the 205 cm^{-1} mode is a forbidden longitudinal mode, which becomes allowed due to the resonance effect.

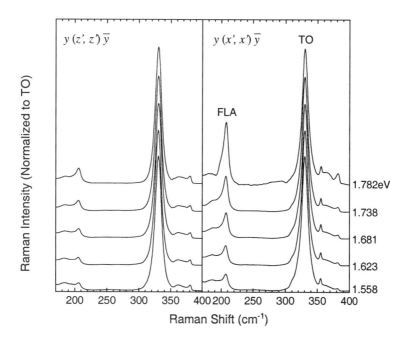

Figure 11. Excitation energy dependence of Raman spectra in (110) backscattering. The excitation energies are indicated on the right.

2.2.4   Anisotropy of the Force Constants

The emergence of a uniaxis upon spontaneous ordering in GaInP$_2$ inevitably introduces anisotropy of force constants, leading to angular dispersion of phonon modes for the angle ϕ between the phonon wave vector q and the ordering axis z'. A set of careful micro-Raman measurements has established that the anisotropy of the short-range interatomic forces (A$_1$–E splitting) is very small compared with the long-range electrostatic forces (LO-TO splitting) [29], unambiguously providing additional confirmation for Cheong *et al.*'s interpretation of the 354 cm^{-1} phonon, [23] discussed in the previous section, as opposed to the assignment proposed by Mintairov *et al.* [30-32]

In the backscattering geometry on the (110) cleaved surface, the phonon wave vector q is perpendicular to the ordering axis; the E$_{y'}$ tensor represents longitudinal modes and the transverse modes are represented by E$_{x'}$ and A$_1$ tensors. According to Table 1, only transverse modes are allowed in any of the four polarization configurations. Furthermore, it is important to note that only the A$_1$(TO) mode is allowed in $y'(z', z')\bar{y}'$ configuration whereas E$_{x'}$(TO) is the only allowed mode in $y'(z', x')\bar{y}'$ configuration. Hence, any difference in the TO phonon frequency between the two polarization configurations is due to the ordering-induced anisotropy (A_1-E splitting). Two Raman spectra with $y'(z', z')\overline{y}'$ and $y'(z', x')\overline{y}'$ are displayed in Figure 12; the A_1(TO) phonon energy is observed at 330.1 cm^{-1}, ~ 2.1 cm^{-1} higher than that of $E_{x'}$(TO). As discussed in the previous section (2.1.1), the bonds along the ordering axis gets stiffer as the order parameter increases [20]. Since A_1(TO) represents a vibration along the ordering axis and $E_{x'}$(TO) represents a vibration perpendicular to it, one would expect the phonon frequency for A_1(TO) to be higher than that for $E_{x'}$(TO). As the angle ϕ changes from 0° to 90°, the A_1(TO) frequency shows a continuous variation from the degenerate value (328.0 cm^{-1}) to the maximally split value (330.1 cm^{-1}) while the $E_{x'}$(TO) mode does not exhibit any change, which is illustrated in the inset of Figure 12. The distinct difference observed in the TO phonon energy between the two Raman spectra provides clear evidence of a small anisotropy induced by spontaneous ordering in the GaInP$_2$ crystal.

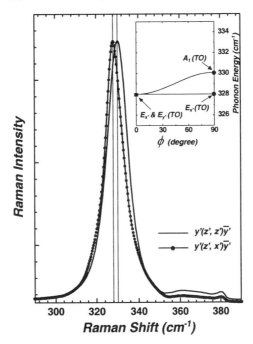

Figure 12. Raman spectra measured in $y'(z', z')\overline{y}'$ (solid line) and $y'(z', x')\overline{y}'$ (dotted solid line) polarization configurations. The $y'(z', x')\overline{y}'$ spectrum is enlarged by a factor of 4 to facilitate the comparison of the TO phonon signatures. Angular dispersion of the TO modes is illustrated in the inset, where the two circular dots indicate the TO modes experimentally observed, the error bars (~0.32 cm^{-1}) are slightly smaller than the size of these dots. The continuous angular variation is indicated with a simple guide line representing $\sqrt{330.1^2 \sin^2 \phi + 328.0^2 \cos^2 \phi}$.

In the $y'(x', x')\bar{y}'$ geometry, both $A_1(TO)$ and $E_{x'}(TO)$ are allowed with Raman efficiencies a^2 and c^2, respectively, making the TO phonon signature in this scattering geometry a superposition of the two TO modes. As shown in Figure 13, the peak position of the TO phonon signature for $y'(x', x')\bar{y}'$ is observed at a Raman shift smaller than that in $y'(z', z')\bar{y}'$ by ~1.3 cm^{-1}, which is another confirmation of the anisotropy of the TO modes manifested in Figure 12. Although LO phonon Raman scattering is forbidden in this (110)-backscattering geometry, small signatures of the LO modes around ~360 cm^{-1} and ~380 cm^{-1} are observed and can be partly due to the geometry of the micro-Raman optics employed in this study as discussed earlier. However, there is a clear directionality; the forbidden modes are distinctly stronger for the $y'(x', x')\bar{y}'$ configuration than for the $y'(z', z')\bar{y}'$ and the 354 cm^{-1}-mode labeled with an asterisk exhibits a striking difference in its selection rule from the LO phonon signature around ~380 cm^{-1}. Since the excitation energy, ~1.69 eV, is not far from the fundamental band gap of the sample, ~1.80 eV, at room temperature this indicates that the Fröhlich interaction is involved, making these forbidden modes partially observable [28]. In addition to the forbidden LO phonon signatures, a shoulder-like feature identified with an arrow, which is absent in the $y'(z', z')\bar{y}'$ configuration, is observed at ~310 cm^{-1} in the $y'(x', x')\bar{y}'$ configuration. Although a point group analysis for a crystal of C_{3v} symmetry yields nine Raman-active modes, i.e. $3A_1(z)+3\{E(x)+E(y)\}$, at the zone-center of the perfectly CuPt-ordered GaInP$_2$, only seven Raman modes have been experimentally identified until recently. Since the "*missing modes*" originate from the zone boundary TO phonon (*L*-point) of a random GaInP$_2$ alloy and become Raman-active due to Brillouin zone folding induced by spontaneous ordering, one would expect that their frequency should be slightly lower than that of the TO mode at ~330 cm^{-1} which evolves from the zone center TO phonon (Γ-point) of a random GaInP$_2$ alloy. Therefore, the mode observed at ~310 cm^{-1} in Figure 13 can be identified with the "*missing modes*" represented by the $E_{x'}$ and $E_{y'}$ tensors.

In order to probe phonons whose wave vector q are along the ordering axis, a right-angle scattering geometry is employed where the incident light is in the $\bar{Z}$ direction and the scattered light in the $\bar{X}$ direction. As discussed in section 2.2.2, in this geometry it is a good approximation to assume that q is parallel to z' and the A_1 tensor represents longitudinal modes and the $E_{x'}$ and $E_{y'}$ tensors transverse modes. The Raman spectrum measured in $\bar{Z}(X, Z)\bar{X}$ configuration, where both A_1 and $E_{x'}$ modes are allowed, is displayed and compared to that measured in the $y'(x', x')\bar{y}'$ configuration in Figure 14. The anisotropy of the LO phonon mode at ~380 cm^{-1} is clearly evidenced by its unmistakable -1.2 cm^{-1} phonon frequency shift as it changes from a $A_1(LO)$ mode in the $\bar{Z}(X, Z)\bar{X}$ configuration to an $E_{y'}(LO)$ mode in the

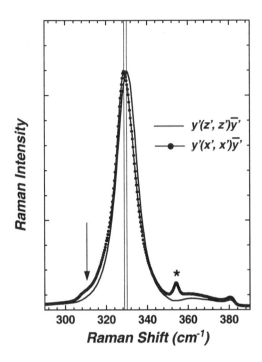

Figure 13. Raman spectra measured in $y'(z', z')\bar{y}'$ (solid line) and $y'(x', x')\bar{y}'$ (dotted solid line) polarization configurations. The $y'(z', z')\bar{y}'$ spectrum is enlarged by a factor of 2 to facilitate the comparison of the TO phonon signatures.

$y'(x', x')\bar{y}'$ configuration. The frequency change of the LO mode at ~380 cm^{-1} can be understood in terms of a stiffening of the bond along the ordering axis as discussed earlier in the case of TO mode at ~330 cm^{-1}. In surprising contrast, the phonon mode at ~354 cm^{-1}, whose origin can be traced to the folding of LO phonon dispersion, exhibits an even larger shift in the opposite direction, + 2.5 cm^{-1}, as the angle ϕ between q and the ordering axis changes from 0° to 90°. The angular dispersions for the two phonon modes were further confirmed by measuring the Raman spectrum in the $X(y', y')\bar{X}$ scattering configuration where $\phi \cong 45°$ and the corresponding Raman frequencies are indicated with circular dots around $\phi \cong 45°$ in the inset of Figure 14.

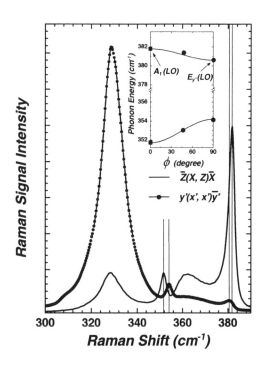

Figure 14. Raman spectra measured in $y'(x', x')\overline{y}'$ (dotted solid line) and $\overline{Z}$ (X, Z) $\overline{X}$ (solid line) scattering geometries. Angular dispersions of the two phonon modes are illustrated in the inset, where circular and square dots represent the experimentally observed modes, using a similar guide line as used in Figure 12.

In summary, comprehensive Raman studies on partially ordered GaInP$_2$ in various scattering configurations including micro-Raman measurements, where the phonon wave vector $\boldsymbol{q}$ is either parallel or perpendicular to the ordering axis z', have revealed four additional distinct vibrational modes, observed at 60, 205, 310, and 354 cm^{-1}, that are absent in the random GaInP$_2$ alloy. The emergence of these new phonon modes and their Raman selection rules can be adequately explained using the ordering-induced crystal symmetry change from T_d (cubic) to C_{3v} (trigonal). A small but unmistakable angular dispersion, originating from the force constant anisotropy induced by ordering, was observed for some of the phonons, providing strong evidence for a small A$_1$-E splitting in comparison to a large LO-TO splitting for these modes.

3.   INFRARED STUDIES

In addition to Raman scattering, infrared (IR) spectroscopy can also be used to investigate the vibrational properties of crystals, often providing complementary information to that obtained by Raman scattering. The IR reflectance is sensitive to the longitudinal A_1 and E vibrations, even when measured in a nonprincipal plane. The IR reflectance spectrum of a $Ga_xIn_{1-x}P$ random alloys exhibits two Restrahlen bands for any alloy composition, and each band can be associated with a longitudinal-transverse (LO-TO) vibrational mode pair belonging to an end-compound [3,7]. On the other hand, since only transverse modes are IR active in an IR transmission spectrum measured in a normal incidence geometry, one may use IR transmission to identify transverse modes by comparing the IR transmission spectrum with the IR reflection and Raman spectra. This section will summarize results of IR reflection[33] and transmission[9] studies on spontaneously ordered $GaInP_2$ alloys.

3.1   IR Reflection

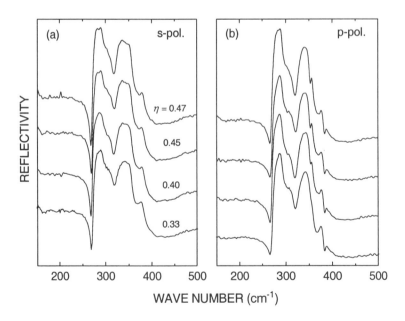

Figure 15. Experimental reflectance spectra of ordered $GaInP_2$ samples with 45° incidence angle, for: (a) *s* polarization and (b) *p* polarization. The estimated order parameter (η) for each sample is indicated.

Figure 15 shows a series of reflectance spectra measured from different samples having increasing values of η, for both s and p [34] light polarizations. Development of a feature at ~355 cm^{-1} with increasing ordering is clearly seen in the p-polarization, but not in the s polarization. The fact that ordering is partial complicates analysis of the data. Zhang *et al.* [35] showed that simple η^2 interpolation scheme, argued by Laks *et al.*[36], is not generally adequate to account for electronic properties of partially ordered alloys. However, since a similar theoretical model calculation has not been done for the lattice dynamics of partially ordered alloys, the dielectric function of partially ordered alloys can be approximated in its simplest form using

$$\frac{\hat{\varepsilon}(\eta)-1}{\hat{\varepsilon}(\eta)+2}=\eta^2\frac{\hat{\varepsilon}_{\text{ord}}-1}{\hat{\varepsilon}_{\text{ord}}+2}+(1-\eta^2)\frac{\varepsilon_{\text{dis}}-1}{\varepsilon_{\text{dis}}+2},$$

where $\hat{\varepsilon}_{\text{ord}}$ is the dielectric tensor for perfectly ordered GaInP$_2$,

$$\hat{\varepsilon}_{\text{ord}}=\begin{pmatrix}\varepsilon_\perp & & \\ & \varepsilon_\perp & \\ & & \varepsilon_\parallel\end{pmatrix},$$

and ε_{dis} is the dielectric function of the disordered alloy. Here, $\varepsilon_\perp$ and $\varepsilon_\parallel$ refer to dielectric tensor components perpendicular and parallel to the ordering axis, respectively. The experimental data can then be simulated by choosing appropriate dielectric functions $\varepsilon_\perp(\eta)$ and $\varepsilon_\parallel(\eta)$. A detailed account of the analysis is omitted here, but interested readers are encouraged to read Ref. [33].

Figure 16 shows experimental and calculated reflectivity for the $\eta=0.45$ sample for the s and p polarizations. The calculated spectra reproduce all the major features of the experimental spectra. Figure 17 shows the dielectric functions used to fit the data in Figure 16. The imaginary part of $\hat{\varepsilon}(\eta)$ reproduce the TO component at ~330 cm^{-1}, and there appears to be a weak TO component near 370 cm^{-1}. This TO mode near 370 cm^{-1} was not observed in Raman spectra, as discussed earlier. The imaginary part of $-1/\hat{\varepsilon}(\eta)$, on the other hand, reproduces the GaP-like and InP-like LO phonons and the additional phonon at 354 cm^{-1}. These results are consistent with the Raman scattering data.

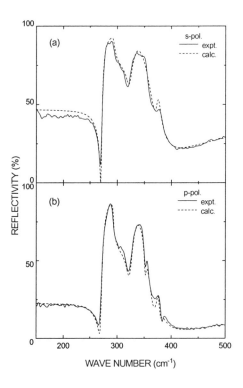

Figure 16. Experimental (solid line) and calculated (dashed line) reflectance spectra of a partially ordered GaInP/GaAs sample with 45° incidence angle, for: (a) s polarization and (b) p polarizations.

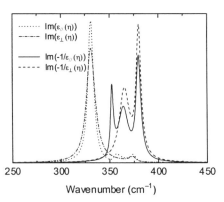

Figure 17. Components of the dielectric functions for a partially ordered GaInP2, where $\epsilon_\perp$ and $\epsilon_\parallel$ are dielectric tensor components perpendicular and parallel to the ordering axis, respectively, obtained from the simultaneous fit analysis of the two polarized spectra shown in Figure 16.

3.2   IR Transmission

Whereas the frequency and strength of the TO modes can be deduced through fitting processes for reflectance data, as illustrated in the preceding section, they can be directly observed in IR transmission spectra. Figure 18 shows IR transmission spectra of a nominally random alloy (disordered) $GaInP_2$ sample, taken in a normal incidence geometry with polarizations along the [110] and [1 $\bar{1}$ 0] directions. At normal incidence, only TO modes are IR active. Since the random alloy $GaInP_2$ has cubic T_d symmetry, the two spectra for the two polarizations are essentially identical. The calculated spectrum reproduces the main features of the experimental spectra very well. The two local minima in transmittance near ~330 cm^{-1} and ~372 cm^{-1} correspond to the two TO phonon modes. Unlike the reflectance studies, no fitting process is necessary to 'detect' the presence of a TO phonon mode in transmission.

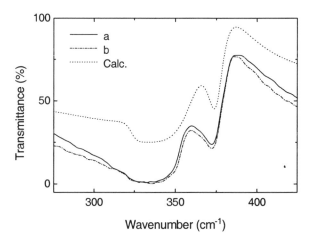

Figure 18. Transmission spectra at normal incidence of disordered $GaInP_2$, for polarization along (a) [110] and (b) [1 $\bar{1}$ 0] directions. The calculated transmission spectrum (dotted curve) is shifted for clarity.

Figure 19 compares transmission spectra of an ordered $GaInP_2$ sample with $\eta = \sim 0.5$ in two polarizations: the *s*-polarization where the polarization is along the [110] direction and the *p*-polarization with the polarization along the [1 $\bar{1}$ 0] direction. In the *s*-polarization, two minima at 332 and 372 cm^{-1} are observed as in the case of the random alloy. In the *p*-polarization, an additional minimum at 354 cm^{-1} is observed. This reflects the mixed LO-TO character of this mode and is consistent with the results of Raman scattering measurements.

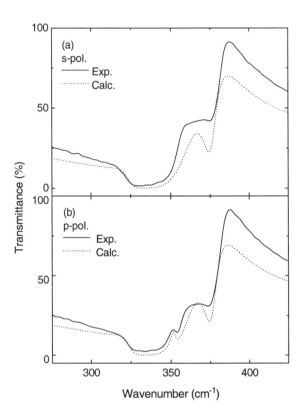

Figure 19. Experimental and calculated transmittance spectra of the ordered GaInP$_2$.

4.   PHONONS IN DOUBLE VARIANT ORDERED GaInP$_2$

Although the effect of single variant ordering on the lattice vibration has been extensively studied for the simple CuPt structure, [18-20] the additional effects which may arise due to the formation of an orientational superlattice (OSL) type structure are less known [37,38]. Three CuPt ordering induced new Raman peaks at ~ 60 cm^{-1}, ~ 205 cm^{-1} and 354 cm^{-1} have been clearly observed in single variant ordered samples [18-20]. In double-variant ordered samples, an additional peak at ~ 219 cm^{-1} has been observed by Kwok *et al*, [37] and explained as a folded LA(X) phonon occurring due to the OSL effect (where the authors referred the ordered domain twins as antiphase domain boundaries or APBs). However, the 354 cm^{-1} peak was not seen in their double-variant ordered samples. Zhang and Mascarenhas [38] recently observed that the 219 cm^{-1} Raman feature exists even in single variant ordered samples, as shown in the Raman spectra of Fig.20 (the phonon energy is slightly changed due to the temperature difference), which suggests that this feature may not relate to the OSL structure but to lateral anti-phase boundaries (APBs) which are always present in ordered alloys whether single or double variant [39,40]. The planar defect referred to as APB consists of an out-of-phase arrangement of Ga-rich and In-rich atomic planes which may exist even in single variant ordered sample. In addition, the 354 cm^{-1} feature is observable even in double variant ordered samples, [38] but is much weaker than in high quality single variant ordered samples. More careful studies are needed to explore possible effects of double variant ordering on the vibrational properties of ordered alloys.

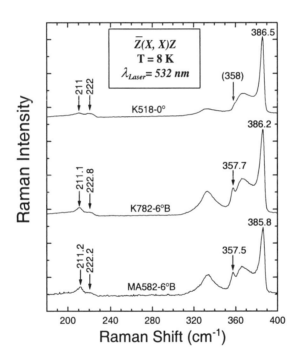

Figure 20. Raman spectra of single-variant (K782-6°B and MA582-6°B) and double-variant (K518-0°) ordered GaInP$_2$ measured in $\bar{Z}$ (X, X)Z backscattering geometry.

5.   CONCLUSION

Our understanding of the phonons in ordered GaInP$_2$ alloys has greatly improved greatly during the past decade. Four additional phonon modes at 60, 205, 310 and 354 cm^{-1}, attributed to spontaneous ordering in GaInP$_2$ have been identified and their origin explained through careful comparison of Raman, micro-Raman, resonant-Raman, IR reflectance, IR transmission data. However, a major obstacle in further enhancing the understanding of this subject is the lack of a theoretical model that can satisfactorily calculate the observed frequencies and selection rules of the phonons in ordered alloys. Although there have been some attempts, [41,42] the discrepancy between the predictions of these calculations and the existing experimental data is still too large. Difficulties arise due to the fact that ordering is only partial. Although the method used by Zhang *et al.* [35] to account for the statistical nature of the alloy distribution has been successful in fitting experimental data for electronic properties of partially ordered alloys, analogous methods need to be developed for the lattice dynamics of ordered alloys.

6.   ACKNOWLEDGEMENT

We would like to acknowledge the invaluable contributions of Jerry Olson, John Geisz, D. Friedman, Mark Hanna, to the research that has been presented. We thank the Office of Science/Basic Energy Sciences/Division of Materials Sciences for their generous support of this research at NREL.

REFERENCES

1.   W. Hayes and R. Loudon. *Scattering of light by crystals* (John Wiley & Sons, New York, 1978).
2.   I. F. Chang and S. S. Mitra, Advances in Physics, **20**, 359 (1971).
3.   G. Lucovsky, M. H. Brodsky, M. F. Chen, R. J. Chicotka, and A. T. Ward, Phys. Rev. B, **4**, 1945 (1971).
4.   R. Beserman, C. Hirlimann, and M. Balkanski, Solid State Commun., **20**, 485 (1976).
5.   T. Kato, T. Matsumoto, and T. Ishida, Jpn. J. Appl. Phys., **27**, 983 (1988).
6.   P. Galtier, J. Chevallier, M. Zigone, and G. Martinez, Phys. Rev. B, **30**, 726 (1984).
7.   E. Jahne, W. Pilz, M. Giehler, and L. Hildish, Phys. Status Solidi B, **91**, 155 (1979).
8.   B. Jusserand and S. Slempkes, Solid State Commun., **49**, 95 (1984).
9.   F. Alsina, J. D. Webb, A. Mascarenhas, J. F. Geisz, J. M. Olson, and A. Duda, Phys. Rev. B, **60**, 1484 (1999).
10. T. Suzuki, A. Gomyo, S. Iijima, K. Kobayashi, S. Kawata, I. Hino, and T. Yuasa, Jpn. J. Appl. Phys., **27**, 2098 (1988).
11. A. Mascarenahs, unpublished.
12. M. Kondow and S. Minagawa, J. Appl. Phys., **64**, 793 (1988).

13. K. Sinha, A. Mascarenhas, G. S. Horner, R. G. Alonso, K. A. Bertness, and J. M. Olson, Phys. Rev. B, **48**, 17591 (1993).

14. K. Sinha, A. Mascarenhas, G. S. Horner, K. A. Bertness, S. R. Kurtz, and J. M. Olson, Phys. Rev. B, **50**, 7509 (1994).

15. K. Uchida, P. Y. Yu, N. Noto, Z. Liliental-Weber, and E. R. Weber, Philos. Mag. B, **70**, 453 (1994).

16. H. M. Cheong, A. Mascarenhas, P. Ernst, and C. Geng, Bulletin of the American Physical Society, **41**, 477 (1996).

17. A. M. Mintairov, B. N. Zvonkov, T. S. Babushkina, I. G. Malkina, and Y. N. Saf'yanov, Phys. Solid. State, **37**, 1985 (1995).

18. F. Alsina, N. Mestres, J. Pascual, C. Geng, P. Ernst, and F. Scholz, Phys. Rev. B, **53**, 12994 (1996).

19. A. Hassine, J. Sapriel, P. Le Berre, M. A. Di Forte-Poisson, F. Alexandre, and M. Quillec, Phys. Rev. B, **54**, 2728 (1996).

20. H. M. Cheong, A. Mascarenhas, P. Ernst, and C. Geng, Phys. Rev. B, **56**, 1882 (1997).

21. H. M. Cheong, S. P. Ahrenkiel, M. C. Hanna, and A. Mascarenhas, Appl. Phys. Lett., **73**, 2648 (1998).

22. M. C. Hanna, H. M. Cheong, and A. Mascarenhas, Appl. Phys. Lett., **76**, 997 (2000).

23. H. M. Cheong, F. Alsina, A. Mascarenhas, J. F. Geisz, and J. M. Olson, Phys. Rev. B, **56**, 1888 (1997).

24. Our OMVPE-grown GaInP$_2$ samples are moderately *n*-type with a background doping density of ~10^{17} cm^{-3}.

25. F. Schäffler and G. Abstreiter, Phys. Rev. B, **34**, 4017 (1986).

26. A. Zunger and S. Mahajan. *Materials, Properties and Preparation*. S. Mahajan. (North-Holland, Amsterdam, 1994), 1399.

27. N. Mestres, F. Alsina, J. Pascual, J. M. Bluet, J. Camassel, C. Geng, and F. Scholz, Phys. Rev. B, **54**, 17754 (1996).

28. H. M. Cheong, A. Mascarenhas, J. F. Geisz, and J. M. Olson, Phys. Rev. B, **62**, 1536 (2000).

29. M. J. Seong, A. Mascarenhas, J. M. Olson, and H. M. Cheong, Phys. Rev. B, **63**, 235205 (2001).

30. A. M. Mintairov, J. L. Merz, A. S. Vlasov, and D. V. Vinokurov, Semicond. Sci. Technol., **13**, 1140 (1998).

31. A. M. Mintairov, J. L. Merz, and A. S. Vlasov, Phys. Rev. B, **63**, 247201 (2001).

32. A. Mascarenhas, H. M. Cheong, F. Alsina, J. F. Geisz, and J. M. Olson, Phys. Rev. B, **63**, 247202 (2001).

33. F. Alsina, H. M. Cheong, J. D. Webb, A. Mascarenhas, J. F. Geisz, and J. M. Olson, Phys. Rev. B, **56**, 13 126 (1997).

34. R. B. Capaz and B. Koiller, Phys. Rev. B, **47**, 4044 (1993).

35. Y. Zhang, A. Mascarenhas, and L.-W. Wang, Phys. Rev. B, **63**, 201312(R) (2001).

36. D. B. Laks, S.-H. Wei, and A. Zunger, Phys. Rev. Lett., **69**, 3766 (1992).

37. S. H. Kwok, P. Y. Yu, and K. Uchida, Phys. Rev. B, **58**, R13 395 (1998).

38. Y. Zhang and A. Mascarenhas, J. Raman Spec., **32**, 831 (2001).

39. J. H. Li, J. Kulik, V. Holy, Z. Zhong, S. C. Moss, Y. Zhang, S. P. Ahrenkiel, A. Mascarenhas and J. Bai, Phys. Rev. B, **63**, 155310 (2001).

40. C. S. Baxter, W. M. Stobbs, and J. H. Wilkie, Journal of Crystal Growth, **112**, 373 (1991).
41. V. Ozolins and A. Zunger, Phys. Rev. B, **57**, R9404 (1998).
42. F. Alsina, N. Mestres, A. Nakhli, and J. Pascual, Phys. Status Solidi B, **215**, 121 (1999).

Chapter 15

Effects of Ordering on Physical Properties of Semiconductor Alloys

Su-Huai Wei
National Renewable Energy Laboratory
Golden, CO 80401, USA

Key words: Theory, Ordering, Strain, Band Structure, Optical Anisotropy, Spin Polarization

Abstract: Many *III-V* semiconductor alloys $A_xB_{1-x}C$ exhibit spontaneous CuPt-like ordering when grown from the vapor phase. This article describes theoretically the ordering induced changes in the structural, electronical, and optical properties of the semiconductor alloys as a function of the long range order parameter η. These include (i) the band gap reduction ΔE_g and the valence band splitting ΔE_{12}, (ii) the appearence of new structural factors, (iii) the change of electric field gradient at the nuclear sites, (iv) the consequence of coexistence of epitaxial strain and chemical ordering on the optical properties, (v) the ordering induced optical anisotropy, and (vi) the spin polarization of photoemitted electrons near the band edge. The effects of stacking fault in order semiconductor alloys are also discussed.

1.   INTRODUCTION

Bulk semiconductor alloys $A_xB_{1-x}C$ grown at high temperatures are nearly perfectly random [1]. On the other hand, during the low temperature organometallic vapor phase epitaxial growth on (001) substrates, surface induced spontaneous CuPt-like ordering of semiconductor alloys has been widely observed in many III-V systems, especially of size-mismatched alloys such as $Ga_xIn_{1-x}P$ [2-4]. The ordered phase consists of alternate cation monolayer planes $A_{x+\frac{\eta}{2}}B_{1-x-\frac{\eta}{2}}$ and $A_{x-\frac{\eta}{2}}B_{1+x+\frac{\eta}{2}}$ stacked along the [111] (or equivalent) directions, where $0 \leq \eta \leq 1$ is the long range order (LRO) parameter. The degree of ordering depends on growth temperature, growth rates, *III/V* ratio, substrate misorientation and doping. Perfect ordering (η=1) corresponds to successive planes of pure *A* followed by pure *B*, etc. (Fig. 1).

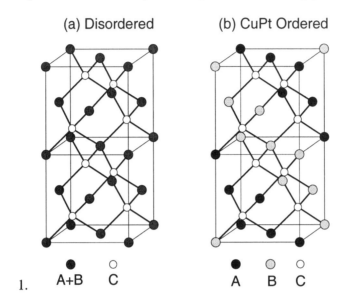

(a) Disordered   (b) CuPt Ordered

1.   ● A+B   ○ C   ● A   ◎ B   ○ C

Figure 1. Crystal structures of (a) disordered zinc-blende alloy and (b) CuPt *ordered alloy.*

When the zinc-blende (ZB) disordered alloy forms the long range ordered CuPt superlattice, the unit cell is doubled, the Brillouin zone is reduced by half, and the point-group symmetry is changed from T_d to C_{3v} [5–6]. Consequently, new X-ray diffraction spots appear at $\{G_{ZB}\}+(\frac{1}{2},\frac{1}{2},\frac{1}{2})$, where $\{G_{ZB}\}$ are zinc-blende reciprocal lattice vectors [7]. Furthermore, in the ordered phase, two zinc-blende **k**-points (and states associated with them) fold into a single **k**-point in the CuPt Brillouin zone. The folded states that have the same superlattice symmetry can interact with each other. This coupling leads to a series of changes in the optical properties of the alloys near the band edge [7-17]. For example, the states at $\overline{\Gamma}$ (we use an overbar to denote superlattice states) are constructed from zinc-blende like states at Γ and L^{111} [5]. The coupling between the $\overline{\Gamma}_{3v}(\Gamma_{15v})$ and $\overline{\Gamma}_{3v}(L_{3v})$ states and the coupling between the $\overline{\Gamma}_{1c}(\Gamma_{1c})$ and $\overline{\Gamma}_{1c}(L_{1c})$ states shifts the energy of the valence band maximum (VBM) upwards and the energy of the conduction band minimum (CBM) downwards, thus causing a splitting at the VBM and a lowering of the band gap relative to the random alloy (Fig. 2) [7]. This article will review recent advances of the theoretical understanding of the ordering induced changes in semiconductor alloys as a function of the long range order parameter η [7-18]. Review of early theoretical work on semiconductor ordering can also be found in this book and in Refs. 2, 19 and 20.

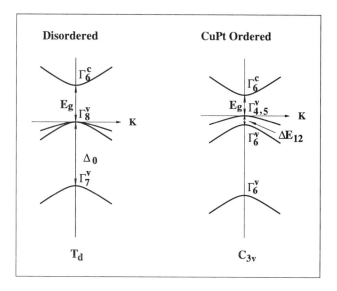

Figure 2. Schematic plot of the band structures near the band edge for (a) disordered zinc-blende alloy and (b) CuPt ordered alloy.

2.   DEPENDENCE ON THE DEGREE OF LRO

Early theoretical studies of semiconductor alloys are focused on either perfectly random or perfectly ordered compound-like structures [5,8,21,22]. However, for *spontaneously* ordered semiconductor alloys, the LRO is never perfect [2-4]. To study the physical property P of a partially ordered alloy, a general theory [23,24] was derived which provides an easy way to deduce the degree of ordering from the analysis of the experimental data.

The theory is based on a statistical description of substitutional $A_{1-x}B_x$ system in terms of a generalized Ising model [25]. In this approach, a particular ordered configuration σ is determined by the occupation of each of the N sites of the lattice with either an A atom or a B atom, and assigning the Ising spin variable of $\hat{S}_i = -1$ if site i is occupied by an A atom, and $\hat{S}_i = +1$ if it is occupied by a B atom. The lattice sites are grouped into figures with k vertices so that k=1,2,3 are single site, pair and triangle figures. For each class F of symmetry-equivalent figures (e.g., all nearest-neighbor pairs) and for each configuration σ, one defines a "correlation function" that is the average of the spin products of all figures in the class:

$$\overline{\Pi}_F(\sigma) = 1/O_F \sum_f \hat{S}_{i_1}(\sigma)\hat{S}_{i_2}(\sigma)\cdots\hat{S}_{i_k}(\sigma) , \tag{1}$$

where f runs over the O_F figures of class F, and the spin product is taken over the k sites of f. The set $\{\overline{\Pi}_F(\sigma)\}$ forms an orthonormal basis; consequently, any lattice property $P(\sigma)$ (e.g., total energy, band gap) can be rigorously expanded as

$$P(\sigma) = N \sum_f D_F p_F \overline{\Pi}_F(\sigma) , \tag{2}$$

where D_F is the number of figures of class F per site ($= O_F/N$), and p_F is the contribution of F to property P. Calculations [25-27] have shown that often there is a hierarchy of figures, i.e., that interactions between widely separated atoms, or between many atoms interacting simultaneously are less important than those between nearby pairs of atoms. In this case, it is possible to calculate the property $P(\sigma)$ of any of the 2^N ordered configurations using a small set of dominant interactions which can be obtained from highly accurate, but computationally demanding first-principles band structure methods. This approach has been widely applied to study the *ordered configurations* of many binary alloys [21,22,25-27].

This approach can be generalized to *partially* ordered structures [23-24]. The degree of η LRO is defined with respect to a particular ordered structure σ whose composition is X_σ. For example, the fully ordered CuPt structure has $X_\sigma = 1/2$ and consists of two sublattices: one occupied by A and the other occupied by B. In the *partially* ordered CuPt structure the A sublattice is occupied in part also by B and vice versa. To study such alloy with composition x (which may differ from X_σ), the *discrete* spin-variables $\hat{S}_i = \pm 1$ on each sublattice is replaced by the *ensemble average* value (assuming that the atom distributions on the A-rich sites and, separately, on the B-rich sites are random). That is,

$$< \hat{S}_i(x,\eta) > = (2x-1) + \eta[\hat{S}_i(\sigma) - (2X_\sigma - 1)]. \tag{3}$$

Here $(2x-1)$ is the site average of spin for the random alloy. To compute the properties $<P>$ of the alloy at any η value, one need to first obtain the correlation functions of Eq. (1) using Eq. (3) and then to insert these into the

Ising model (2). This provides a general Ising expansion for any lattice property P at an arbitrary η.

Note that η is limited by the requirement that $|< \hat{S}_i(x,\eta) >| \leq 1$, which implies that

$$\eta \leq \min\left[x / X_\sigma, (1-x)/(1-X_\sigma)\right].\tag{4}$$

Thus, perfect ordering ($\eta = 1$) is possible only when $x = X_\sigma$. In many practical situations, the pair interactions are dominant. We can derive correlation functions using Eqs. (1) and (3) for F=pair figures:

$$\overline{\Pi}_F(x,\eta) = (2x-1)^2 + \eta^2 \left[\overline{\Pi}_F(\sigma) - (2X_\sigma - 1)^2\right].\tag{5}$$

Using Eq. (2), Eq. (5) implies that for property P:

$$P(x,\eta) = P(x,0) + \eta^2 \left[P(X_\sigma,1) - P(X_\sigma,0)\right],\tag{6}$$

provided that the property can be well expressed in terms of single site and pair interactions and that the lattice property is independent of the choice of the origin of the system of coordinates. This equation relates the property P at any degree of LRO to the corresponding properties in (i) the perfectly random alloy at compositions x and X_σ and (ii) the perfectly ordered structure at composition X_σ. Thus, if we know the properties $P(x,0)$ and $P(X_\sigma,1)$, we can easily derive the ordering parameter η from experimental measurement of $P(x,\eta)$ and Eq. (6).

3. ENERGY LEVEL SHIFT AND SPLITTING

The optical fingerprints [6-9] of ordering include the band gap reduction relative to the random alloy

$$\Delta E_g(\eta) = E_g(\eta) - E_g(0),\tag{7}$$

as well as the two valence band splittings (Fig. 2)

$$\Delta E_{12}(\eta) = E_1\left(\overline{\Gamma}_{4,5v}\right) - E_2\left(\overline{\Gamma}_{6v}^{(1)}\right)$$

$$\Delta E_{13}(\eta) = E_1\left(\overline{\Gamma}_{4,5v}\right) - E_3\left(\overline{\Gamma}_{6v}^{(2)}\right). \tag{8}$$

Here $|1\rangle = \overline{\Gamma}_{4,5v}$ is a heavy hole $|\frac{3}{2},\frac{3}{2}\rangle$ state, whereas the two $|2\rangle = \overline{\Gamma}_{6v}^{(1)}$ and $|3\rangle = \overline{\Gamma}_{6v}^{(2)}$ states are mixtures of the light hole $|\frac{3}{2},\frac{1}{2}\rangle$ and split-off $|\frac{1}{2},\frac{1}{2}\rangle$ states. The conduction band minimum at Γ is the $\overline{\Gamma}_{6c}|\frac{1}{2},\frac{1}{2}\rangle$ state. Using the quasicubic model [28] the valence band splittings at the top of the valence band for CuPt ordering are given by

$$\Delta E_{12}(\eta) = \frac{1}{2}[\Delta_{SO} + \Delta_{CF}] - \frac{1}{2}\left[(\Delta_{SO} + \Delta_{CF})^2 - \frac{8}{3}\Delta_{SO}\,\Delta_{CF}\right]^{1/2}$$

$$\Delta E_{13}(\eta) = \frac{1}{2}[\Delta_{SO} + \Delta_{CF}] - \frac{1}{2}\left[(\Delta_{SO} + \Delta_{CF})^2 - \frac{8}{3}\Delta_{SO}\,\Delta_{CF}\right]^{1/2} \tag{9}$$

where Δ_{SO} is the spin-orbit splitting and $\Delta_{CF} = \overline{\Gamma}_{3v} - \overline{\Gamma}_{1v}$ is the ordering-induced crystal field splitting in the absence of spin-orbit coupling. It is caused by (a) an upward shift of the $\overline{\Gamma}_{3v}$ (Γ_{15v}) which couples with the state ($\overline{\Gamma}_{3v}$ (Γ_{3v})   below, and by (b) the downward shift of the ($\overline{\Gamma}_{1v}$ (Γ_{15v}) which couples with the $\overline{\Gamma}_{1c}$(Γ_{1c}) state above.

The quantities that are accessible experimentally are $\Delta E_{12}(\eta)$, $\Delta E_{13}(\eta)$, and $E_g(\eta)$ for partially ordered alloys and for random alloys. These values can be used to derive

$$\Delta_{SO}(\eta) - \Delta_{SO}(0) = [\Delta_{SO}(1) - \Delta_{SO}(0)]\eta^2$$
$$\Delta_{CF}(\eta) = \Delta_{CF}(1)\eta^2 \tag{10}$$
$$\Delta E_g(\eta) = \Delta E_g(1)\eta^2$$

using Eqs. (7), (9) and (6) [note that $\Delta_{CF}(0) = 0$]. However, since (i) perfectly ordered ($\eta = 1$) samples are unavailable and (ii) the degree of LRO η of a given sample usually is not known independently, one can not find $\Delta_{SO}(1)$, $\Delta_{CF}(1)$, and $\Delta E_g(1)$ by this fitting procedure [29]. In fact, only the ratio

$$\xi = -\Delta E_g(1)/[\Delta_{SO}(1) - \Delta_{SO}(0)]$$

$$\zeta = -\Delta E_g(1)/\Delta_{CF}(1) \tag{11}$$

can be determined from experimental measurement of $\Delta E_{12}(\eta)$, $\Delta E_{13}(\eta)$, and $E_g(\eta)$. On the other hand, if $[\Delta_{SO}(1) - \Delta_{SO}(0)]$, $\Delta_{CF}(1)$, and $\Delta E_g(1)$ were known independently (e.g., from theoretical calculation [7]), then experimental measurements [30-32] of $\Delta E_{12}(\eta)$, $\Delta E_{13}(\eta)$, and $E_g(\eta)$ could be used to derive the ordering parameter η from the above equations.

Table I. Calculated, ordering induced changes in the spin-orbit splitting $[\Delta_{so}(1) - \Delta_{so}(0)]$, the crystal-field splitting $\Delta_{CF}(1)$, the band gap reduction $\Delta E_g(1)$, and the VBM energy level $\Delta E_v(1) = E_v(1) - E_{v(0)}$ (in eV) for the four III-V alloys. The calculated ratios ζ are also given. An LDA-corrected formalism [7] is included.

	$Al_{0.5}In_{0.5}P$	$Al_{0.5}In_{0.5}As$	$Ga_{0.5}In_{0.5}P$	$Ga_{0.5}In_{0.5}As$
$[\Delta_{SO}(1) - \Delta_{SO}(0)]$	0.02	0.01	0.01	0.00
$\Delta_{CF}(1)$	0.24	0.21	0.16	0.13
$\Delta E_g(1)$	−0.27	−0.18	−0.43	−0.25
$\Delta E_v(1)$	0.13	0.11	0.10	0.10
$\zeta = -\Delta E_g(1)/\Delta_{CF}(1)$	1.13	0.86	2.69	1.92

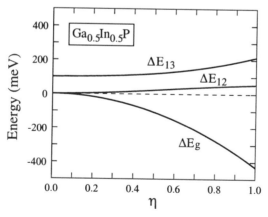

Figure 3. Calculated valence band splitting ΔE_{12}, ΔE_{13} and band gap reduction energies ΔE_g as functions of the LRO parameter η for $Ga_{0.5}In_{0.5}P$ (solid lines).

Table I lists the calculated, ordering induced changes $[\Delta_{SO}(1) - \Delta_{SO}(0)]$, $\Delta_{CF}(1)$, $\Delta E_g(1)$, and the ratio $\zeta = -\Delta E_g(1) / \Delta_{CF}(1)$ for ordered $Al_{0.5}In_{0.5}P$, $Al_{0.5}In_{0.5}As$, $Ga_{0.5}In_{0.5}P$, and $Ga_{0.5}In_{0.5}As$. It also gives the calculated VBM state shift $\Delta E_v (1) = E_v (1) - E_v(0)$. Figure 3 shows the calculated results of ΔE_{12}, ΔE_{13}, and ΔE_g as function of η for the ordered $Ga_{0.5}In_{0.5}P$ alloy using the data given in Table I. The calculation was done using the local density approximation (LDA) [33,34] as implemented by the self-consistent linearized augmented plan wave (LAPW) method [35,36]. Since LDA underestimates the band gap, an LDA-corrected formalism [7,37] is included. Good agreement between the calculated values and experimental data are obtained. For example, using a pump-probe exciton absorption/bleaching method, the measured [31] ratio ζ for $Ga_xIn_{1-x}P$ is $\zeta = 2.66 \pm 0.15$, which is very close to the LDA-corrected value of $\zeta = 2.69$ (Table I). For $Ga_xIn_{1-x}As$, using low temperature absorption and photoluminescence, the measured [30] value is $\zeta = 1.8 \pm 0.4$, also very close to the calculated value $\zeta = 1.92$.

The results shown in Table I indicate that:

(i) CuPt ordering induces large crystal field splitting $\Delta_{CF}(1)$ and band gap reduction $\Delta E_g(1)$ in all four alloys which have large lattice mismatch between their binary constituents.

(ii) $Al_{0.5}In_{0.5}X$ alloys have larger crystal field splitting than $Ga_{0.5}In_{0.5}X$ alloys. Part of the reason is because the valence band offset [38] between AlX/InX (~ 0.6 eV) is much larger than the one for GaX/InX (~ 0.1 eV). Thus, there is a larger perturbation in the valence band of $Al_{0.5}In_{0.5}X$ than in $Ga_{0.5}In_{0.5}X$.

(iii) $Al_{0.5}In_{0.5}X$ alloys have a smaller band gap reduction $\Delta E_g(1)$ than $Ga_{0.5}In_{0.5}X$. This is because (1) the atomic s orbital energies of Al, Ga, and In are non-monotonic, namely -7.9, -9.3, and -8.6 eV, respectively, and (2) atomic relaxation in lattice mismatch common-anion alloys tends to shift the charge from the long bond (In-X) to the short bonds (Ga-X in $Ga_{0.5}In_{0.5}X$ and Al-X in $Al_{0.5}In_{0.5}X$) [39]. Consequently, the band gap reduction due to atomic relaxation is larger in $Ga_{0.5}In_{0.5}X$ (since Ga receives charge, and its s orbital is deeper in energy than that for In), but smaller in $Al_{0.5}In_{0.5}X$ (since Al receives its charge, and its s orbital is shallower in energy than that for In).

(iv) Relative to the random alloy, the VBM wavefunction of the ordered compounds is more localized on the cation having a larger atomic number [40]. Thus, $[\Delta_{SO}(1) - \Delta_{SO}(0)] > 0$. However, for common-anion systems Δ_{SO} of the two binary constituents are similar, thus the ordering induced increase $[\Delta_{SO}(1) - \Delta_{SO}(0)]$ is rather small (0.00 ~ 0.02 eV). The increase is slightly greater for $Al_{0.5}In_{0.5}X$ than for $Ga_{0.5}In_{0.5}X$, because of the larger atomic number difference between Al and In.

Note that the ordering induced energy level shifts and splittings also occur at higher energy transitions [13]. In combination with theoretical calculations, measurement of these high energy transitions [41,42] can also be used to derive the ordering parameter. Furthermore, ordering induced band gap reduction ΔE_g and the crystal field splitting Δ_{CF} at the valence band maximum are not restricted only to CuPt-like structures. These phenomena can occur also in other sponteneous or artificially ordered superlattice structures [5,8]. Table II gives the calculated ΔE_g and Δ_{CF} for the four alloys ordered in the CuAu structure [a (1,1) superlattice] and Z2 structure [a (2,2) superlattice] along the [001] direction. We find that ΔE_g and Δ_{CF} are smaller in the CuAu and Z2 structures than in the CuPt structure because the coupling in the CuAu and Z2 structures (between Γ and folded X and Δ states) are weaker than the coupling in the CuPt structure (between Γ and the folded L states).

Table II. Calculated ordering induced changes in the crystal-field splitting $\Delta_{CF}(1)$ and band gap reduction $\Delta E_g(1)$ (in eV) for four III-V alloys in the CuAu and Z2 structures. An LDA-corrected formalism is included

	$Al_{0.5}In_{0.5}P$	$Al_{0.5}In_{0.5}As$	$Ga_{0.5}In_{0.5}P$	$Ga_{0.5}In_{0.5}As$
		CuAu ordering		
$\Delta_{CF}(1)$	0.21	0.18	0.10	0.08
$\Delta E_g(1)$	-0.13	-0.06	-0.07	-0.03
		Z2 ordering		
$\Delta_{CF}(1)$	0.21	0.17	0.11	0.09
$\Delta E_g(1)$	-0.20	-0.16	-0.11	-0.05

4.   X-RAY STRUCTURE FACTOR

Another fingerprint of CuPt ordering is the appearance of new X-ray diffraction spots that do not exist in zinc-blende material [7, 43]. Observation of these new superstructure spots would be one of the strongest indications of the existence of the ordered phase. Since the intensity of the new spots is proportional to η^2, accurate measurement of the intensity can, in principle, be used to derive the degree of order η by comparing it with the calculated values for perfectly ordered systems. This provides an independent verification to the ordering parameter measured from optical experiment.

Table III gives the calculated [7] static X-ray structure factors $\rho(\mathbf{G})$ for the fully ordered $AlInP_2$, $AlInAs_2$, $GaInP_2$, and $GaInAs_2$. The structure

Spontaneous Ordering in Semiconductor Alloys

factors $\rho(\mathbf{G})$ are the Fourier transform of the electron charge density $\rho(\mathbf{r})$, i.e.,

$$\rho(\mathbf{G}) = \frac{1}{\Omega} \int_\Omega \rho(\mathbf{r}) e^{i\mathbf{G}\cdot\mathbf{r}} d\mathbf{r}. \tag{12}$$

Here $\mathbf{G}$ is the reciprocal lattice vector and Ω is the unit cell volume. The diffraction intensity I is proportional to $|\rho(\mathbf{G})|^2$.

Table III. Calculated structure factors $|\rho(\mathbf{G})|$ of fully CuPt ordered $AlInP_2$, $AlInAs_2$, $GaInP_2$, and $GaInAs_2$ (in electrons per atom). Note that for $\mathbf{G}$ vectors in the same star [(x,y,z) and its cyclical permutations] the structural factors are the same. Here, the reciprocal lattice vector $\mathbf{G}$ is in units of $\frac{2\pi}{a}$, where a is the cubic lattice constant. An asterisk next to a $\mathbf{G}$ vector indicates that it is a superstructure spot.

G	$AlInP_2$	$AlInAs_2$	$GaInP_2$	$GaInAs_2$
0 0, 0	23.00	32.00	27.50	36.50
*$\frac{1}{2},\frac{1}{2},\frac{1}{2}$	8.26	7.79	3.83	3.39
*$\frac{1}{2},\frac{1}{2},\frac{3}{2}$	8.02	7.91	3.83	3.70
1, 1, 1	14.43	19.53	18.28	22.55
1, 1, 1	14.20	19.11	18.12	22.23
0, 0, 2	7.16	0.94	11.21	3.17
*$\frac{1}{2},\frac{3}{2},\frac{3}{2}$	8.30	9.25	4.34	5.33
*$\frac{1}{2},\frac{1}{2},\frac{5}{2}$	8.16	9.52	4.51	6.00
*$\frac{3}{2},\frac{3}{2},\frac{3}{2}$	6.48	5.74	3.01	3.26
2, 0, 2	15.94	23.73	19.58	27.43
2, 2, 0	15.77	23.36	19.40	27.04

The calculated results show that (i) the structure factors for the ordered alloy taken at the ZB allowed $\{\mathbf{G}_{ZB}\}$ are very similar to those of the random alloys (not shown), except for some small splittings due to the lower symmetry of the ordered alloy. However, (ii) new structure factors (marked with * in Table III) appear at $\{\mathbf{G}_{ZB}\} + (\frac{1}{2},\frac{1}{2},\frac{1}{2})$ in the ordered alloy which do not exist in perfectly random alloy.

In an actual experimental measurement at finite temperature, the measured intensity is reduced by the thermal vibration of the lattice. The dynamic (temperature) effect is often approximated by the Debye-Waller factors [44]. In this approximation the relation between the measured dynamic structure factor $\rho_{expt}(\mathbf{G}, \eta)$ and the calculated static structure factor $\rho_{calc}(\mathbf{G}, \eta)$ is

$$\rho_{expt}(\mathbf{G},\eta) = \rho_{calc}(\mathbf{G},\eta)e^{-B(T)G^2} , \tag{13}$$

where $B(T)$ is a temperature dependent constant. Since $\rho(\mathbf{G}_{ZB}, \eta)$ is essentially ordering independent for the zinc-blende allowed $\mathbf{G}_{ZB}$ vectors, measuring $\rho_{expt}(\mathbf{G}_{ZB})$ can be used to derive the value $B(T)$ from Eq. (13) and Table III. This $B(T)$ can in turn be used in Eq. (13) to calculate $\rho_{calc}(\mathbf{G}, \eta)$ from measured $\rho_{expt}(\mathbf{G})$ for the superstructure sports. Finally, the obtained $\rho_{calc}(\mathbf{G}, \eta)$ can be used to derive the ordering parameters η using Eq. (6) and the values given in Table III.

5.   ELECTRIC FIELD GRADIENT

Besides the direct observation of LRO through X-Ray diffraction techniques [7,43], or indirect observation through optical fingerprints [7,29-32], other techniques which are sensitive to the symmetry lowering LRO can also be used to derive the ordering parameters η. One of the techniques suggested recently [45] is to measure the quadrupole interactions between the nuclei and the local electric field gradient (EFG) using nuclear magnetic resonance (NMR).

The EFG of a LRO alloy can be calculated using the variational potential derived from the all-electron full-potential LAPW method [18]. In the LAPW approach, space is divided into two regions: the (non-overlapping) muffin-tin (MT) spheres (with radii R_{MT} centered about each atom, and the interstitial region between the atoms). The Coulomb potential of the crystal is expanded in terms of spherical harmonics Y_{lm} inside the MT region and in terms of plane waves in the interstitials. That is

$$V(\mathbf{r}) = \begin{cases} \sum_{l,m} V_{lm}(\mathbf{r})Y_{lm}(\hat{r}) & \text{Inside the MT spheres,} \\ \sum_G V_G e^{i\mathbf{G}\cdot\mathbf{r}} & \text{In the interstitial region.} \end{cases} \tag{14}$$

Given the Coulomb potential $V(\mathbf{r})$, the electric field gradient (EFG) tensor V_{ij}^{α} at the αth nuclear site is defined as

$$V^{\alpha}_{ij} = \frac{\partial^2 V^{\alpha}}{\partial x_i \partial x_j}\bigg|_{r_\alpha = 0}. \tag{15}$$

where i and j are Cartesian coordinates. (For simplicity, the index α will be ignored below.) Since the potential of Eq. (14) near the nucleus at $r = 0$ has the asymptotic form [46]

$$V(r \to 0) = \sum_{l,m} r^l \, \Phi_{lm} \left[\frac{4\pi}{2l+1}\right]^{1/2} Y_{lm}(\hat{r}), \tag{16}$$

Only the $l = 2$ ($m = -2$ to 2) terms give non-zero EFG. Specifically, the EFG at a nuclear site are

$$\begin{aligned}
V_{xx} &= \left(\tfrac{3}{2}\right)^{1/2} \left(\Phi_{2,-2} + \Phi_{2,2}\right) - \Phi_{2,0} \\
V_{yy} &= -\left(\tfrac{3}{2}\right)^{1/2} \left(\Phi_{2,-2} + \Phi_{2,2}\right) - \Phi_{2,0} \\
V_{xy} &= -i \left(\tfrac{3}{2}\right)^{1/2} \left(\Phi_{2,-2} + \Phi_{2,2}\right) \\
V_{yz} &= -i \left(\tfrac{3}{2}\right)^{1/2} \left(\Phi_{2,-1} + \Phi_{2,1}\right) \\
V_{zx} &= \left(\tfrac{3}{2}\right)^{1/2} \left(\Phi_{2,-1} - \Phi_{2,1}\right).
\end{aligned} \tag{17}$$

Here,

$$\Phi_{2m} = \Phi^{*}_{2,-m} = \lim_{r \to 0}\left[\frac{5}{4\pi}\right]^{1/2} \frac{V_{2m}(r)}{r^2}, \tag{18}$$

and the traceless condition of the EFG requires

$$V_{zz} = -(V_{xx} + V_{yy}) = 2\Phi_{2,0}. \tag{19}$$

These results show that the EFG can be easily obtained once the self-consistent Coloumb potential is known. Further, one can always find principal axes (x', y', z') of the potential V such that $V_{ij} = 0$ if $i \neq j$. Conventionally, one orders the Cartesian components according to their magnitude so that $|V_{z'z'}| > |V_{y'y'}| > |V_{x'x'}|$. Thus, the EFG are usually specified by just two parameters: the principal component $V_{z'z'}$ and the anisotropy parameter

$$\lambda = (V_{x'x'} - V_{y'y'})/V_{z'z'}. \tag{20}$$

Table IV. Calculated Cartesian coordinates $(x, y, z)a$ ($a = 5.671$ Å) of atoms in the relaxed CuPt ordered GaInP$_2$ and the principle components of the electric field gradient $V_{z'z'}$ (in Ry/Bohr^2). Here, z' is along the ordering [111] direction. The two types of P atoms are denoted P(Ga$_3$In) and P(GaIn$_3$), respectively.

	x	y	z	EFG
Ga	0.00000	0.00000	0.00000	-0.24162
In	1.00325	1.00325	1.00325	0.13746
P(Ga$_3$In)	1.26231	1.26321	1.26321	-0.24958
P(GaIn$_3$)	0.23993	0.23993	0.23993	0.27366

This method has been used to calculate the EFG for CuPt ordered GaInP$_2$. The equilibrium lattice parameters and cell-internal crystallographic parameters of the ordered GaInP$_2$ were determined by minimization of the total energies. The calculated structural parameters of CuPt ordered GaInP$_2$ are given in Table IV. In this ordered structure there are two chemical types of cations (Ga and In) and two crystallographic types of anions inside the trigonal primitive unit cell. Each cation is surrounded by four nearest neighbor P atoms. One anion P is surrounded by three Ga and one In atoms [denoted as P(Ga$_3$In)], while the second P is surrounded by one Ga and three In atoms [denoted as P(GaIn$_3$)]. All atoms have locally a trigonal (C$_{3v}$) symmetry, thus the EFGs at all sites are axially symmetric ($\lambda = 0$) with the principal components $V_{z'z'}$ oriented along the ordering [111] direction. The last column of Table IV lists our calculated $V_{z'z'}$ (denoted as *EFG*) for the four atoms in perfectly ordered ($\eta = 1$) GaInP$_2$.

To understand the different contributions to the EFG of GaInP$_2$, the $V_{z'z'}$ has been decomposed into two parts: (i) $V_{z'z'}^{\text{sphere}}(R)$ from the anisotropic charge distribution *inside* a sphere of radius R centered at the nuclear site and (ii) $V_{z'z'}^{latt}(R)$ from charges *elsewhere in the lattice*. The sphere's contribution $V_{z'z'}^{\text{sphere}}(R)$ is given by

$$V_{z'z'}^{\text{sphere}}(R) = \left[\frac{4\pi}{5}\right]^{1/2} \int_0^R \frac{\rho_{z'z'}}{r^3} r^2 \, dr. \tag{21}$$

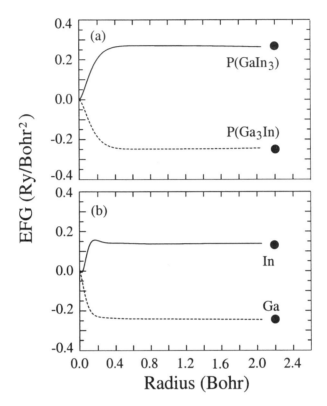

Figure 4. Calculated EFG of CuPt ordered GaInP$_2$. The lines show the contributions to EFG from charges inside a sphere as a function of the sphere's radius R [Eq. (21)]. The solid dots on the right hand side of the figures give the total EFG—(a) for the two types of P atoms, and (b) for Ga and In.

Figure 4 plots $V_{z'z'}^{\text{sphere}}(R)$ as a function of the sphere radius R for the four atom types in the GaInP$_2$ unit cell. The values of the *total* $V_{z'z'}$ are shown on the right hand side of Fig. 4 as solid dots. The difference between the total $V_{z'z'}$ and $V_{z'z'}^{\text{sphere}}(R)$ is $V_{z'z'}^{\text{latt}}(R)$. The calculated results show that *most of the EFG comes from the anisotropic charge distribution of the electron inside a sphere radius R_0 of about 0.4 Bohr*. The contribution to the EFG of charges *outside R_0 is very small* because of the $1/r^3$ dependence of the EFG to a point charge [Eq. 21]. These results indicate that to model the LRO induced change in NMR measurement [45], it is crucial to have correct charge distribution in a region very close to the nuclei.

The calculated $V_{z'z'}$ can be used to compute the nuclear quadrupole coupling constant [47] $Q_{cc} = e^2 qQ$, where $eq = V_{z'z'}$, and Q is the quadrupole moment of the nucleus. Using the values [48] of $Q = 0.17$, 0.10 and 0.81

barn for ^{69}Ga, ^{71}Ga, and ^{115}In, respectively, the calculated quadrupole coupling constants

$$Q_{cc}(^{69}Ga) = -4.83\,\text{MHZ}$$
$$Q_{cc}(^{71}Ga) = -2.84\,\text{MHZ} \tag{22}$$
$$Q_{cc}(^{115}In) = 13.08\,\text{MHZ}$$

Since ^{31}P has no quadrupole moment, $Q_{cc}(^{31}\text{P})$ is zero.

6.   COUPLING OF ORDERING WITH STRAIN

Since the ordered samples are grown on a substrate with nominally (001) orientation, strain caused by lattice mismatch between the film and the substrate can exist [4]. Ordering of semiconductor alloys leads to lowering of the band gap and to splitting of the VBM states. On the other hand, strain produced by misfit between the substrate and the alloy can also lead to valence band splitting and to band gap changes [9,10,32]. Unlike the case of *artificially* grown strain layer superlattices in which the direction of layer modulation coincides with the direction of strain, so these effects add up co-linearly, the coexistence of chemical ordering in the direction $\mathbf{G}_{ord} = (111)$ with epitaxial strain in the direction $\mathbf{G}_{substrate} = (001)$ leads to a non-co-linear, "vector addition" of ordering and strain.

Assuming small deviation of x from the lattice matched composition x_0 (i.e., small strain), and using the six valence states $(x\uparrow, y\uparrow, z\uparrow, x\downarrow, y\downarrow, z\downarrow)$ as a basis, where $\hat{\sigma} = \uparrow$ and $\downarrow$ are the spinors parallel or antiparallel to the z direction, respectively, the effects of coupling between the (111) ordering and (001) strain at $\mathbf{k} = \mathbf{0}$ VBM can be described [9] by the Hamiltonian

$$H_v = \frac{1}{3} \begin{pmatrix} \Delta^S & -\Delta^O - \Delta^{SO} & -\Delta^O & 0 & 0 & \Delta^{SO} \\ -\Delta^O + i\Delta^{SO} & \Delta^S & -\Delta^O & 0 & 0 & -i\Delta^{SO} \\ -\Delta^O & -\Delta^O & -2\Delta^S & -\Delta^{SO} & i\Delta^{SO} & 0 \\ 0 & 0 & -\Delta^{SO} & \Delta^S & -\Delta^O + i\Delta^{SO} & -\Delta^O \\ 0 & 0 & -i\Delta^{SO} & -\Delta^O - i\Delta^{SO} & \Delta^S & -\Delta^O \\ \Delta^{SO} & i\Delta^{SO} & 0 & -\Delta^O & -\Delta^O & -2\Delta^S \end{pmatrix} \tag{23}$$

Here Δ^{SO} is the spin-orbit splitting at the VBM. $\Delta^S(\varepsilon) = 3b\frac{C_{11}+2C_{12}}{C_{11}}\varepsilon$ is the strain-induced valence band crystal field splitting, where b is the tetragonal deformation potential and C_{ij} are the elastic constants. The epitaxial strain is given by $\varepsilon(x) = [a_s - a_f(x)]/a_f(x)$, where $a_f(x)$ and a_s are the lattice constants of the film and the substrate, respectively. $\Delta^O(\eta) = \Delta^O(1)\eta^2$ is the ordering induced valence band crystal field splitting. Solving Eq. (23) gives the valence band levels (in decreasing order) E_1, E_2, and E_3 and their eigenstates as functions of composition x (or strain ε) and degree of ordering η.

Figure 5 depicts the valence band splitting $\Delta E_{12}[\eta, \varepsilon(x)]$ for the $Ga_xIn_{1-x}P$ alloy grown on a GaAs substrate as a function of the film composition x. The following are the important features: (i) Any non-zero strain leads to a splitting of the VBM. (ii) The ΔE_{12} vs. x curve has a cusp at $x = x_0$ for the random alloy, reflecting the change of the VBM from heavy hole when $x < x_0$ to light hole when $x > x_0$. (iii) Chemical order universally leads to a valence band splitting for any $\eta > 0$. For the valence band splitting, chemical ordering is analogous to in-plane *compressive* strain in that both yield a heavy-hole state at the top of the valence band. (iv) The coexistence of ordering ($\eta \neq 0$) with strain ($x \neq x_0$) is predicted (a) to remove the cusp in the ΔE_{12} vs. x curves. This is so since any amount of chemical order will mix equal amounts of light with heavy hole states at $x = x_0$. (b) Chemical ordering is also predicted to reduce the dependence of ΔE_{12} on x, as illustrated by the flattening of the $\eta = 0.5$ curve near $x = x_0$ in Figure 5. (c) (001) strain increases the valence band splitting produced by pure ordering, and vise versa [49]. This reflects the fact that the ordering-induced splitting Δ^O and the strain-induced splitting Δ^S are complementary, so the combined effect is larger than the individual ones.

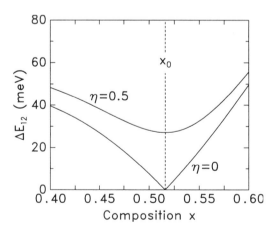

Figure 5. Valence band splitting ΔE_{12} of the $Ga_xIn_{1-x}P$ alloy strained on GaAs as a function of composition x at $\eta = 0$ and $\eta = 0.5$. The lattice matching composition is at $x_0 = 0.516$.

7.   OPTICAL ANISOTROPY IN ORDERED ALLOYS

The splitting of the VBM for the ordered alloy induces an anisotropy in the intensities of the transitions between these split VBM components and the conduction band minimum [6,9,11]. Quantitative analysis of experimental data [29-32,50-54] requires knowledge of *optical transition rate* between the split VBM components and the conduction band minimum (CBM) as a function of the polarization of the light $\hat{e}$ and the degree of long range order η.

The valence eigenstates Ψ_v can be expanded in terms of the six basis functions $\{ p_v \hat{\sigma} \}$ and can be obtained by diagonalizing the corresponding 6x6 Hamiltonian of Eq. (23). The two degenerate conduction states Ψ_v can be described by the two s-like states ($s \uparrow, s \downarrow$). The transition intensity between Ψ_c and Ψ_v is proportional to the matrix element squared $I_{c,v} = | < \Psi_c | H_{int} | \Psi_v > |^2$, where H_{int} is the interacting Hamiltonian. For *linearly* polarized light along the [l,m,n] direction we have $H_{int} \propto lx + my + nz$, while for *circularly* polarized light $\sigma^{\pm}$ with angular momentum parallel and antiparallel to z' we have $H_{int} \propto x' \pm iy'$. The transition matrix elements can be calculated by writing the orbital wavefunctions and H_{int} in terms of the spherical harmonics Y_{lm} and by noticing that the allowed dipole transitions are for $\Delta m = \pm 1$. This gives the simple selection rule

$$<s\hat{\sigma} \,|\, x_\mu \, \rho_v \hat{\sigma}' > \,=\, \gamma \, \delta_{\mu,v} \delta_{\hat{\sigma},\hat{\sigma}'} \,, \tag{24}$$

where (μ, $v = x, y, z$) and γ is a normalization parameter.

Using the calculated eigenstates of Eq. (23) and the selection rule of Eq. (24), the transition intensities $I_{c,v}$ between the valence states $| 1 >$ and $| 2 >$ and the conduction state in pure (111) ordered $Ga_{0.5}In_{0.5}P$ as a function of η have been calculated. The linearly polarized light with polarization $\hat{e} \parallel [110]$ (defined as $\Theta = 0^o$); $\hat{e} \parallel [1\bar{1}0]$ (defined as $\Theta = 90^o$) and in-between are considered. For the random alloy the transition intensities $I_{c,v}$ are independent of Θ.

Figure 6 depicts the calculated normalized intensities as function of polarization angle Θ. We find that in this case the intensity can be described by

$$I(\Theta) = I_{1\bar{1}0} \, sin^2 \, \Theta + I_{110} cos^2 \, \Theta \,. \tag{25}$$

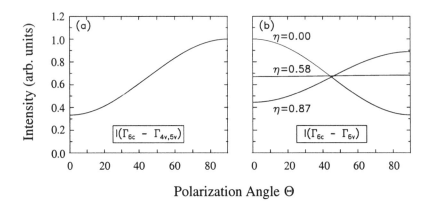

Figure 6. Calculated transition intensity of (111) ordered $Ga_{0.5}In_{0.5}P$ as a function of polarization angle Θ. (a) The Γ_{6c} - $\Gamma_{4v,5v}$ transition. The intensity is independent of η. (b) The Γ_{6c} - Γ_{6v} transition at $\eta = 0.00$, 0.58, and 0.87.

For the Γ_{6c} - $\Gamma_{4v, 5v}$ transition (Fig. 6a) the intensity is independent of η, since there is no coupling between $\Gamma_{4v, 5v}$ and the other two Γ_{6v} valence states. For the Γ_{6c} - Γ_{6v} transition, however, we see a strong dependence on the ordering parameter η. This is due to the coupling between the two Γ_{6v} states. $I(\Theta)$ can be either an increasing function (at large η) or a decreasing function (at small η) of the polarization angle Θ. Thus, by measuring the intensity $I(\Gamma_{6c}$ - $\Gamma_{6v})$ as a function of the polarization angle Θ, and fitting the results to theoretically calculated curves of Eq. (25), it is possible to derive the LRO parameter η. Similarly, the optical anisotropy in the transitions intensities can be detected using the reflectance-difference spectroscopy (RDS). General formulae relating the ordering-induced bulk RDS intensity with the degree of LRO have been derived and discussed in Refs. 12, 53, and 54.

8.   SPIN POLARIZATION IN ORDERED ALLOYS

Ordering also induces changes in spin polarization of emitted photoelectrons [9,11,20,55,56] near the band edge. For this purpose circularly polarized light σ^+ with its angular momentum along the ordering direction $z' = [111]$ is used. That is, $H_{int} \propto x' + iy'$, where $x' = \frac{1}{\sqrt{2}} (-x + y)$ and $y' = \frac{1}{\sqrt{6}} (x + y - 2z)$. The spinors parallel and anti parallel to the [111] ordering direction are given by

$$\uparrow' = cos\frac{\theta}{2}e^{-i\frac{\varphi}{2}}\uparrow + sin\frac{\theta}{2}e^{i\frac{\varphi}{2}}\downarrow$$
$$\downarrow' = sin\frac{\theta}{2}e^{-i\frac{\varphi}{2}}\uparrow + cos\frac{\theta}{2}e^{i\frac{\varphi}{2}}\downarrow,$$

(26)

where the angles θ and φ are determined by the equation [$sin\theta cos\varphi$, $sin\theta sin\varphi$, $cos\theta$] = $\frac{1}{\sqrt{3}}$ [111].

Figure 7 shows schematically how the spin-polarized electrons are generated from zinc-blende derived systems. Without the spin-orbit interaction ($\Delta^{SO} = 0$) and the symmetry lowering crystal field interaction (Fig. 7a), the six Γ_{15v} valence states are degenerate. In this case the spin polarization P defined as

$$P = \frac{I\uparrow' - I\downarrow'}{I\uparrow' + I\downarrow'},$$

(27)

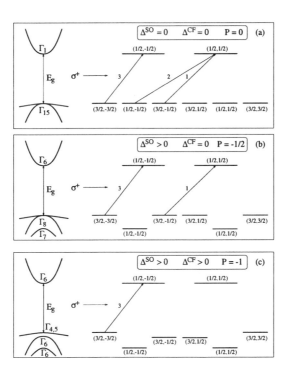

Figure 7. Schematic plot of energy levels at Γ and transition probabilities for zinc-blende systems. (a) for $\Delta^{SO} = 0$ and $\Delta^{CF} = 0$, (b) $\Delta^{SO} > 0$ and $\Delta^{CF} = 0$ (e.g., GaAs), and (c) both $\Delta^{SO} > 0$ and $\Delta^{CF} > 0$ (e.g., CuPt ordered Ga$_{0.5}$In$_{0.5}$P).

will be zero. Here, $I \uparrow'$ and $I \downarrow'$ are the transition intensities for $\uparrow'$ spin and $\downarrow'$ spin, respectively. When the spin-orbit interaction is switched on (Fig. 7b), the four Γ_{8v} ($j = 3/2$) valence states are separated from the two Γ_{7v} ($j = 1/2$) valence states by an energy of the spin-orbit splitting Δ^{SO}. In this case (e.g., GaAs and random alloy), if we can control the optical pumping so that the transition to the conduction band results only from the highest occupied $j = 3/2$ state, the generated conduction electron will have a spin polarization of 50% (Fig. 7b). For CuPt ordered semiconductor alloys, the large, positive crystal field splitting further removes the degeneracy at the VBM, and puts the heavy hole $| 1 > = \Gamma_{4v,5v}$ ($j = 3/2$, $m_j = \pm 3/2$) states above the light hole $| 2 > = \Gamma_{6v}$ ($j = 3/2$, $m_j = \pm 1/2$) states. Figure 8 shows the calculated spin intensities $I(\Gamma_{6c} - \Gamma_{4v,5v})$ and $I(\Gamma_{6v} - \Gamma_{6v})$ for ordered $Ga_{0.5}In_{0.5}P$ alloys as a function of the valence band splitting ΔE_{12}, which increases as the ordering parameter increases. In both cases the photoelectrons generated from these transitions are *fully polarized*. The transition intensity $I(\Gamma_{6c} - \Gamma_{4v,5v})$ is independent of the ordering parameter η, while $I(\Gamma_{6c} - \Gamma_{4v,5v})$ decreases as η increases. Therefore, if the splitting ΔE_{12} is large enough to allow optical pumping only from the highest $\Gamma_{4v,5v}$ state, the generated photoelectrons can be 100% spin polarized. Because these ordered alloys have all the advantages associated with III-V semiconductors, they have been proposed [20] as strong candidates for next generation high quality spin-polarized electron source with high spin-polarization, high quantum efficiency, and high reliability [57-59].

As discussed in Sec. 6, the coexistence of (001) strain and (111) ordering will mix the heavy hole states with the light hole states, thus reducing the spin polarization [9,11]. In this case the maximum spin-polarization (< 100%) can be obtained in a direction parallel to one of the principle axis defined by the Hamiltonian of Eq. (23).

Note that despite the identical optical response with respect to the linearly polarized light along [110] and [$1\bar{1}0$] of the two $CuPt_A$ [(111) and ($1\bar{1}1$)] subvariants, their response to the circularly polarized light is predicted to be different. Using the σ^+ light noted above but for ($1\bar{1}1$) ordering, we find that the spin polarization P for the transition from the top $\Gamma_{4v,5v}$ state is only 20% and the total intensity $I \downarrow' + I \uparrow'$ is reduced to 55% of the intensity for (111) ordering. This difference can be used to distinguish (111) ordering from ($1\bar{1}1$) ordering, which is not possible using the linearly polarized light. This also indicates that in order to obtain the highest efficiency in generating spin polarized electrons, single variant crystals are required.

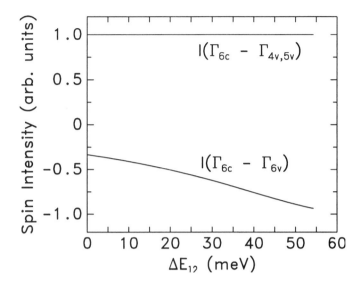

Figure 8. Calculated spin intensity $I(\Gamma_{6c} - \Gamma_{4v,5v})$ and $I(\Gamma_{6c} - \Gamma_{6v})$ (in arbitrary units) of ordered $Ga_{0.5}In_{0.5}P$ as a function of the valence band splitting ΔE_{12}. The angular momentum of the circularly polarized light is in the same direction as the ordering vector. The generated photoelectron from each band is fully polarized.

9.   EFFECTS OF STACKING FAULTS

Unlike classic cases of long-range order in metallurgical systems, semiconductors often show surprisingly a coexistence of domains of a few types of ordered structures separated by stacking faults in the same sample. For example, nominally CuPt-like $GaInP_2$ samples are suggested to contain a Y2-like phase [60] (Fig. 9d), anti-phase boundaries (APB) on the (001) planes [61], and "orientational superlattices" (Fig. 9c) with periodically alternating (111) and (11$\bar{1}$) ordered subvariants (Figs. 9a and 9b) [62-65]. Such samples with mixed ordering domains often exhibit interesting optical effects such as spatially indirect interband transitions and excitation intensity dependent emission energies [66-70]. Therefore, one needs to identify the microstructures that lead to these highly unusual effects.

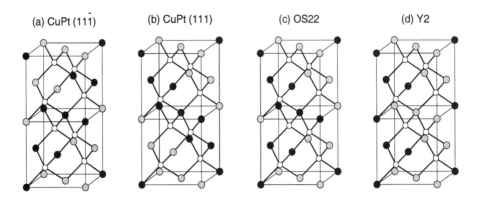

Figure 9. Crystal structures of (a) CuPt with (11$\bar{1}$) ordering plane, (b) CuPt with (111) ordering plane, (c) (2,2) orientation superlattice, and (d) the Y2.

Because of the complexity of the mixed phase systems, the unit cell size of the mixed phase structure is often too large to be calculated directly using currently amenable first-principles method. This problem is overcome [16] by noticing that these multiphase ordered structures are formed by different stacking of a basic (001) atomic plane, which has alternating A lines and B lines along the [$\bar{1}$10] direction (Fig. 9). That is, they are zinc-blende based *polytypes*. Thus, the physical property P of the polytypes can be described in a genralized one dimensional Ising-like model [16,71]. In this approach, a spin variable S_i is assigned to each plane i. A reference structure (CuPt) is chosen for which $S_i \equiv 1$ for all i. For other polytypes, if a plane i is shifted by $\tau = (1/2, 1/2, 0)a$ relative to the same plane in the reference structure, then $S_i = -1$. Using this description, Fig. 10 shows the schematic diagrams of the atomic arrangements, the spin variables $\{S_i\}$ in a conventional unit cell, and the associated Zhdanov notation [72] for some of the polytypes. In this notation the CuPt structure is denoted as $< \infty >$, while the Y2 structure is denoted as $< 2 >$. $< nm >$ indicates a periodic (001) APB structure where n unshifted CuPt (111) ordered planes are followed by m shifted CuPt (111) ordered planes [$< n > \equiv < nn >$]. Note that $< \infty >$ and $< 1 >$ are two subvariants of the same bulk ordered structure with different orientations, while a periodic orientational superlattice (OS) with n layer CuPt (111) ordered plane followed by m layer CuPt (11$\bar{1}$) ordered plane is denoted as OSnm and described by n spin 1 and $m/2$ spin $1\bar{1}$ pairs. For example, OS44=[1111$\bar{1}\bar{1}\bar{1}\bar{1}$]=$< 5111 >$.

The physical properties P of any polytype configuration σ are expanded in terms of a small set of effective, kth neighbor layer-layer interaction energies $\{J_k\}$, which can be determined by mapping first-principles calculated properties $P(\sigma)$ of a few (small cell) polytypes σ onto this formal expansion. Specifically, we define

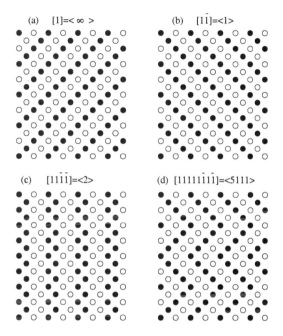

Figure 10. Schematic diagrams of atomic arrangement and the notations of four pseudobinary polytype semiconductors.

$$P(\sigma) = J_0 + \sum_{k=\text{even}}^{\text{pairs}} \overline{\Pi}_k(\sigma) J_k + O(> pairs) . \qquad (28)$$

Here, $0 \leq \overline{\Pi}_k(\sigma) \leq 1$ is the average of the kth neighbor layer pair correlation function, given by

$$\overline{\Pi}_k(\sigma) = \frac{1}{2N} \sum_{i=1}^{N} [1 + S_i(\sigma) S_{i+k}(\sigma)] , \qquad (29)$$

where N is the layer period of the polytype and $O(> pairs)$ contain high order interactions. Since $P(< \infty >) = P(< 1 >)$, only $k = $ even pair interactions are not zero. Table V gives $\overline{\Pi}_{k=\text{even}}$ for some of the polytype structures.

This method has been used to study the band structure of GaInP$_2$ polytypes. Figure 11 plots the band gap energy and the VBM and CBM energies as a function of the layer thickness n of the APB superlattice $< n >$. In this case, the calculated band gap energies and the conduction band energies show an odd-even oscillation as a function of the APB period $< n >$, especially when n is small. This behavior is due to different band-folding and resulting level-repulsion and wave function localization of the CBM in this polytype: when n is odd, the zinc-blende L state folds into Γ, while when n is even, the zinc-blende X state folds into Γ. Since $\Gamma - L$ coupling (and thus level repulsion) is larger than the $\Gamma - X$ coupling [5], the band gap and the CBM energies tend to be lower when n is odd. The oscillation of the CBM states also leads to type-II band alignment between some polytype pairs, especially for polytypes with small domain sizes. This is the case, e.g., between $< 3 >$ and $< 4 >$ (Fig. 11). This finding provides a possible explanation to the experimentally observed type-II band alignment behavior in some multidomain samples [61,67-69], especially those having small domain sizes.

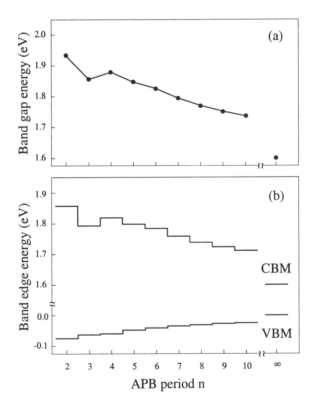

Figure 11. Band gap energies and VBM and CBM energy levels of GaInP$_2$ as function of layer thickness n of APB polytypes $< n >$.

Table V. Correlation functions for some of the zinc-blende based polytype structures. The Zhdanov notation for these polytype structures is also given.

Structure	Notation	$\overline{\Pi}_2$	$\overline{\Pi}_4$	$\overline{\Pi}_6$	$\overline{\Pi}_8$
CuPt	$<\infty>$	1	1	1	1
Y2	$<2>$	0	1	0	1
OS21	$<3>$	1/3	1/3	1	1/3
OS31	$<4>$	1/2	0	1/2	1
OS22	$<31>$	1/2	1	1/2	1
OS33	$<411>$	2/3	1/3	0	1/3
OS44	$<5111>$	3/4	1/2	3/4	1
OS5111	$<62>$	1/2	1/2	1/2	1

The electronic consequences of layer thickness fluctuations along the ordering [111] direction in CuPt-ordered $GaInP_2$ alloy (layer sequence Ga-In-Ga-In...) were are also investigated recently using plane-wave pseudopotential calculations [17]. These calculations show that the formation of a "sequence mutated" Ga-In-In-Ga... region (Fig. 12a) creates a hole state $h1$ localized in the In-In double layer, while the lowest electron state $e1$ is localized in the CuPt region. Thus, the system exhibits electron-hole *charge separation* ($e1 - h1$ transition is spatially indirect) in addition to spatial localization. This physical picture is preserved [17] when the dimension of the mutated segment is reduced from 2D to 0D (Fig. 12b),

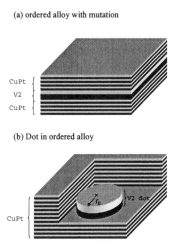

(a) ordered alloy with mutation

(b) Dot in ordered alloy

Figure 12. Structure of sequence mutation in ordered $GaInP_2$: one period of Ga-Ga-In-In (V2) inside the Ga-In-Ga-In (CuPt) structure. In part (a) the V2 region forms a 2D quantum well, while in part (b) V2 is disk-like (0D quantum dot with radius r_0).

resulting in disk-like dot structures. The calculated transition energies are in good agreement with experiment, and provide strong evidence that the experimentally seen peculiar luminescence properties [66-70] of ordered GaInP$_2$ are a consequence of quantum-disk like microstructures formed due to sequence mutations in [111] superlattices.

10.   SUMMARY

Ordering-induced changes in the structural, electronical, and optical properties of the semiconductor alloys are described theoretically using symmetry arguments and first-principles band structure calculations. These include (i) the change of the band gap ΔE_g and the valence band splitting ΔE_{12}, (ii) the appearance of new structural factors, (iii) the change in the nuclear magnetic resonance signals, (iv) the consequence of coexistence of (001) epitaxial strain and (111) chemical ordering on the optical properties, (v) the ordering induced optical anisotropy and (vi) the spin polarization of photoemitted electrons near the band edge. The effects of stacking fault in order semiconductor alloys are also discussed. Specific, experimentally testable predictions are given.

ACKNOWLEDGMENT

I would like to thank Drs. A. Franceschetti, D. B. Laks, T. Mattila, S. B. Zhang, and A. Zunger for their collaboration in this work and to thank Y. Zhang and A. Mascarenhas for helpful discussion. This work was supported by the U. S. Department of Energy, under contract No. DE-AC36-99-GO10337.

REFERENCES

1.   J. C. Woolley in Compound Semiconductors, edited by R. K. Willardson and H. L. Goering (Reinhold, New York, 1962), p3.
2.   A. Zunger and S. Mahajan, in Handbook of Semiconductors, 2nd ed., edited by S. Mahajan (Elsevier, Amsterdam, 1994), Vol. **3**, p. 1439.
3.   See papers on Compositional Modulation and Ordering in Semiconductors, edited by T. P. Pearsall and G. B. Stringfellow, MRS Bull. **22**, No. 7 (1997).
4.   G. B. Stringfellow, in Thin Films: Heteroepitaxial Systems, edited by M. Santos and W. K. Liu (World Scientific, Singapore, 1998), p.64.
5.   S.-H. Wei and A. Zunger, Phys. Rev. B **39**, 3279 (1989).
6.   A. Mascarenhas, S. Kurtz, A. Kibbler, and J.M. Olson, Phys. Rev. Lett. **63**, 2108 (1989).
7.   S.-H. Wei and A. Zunger, Phys. Rev. B. **57**, 8983 (1998).

8. S.-H. Wei and A. Zunger, Appl. Phys. Lett. **56**, 662 (1990).
9. S.-H. Wei and A. Zunger, Phys. Rev. B **49**, 14337 (1994).
10. S.-H. Wei and A. Zunger, Appl. Phys. Lett. **64**, 757 (1994).
11. S.-H. Wei and A. Zunger, Appl. Phys. Lett. **64**, 1676 (1994).
12. S.-H. Wei and A. Zunger, Phys. Rev. B **51**, 14110 (1995).
13. S.-H. Wei, A. Franceschetti, and A. Zunger, Phys. Rev. B **51**, 13097 (1995).
14. A. Franceschetti, S.-H. Wei and A. Zunger, Phys. Rev. B **52**, 13992 (1995).
15. A. Franceschetti and A. Zunger, Appl. Phys. Lett. **65**, 2990 (1994).
16. S.H. Wei, S.B. Zhang and A. Zunger, Phys. Rev. B. **59**, R2478 (1999).
17. T. Mattila, S.-H. Wei and A. Zunger, Phys. Rev. Lett. **83**, 2010 (1999).
18. S.-H. Wei and A. Zunger, J. Phys. Chem. **107**, 1931 (1997).
19. S.-H. Wei, A. Franceschetti, and A. Zunger, MRS Symp. Proc. **417**, edited by E. D. Jones, A. Mascarenhas, and P. Petroff, (MRS, Pittsburgh, 1996), p. 3.
20. S.-H. Wei, in Proceedings of the 7th International Workshop on Polarized Gas Targets and Polarized Beams, edited by R. J. Holt and M. A. Miller, (AIP, New York, 1998), p. 284.
21. A. Zunger, S.-H. Wei, L. G. Ferreira, and J. E. Bernard, Phys. Rev. Lett. **65**, 353 (1990).
22. S.-H. Wei, L. G. Ferreira, J. E. Bernard, and A. Zunger, Phys. Rev. B **42**, 9622 (1990).
23. S.-H. Wei, D. B. Laks, and A. Zunger, Appl. Phys. Lett. **62**, 1937 (1993).
24. D.B. Laks, S.-H. Wei, and A. Zunger, Phys. Rev. Lett. **69**, 3766 (1992).
25. L. Ferreira, S.-H. Wei, and A. Zunger, Phys. Rev. B 40, 3197 (1989).
26. R. Magri and A. Zunger, Phys. Rev. B **44**, 8672 (1991)
27. R. Magri, J. E. Bernard and A. Zunger, Phys. Rev. B **43**, 1593 (1991).
28. J. J. Hopfield, J. Phys. Chem. Solids **15**, 97 (1960).
29. P. Ernst, C. Geng, F. Scholz, H. Schweizer, Y. Zhang and A. Mascarenhas, Appl. Phys. Lett. **67**, 2347 (1995).
30. R. Wirth, H. Seitz, M. Geiger, F. Scholz, A. Hangleiter, A. Muhe, and F. Phillipp, Appl. Phys. Lett. **71**, 2127 (1997).
31. B. Fluegel, Y. Zhang, H. M. Cheong, A. Mascarenhas, J. F. Geisz, J. M. Olson, and A. Duda. Phys. Rev. B **55**, 13647 (1999).
32. Y. Zhang, A. Mascarenhas, P. Ernst, F. Driessen, D. J. Friedman, K. A. Bertness, J. M. Olson, C. Geng, F. Scholz, and H. Schweizer, J. of Appl. Phys. **81**, 6365 (1997).
33. P. Hohenberg and W. Kohn, Phys. Rev. **136**, B864 (1964).
34. W. Kohn and L. J. Sham, ibid. **140**, A1133 (1965).
35. S.-H. Wei and H. Krakauer, Phys. Rev. Lett. **55**, 1200 (1985).
36. D. J. Singh, Planewaves, pseudopotentials and the LAPW method, (Kluwer, Boston, 1994), and references therein.
37. N. E. Christensen, Phys. Rev. B **30**, 5753 (1984).
38. S.-H. Wei and A. Zunger, Appl. Phys. Lett. **72,** 2011 (1998).
39. S.-H. Wei and A. Zunger, Phys. Rev. Lett. **76**, 664 (1996).
40. S.-H. Wei and A. Zunger, Phys. Rev. B **39**, 6279 (1989).
41. F. Alsina, M. Garriga, M. I. Alonso, J. Pascual, J. Camassel, and R. W. Glew, in Proceedings of the 22nd International Conference on the Physics of Semiconductors, edited by D. J. Lockwoθd, (World Scientific, Singapore, 1995), p. 253.
42. T. Kita, K. Yamashita, and T. Nishino, Phys. Rev. B **54**, 16714 (1996).
43. R L. Forrest, T. D. Golding, S. C. Moss, Y. Zhang, J. F. Geisz, J. M. Olson, A. Mascarenhas, P. Ernst, and C. Geng, Phys. Rev. B **58**, 15355 (1998).
44. Z. W. Lu, A. Zunger, and M. Deutsch, Phys. Rev. B **47**, 9385 (1993).
45. D. Mao, P. C. Taylor, S. R. Kurtz, M. C. Wu, W. A. Harrison, Phys. Rev. Lett. **76**, 4769 (1996).
46. P. Blaha, K. Schwarz, and P. Herzig, Phys. Rev. Lett. **54**, 1192 (1985).
47. C. P. Slichter, Principles of Magnetic Resonance (Springer-Verlag, New York, 1978).

48. CRC Handbook of Chemistry and Physics, 76th edition, edited by A. R. Lide and H. P. R. Frederikse, (CRC Press, New York, 1996).

49. A. Eyal, R. Beserman, S.-H. Wei, A. Zunger, E. Maayan, O. Kreinin, J. Salzman, R. Westphalen, and K. Heime, Jpn. J. Appl. Phys. Suppl. **32-3**, 716 (1993).

50. R.G. Alonso, A. Mascarenhas, G.S. Horner, K.A. Bertness, S.R. Kurtz, and J.M. Olson, Phys. Rev. B **48**, 11833 (1993).

51. T. Kanata, M. Nishimoto, H. Nakayama, and T. Nishino, Phys. Rev. B **45**, 6637 (1992).

52. T. Kanata, M. Nishimoto, H. Nakayama, and T. Nishino, Appl. Phys. Lett. **63**, 26 (1993).

53. J. S. Luo, J. M. Olson, K. A. Bertness, M. E. Raikh, and E. V. Tsiper, J. Vac. Sci. Technol. **B12**, 2552 (1994).

54. J. S. Luo, J. M. Olson, Y. Zhang, and A. Mascarenhas, Phys. Rev. B **55**, 16385 (1997).

55. T. Kita, M. Sakurai, K. Bhattacharya, K. Yamashita, T. Nishino, C. Geng, F. Scholz, and H. Schweizer, Phys. Rev. B **57**, R15044 (1998).

56. B. Fluegel, Y. Zhang, A. Mascarenhas, J. F. Geisz, J. M. Olson, and A. Duda. Phys. Rev. B **60**, R11261 (1999).

57. J. Kessler, Polarized Electrons, 2nd edition (Springer-Verlag, Berlin, 1985).

58. D. T. Pierce and F. Meier, Phys. Rev. B **13**, 5484 (1976).

59. F. Meier, J. C. Grobli, D. Guarisco, and A. Vaterlaus, Physica Scripta **T49**, 574 (1993).

60. S. R. Kurtz, J. Appl. Phys. **74**, 4130 (1993).

61. D. Munzar, E. Dobrocka, I. Vavra, R. Kudela, M. Harvanka, and N. E. Christensen, Phys. Rev. B **57**, 4642 (1998).

62. A. Mascarenhas, Y. Zhang, R. Alonso, and S. Froyen, Solid State Commun. **100**, 47 (1996)

63. Y. Zhang and A. Mascarenhas, Phys. Rev. B **55**, 13100 (1997).

64. E. Morita, M. Ikeda, O. Kumagai, and K. Kaneko, Appl. Phys. Lett. **53**, 2164 (1988).

65. C. S. Baxter, W. M. Stobbs, and J. H. Wilkie, J. Cryst. Growth **112**, 373 (1991).

66. U. Dorr, H. Kalt, D. J. Mowbray, and C. C. Button, Appl. Phys. Lett. **72**, 821 (1998).

67. E. D. Jones, D. M. Follstaedt, H. Lee, J. S. Nelson, R. P. Schneider, Jr., R. G. Alonso, G. S. Horner, J. Machol, and A. Mascarenhas, in 22nd Intl. Conf. on the Phys. of Semiconductors, edited by D. J. Lockwood (World Scientific, Singapore, 1995), p. 293.

68. M. C. Delong, W. D. Ohlsen, I. Viohl, P. C. Taylor, and J. M. Olson, J. Appl. Phys. **70**, 2780 (1991).

69. J. E. Fouquet, M. S. Minsky, and S. J. Rosner, Appl. Phys. Lett. **63**, 3212 (1993).

70. H. M. Cheong, A. Mascarenhas, J. F. Geisz, J. M. Olson, M. W. Keller and J. R. Wendt, Phys. Rev. B **57**, R9400 (1998).

71. W. Selke, Phys. Report **170**, 213 (1988).

72. G. S. Zhdanov, C. R. Acad. Sci. USSR **48**, 43 (1945).

Chapter 16

Polarization Charges at Spontaneously Ordered (In,Ga)P/GaAs Interfaces

Peter Krispin

Paul-Drude-Institut für Festkörperelektronik, Hausvogteiplatz 5-7, 10117 Berlin, Germany

Key words: (In,Ga)P ordering, polarization charges, InGaP/GaAs interfaces

Abstract: In order to determine band offsets as well as the density and polarity of interfacial charges, depth-resolved electrical characteristics of $n-$ and p–type GaAs/(In,Ga)P/GaAs heterojunctions are discussed as obtained from capacitance-voltage measurements. The two interfaces are not equivalent. Positive and negative sheet charges are observed at the (In,Ga)P-on-GaAs and the GaAs-on-(In,Ga)P interfaces, respectively. The density of these interfacial charges increases with increasing degree of order in (In,Ga)P. Spontaneous ordering induces piezoelectric polarization charges at (In,Ga)P/GaAs interfaces, which lead to a redistribution of free carriers. The carrier distribution at single interfaces is compared with calculations based on the solution of Poisson´s equation. The discontinuities in the conduction and the valence band are independently determined. For spontaneously ordered (In,Ga)P, the (In,Ga)P/GaAs interfaces are found to be of type I.

1.   BACKGROUND

The properties of heterointerfaces between different semiconductors have become more and more important for the performance of advanced electronic devices. In particular, for the design of heterojunction bipolar transistors,[1-4] light-emitting and laser diodes,[5,6] photodetectors,[7-9] solar cells,[10,11] and modulation-doped field-effect transistors,[12,13] the lattice-matched (In,Ga)P/GaAs heterojunction is a real alternative to the (Al,Ga)As/GaAs system, because it is Al-free. Despite a large number of investigations, the features of (In,Ga)P/GaAs interfaces are still

451

controversial. In particular, it is not clear yet, how the interfacial properties are modified by spontaneous ordering of (In,Ga)P.

1.1   Band Offsets

Concerning the band discontinuities between (In,Ga)P and GaAs, the experimental values obtained for the conduction band offset ΔE_C vary between −0.05 and 0.25 eV.[2,7-9,14-27] The valence band offset ΔE_V has only been determined a few times directly.[14,16,27-29] It ranges from −0.24 to −0.41 eV. Although it has been known from previous studies[30] that (In,Ga)P has a tendency to order, most of the above experimental results have been published without discussing the effect of ordering. It is therefore still an open question, how spontaneous ordering of (In,Ga)P affects the band offsets at interfaces to GaAs. Since the band gap of (In,Ga)P decreases with increasing order, some of the rather large variations in the offset data are most likely due to different degrees of order.

From recent calculations of the offsets between GaAs and (In,Ga)P, ΔE_C has been found to be 0.12 eV for disordered and −0.13 eV for fully ordered (In,Ga)P.[31] The discontinuity in the valence band ΔE_V changes correspondingly with increasing order from −0.37 to −0.27 eV. Thus, the interfacial character is expected to switch from type I to type II, when (In,Ga)P becomes more ordered. However, experimental evidence for this transition has not been reported so far. Earlier calculations for the interface between disordered (In,Ga)P and GaAs have led to ΔE_V values between −0.26 and −0.45 eV.[32-35]

1.2   Piezoelectric Polarization of (In,Ga)P

According to theoretical predictions, ordered (In,Ga)P reveals a strong piezoelectric polarization, which may largely modify the electronic characteristics of (In,Ga)P/GaAs interfaces.[31] Sheet charges and related electric fields are therefore expected, for example, at interfaces between GaAs (no polarization) and ordered (In,Ga)P. Electric fields due to polarization charges have been observed between two differently ordered variants of (In,Ga)P by scanning capacitance microscopy.[36] Recently, it has also been demonstrated by capacitance versus voltage (C–V) measurements that (In,Ga)P ordering induces polarization charges at the interface with GaAs, the density of which increases for increasing degree of order.[29]

The presence of interfacial charges at the (In,Ga)P/GaAs interface may substantially alter the offset values as determined by different experimental techniques. Such charges change the diffusion voltage on the (In,Ga)P side of the heterojunction and lead to a carrier redistribution at the interface. In particular, photoluminescence measurements (cf., e.g., Refs. 24 and 28) on GaAs quantum wells in (In,Ga)P are not suitable to determine band offsets, because the (In,Ga)P-on-GaAs (normal) and the GaAs-on-(In,Ga)P (inverted) interfaces, which may reveal opposite polarization charges, cannot be discriminated. In addition, the conduction and valence band offsets cannot be determined independently.

Other methods such as ballistic-electron emission microscopy,[22] photoemission and current-voltage characteristics of various (In,Ga)P/GaAs heterojunctions,[2,8,9,18-20,25] which in principle permit the study of single interfaces, rely crucially on the assumption that the band diagram at the heterojunction is not influenced by interfacial charges. The ΔE_C offset values obtained from these experiments are therefore not always reliable.

1.3   *C–V* **Experiments**

Capacitance versus voltage studies on isotype heterojunctions are known to provide values for the band offsets and the sheet charge densities simultaneously.[37,38] The offset data obtained from this technique are therefore more reliable. Using the *C–V* method different authors have determined ΔE_C for *n*–type (In,Ga)P-on-GaAs interfaces grown by molecular beam epitaxy or metalorganic vapor phase epitaxy (MOVPE). Their values are very close to each other around 0.20 eV.[14,15,27] High concentrations of negative sheet charges make the ΔE_C measurement, however, difficult for the *n*–type GaAs-on-ordered (In,Ga)P interface.[23,26,27] These distinct charges at the inverted interface might be a further reason that ΔE_C values, which have been determined at this interface by other than *C–V* techniques,[8,22,25] largely deviate from the average value of about 0.20 eV. Electron traps and internal strain at the interfaces have been made responsible for the strong electron depletion at the *n*–type GaAs-on-(In,Ga)P interfaces.[23,26]

The valence band offset ΔE_V can be independently determined by *C–V* measurements on *p*–type heterointerfaces.[14,27] Values of −0.24 and −0.27 eV have been assessed at normal and inverted junctions in Refs. 14 and 27, respectively. The valence band discontinuity ΔE_V can be also determined, if experimental values of deep-level energies of transition-metal impurities in GaAs, InP, and GaP are compared.[39] By this method, which is related to bulk properties and therefore not dependent on interfacial charges, ΔE_V is found to be about −0.25 eV, in reasonable agreement with the results of the *C–V* technique. It should be noted that the discontinuities ΔE_C and ΔE_V as determined by the *C–V* method are also consistent with the results of deep-level transient spectroscopy (DLTS) on GaAs quantum wells in (In,Ga)P.[16]

Combining current–voltage and *C–V* measurements on junctions between insulating (In,Ga)P and *n*–type GaAs, positive and negative sheet charges with densities in the 10^{11} cm^{-2} range have been detected at the normal and inverted interfaces, respectively.[20,25] The origin of these mirror charges at opposite interfaces has, however, not been clarified yet. They are most likely induced by the piezoelectric polarization of ordered (In,Ga)P described in this chapter.

1.4   Outline

When discussing the electronic properties of (In,Ga)P/GaAs interfaces, spontaneous ordering and the presence of polarization charges in (In,Ga)P have not been adequately taken into account up to now. In this chapter, we will therefore first discuss depth profiles of carrier densities in isotype heterojunctions obtained from *C–V* measurements. Next, the effect of spontaneous ordering on the redistribution of free carriers at inverted and normal (In,Ga)P/GaAs interfaces will be demonstrated. Furthermore, it is shown that the band offsets ΔE_C and ΔE_V as well as the interfacial sheet charge densities can be separately determined by the simulation of the measured carrier distributions. Experimental evidence is presented that spontaneous (In,Ga)P ordering gives rise to opposite polarization charges at the two (In,Ga)P/GaAs interfaces, the density of which increases with increasing degree of order. Finally, we discuss the influence of the polarization charges in ordered (In,Ga)P on physical investigations and device characteristics.

2.   CARRIER DISTRIBUTION AT INTERFACES

2.1   *C–V* Method

Using rectifying metal-semiconductor (MS) contacts on isotype (In,Ga)P/GaAs heterojunctions, carrier distributions can be measured by the conventional *C–V* method.[37] The free carrier concentration *n* or *p* is obtained from the expression[40]

$$n(W) \; or \; p(W) = \frac{2}{A^2 q \varepsilon \varepsilon_0} \left[\frac{d}{dV} \left(\frac{1}{C^2} \right) \right]^{-1}, \tag{1}$$

where W denotes the thickness of the space-charge layer below the MS contact, A the contact area, q the elementary charge, and $\varepsilon\varepsilon_0$ the dielectric constant. In general, the values $n(W)$ or $p(W)$ measured by the C-V technique stand for the free carrier density at the edge of the space-charge layer.[40] The depth W is calculated from the depletion capacitance C using

$$W(V) = \varepsilon\varepsilon_0 A / C(V).\qquad(2)$$

In order to scan the complete carrier concentration versus depth profile of isotype heterojunctions, the layer sequence is usually recess-etched a few times. The distance between the MS contact and the interfaces is thus varied. By changing the dc bias in reverse direction, the edge of the space-charge layer below the MS contact is then shifted across a single interface of the GaAs/(In,Ga)P/GaAs heterostructure studied. However, the actual depth profile of the carrier concentration can only be determined, when the capacitance C does not depend on the measuring frequency.[40]

2.2   *n*–Type GaAs/(In,Ga)P/GaAs Heterojunctions

For a GaAs/(In,Ga)P/GaAs heterojunction grown by MOVPE on a GaAs(001) substrate, the complete depth profile of the electron concentration n is plotted in Fig. 1.

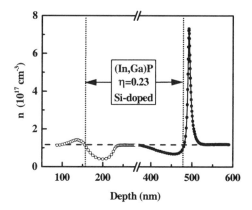

Figure 1. Complete depth profile of the electron density n measured at 1 MHz and 300 K on a GaAs/(In,Ga)P/GaAs heterojunction with weakly ordered (In,Ga)P. The interface positions are indicated by vertical dotted lines. The doping level in the layers is marked by the horizontal dashed line. Measurements on as-grown and recess-etched samples are distinguished using open and full circles, respectively.[27]

The order parameter η has been determined by photoluminescence (PL) measurements at 10 K on the same sample. The position of the PL peaks can be used as a measure for the band gap of the differently ordered (In,Ga)P layers. The degree of order η defined as the average composition of alternating monolayer ordering along a [111] direction is obtained from the relation

$$E_G(0) - E_G(\eta) = 0.471\eta^2 \quad (eV),$$ (3)

which has been experimentally confirmed.[41,42] $E_G(\eta)$ denotes the band gap of (In,Ga)P with order parameter η. The degree of order varies from $\eta = 0$ for complete disordering to $\eta = 1$ for perfect ordering. $E_G(0)$ is found to be 1.99 eV.[42,29]

At both interfaces, the typical peak/valley structure of isotype heterojunctions is established in Fig. 1.[37] Carrier accumulation and depletion are observed on the GaAs and (In,Ga)P sides, respectively. It is therefore evident that the conduction band edge in (In,Ga)P is higher than in GaAs. The conduction band offset ΔE_C is positive.

The properties of the two interfaces in Fig. 1 are apparently not the same. The peak heights on the GaAs side are very different for the two interfaces. Integrating the electron density across each interface with regard to the doping level, a carrier deficit of about 3×10^{11} cm^{-2} and a carrier excess of about 2.5×10^{11} cm^{-2} are determined for the GaAs-on-(In,Ga)P and (In,Ga)P-on-GaAs interfaces, respectively. Such a strong electron depletion at the inverted heterointerface has also been realized by other authors.[23,26] That has been associated with the presence of interfacial electron traps or strain. Additional sources could be unidentified negative charges or variations of the doping level.

The depth profiles of the Si concentration obtained from secondary ion mass spectrometry (SIMS) on the same heterojunction as in Fig. 1 show no reduction of the doping level at the inverted interface. In order to search for electron traps at the GaAs-on-(In,Ga)P interface, deep-level spectra have been measured by DLTS. Figure 2 reveals a sequence of temperature scans for the inverted interface of the above sample. With increasing reverse bias voltage, the edge of the space-charge layer below the MS contact crosses the inverted interface from GaAs to (In,Ga)P.

For a bias of -1 V, the peak observed at a temperature of about 340 K is related to the well-known bulk level EL2 in MOVPE-grown GaAs.[43] Increasing the reverse bias leads to a reduction of the EL2 signal. On the (In,Ga)P side, the EL2 level is missing (curve at -3 V in Fig. 2). Simultaneously, the evolution of a weak broad band is detected at a temperature of about 100 K, which is typically observed in (In,Ga)P layers and most likely linked with residual impurities in the MOVPE precursors.[44]

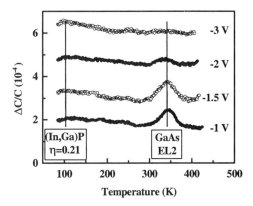

Figure 2. DLTS signals $\Delta C/C$ close to the inverted interface of the heterojunction investigated in Fig. 1. The peak positions of the main bulk levels in (In,Ga)P and GaAs are indicated.[27]

DLTS responses, which may correspond to interfacial levels, are practically absent. The estimated density of interfacial states is in the 10^9 cm^{-2} range and therefore too low to explain the carrier deficit of about 3×10^{11} cm^{-2} measured by the $C-V$ method on the same heterointerface.

2.3   *p*–Type GaAs/(In,Ga)P/GaAs Heterojunctions

For a heterojunction with weakly ordered (In,Ga)P grown by MOVPE on a GaAs(001) substrate, the complete depth profile of the hole concentration *p* is shown in Fig. 3.

As for the corresponding *n*–type sample, carrier accumulation and depletion are observed on the GaAs and (In,Ga)P sides of the interfaces, respectively. The valence band edge in (In,Ga)P is obviously lower than that in GaAs, i.e., the offset ΔE_V is negative. Together with the results on *n*–type heterojunctions grown under similar conditions (see Fig. 1), it is evident that (In,Ga)P/GaAs interfaces with weakly ordered (In,Ga)P are of type I.

The two interfaces in *p*–type heterostructures are also not identical. The peak of the hole density at the normal interface is lower than that at the inverted junction. By integrating the hole distributions at each interface, a carrier deficit and a surplus are found for the normal and inverted interfaces, respectively. The deficit at the normal interface does not originate from interfacial hole traps.[27] To explain the distinct differences between the two interfaces, local changes in the doping level can be excluded using the results of SIMS measurements.

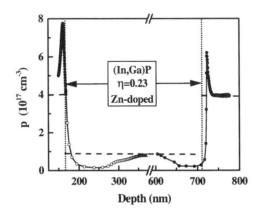

Figure 3. Complete depth profile of the hole concentration p measured at 1 MHz and 300 K on a GaAs/(In,Ga)P/GaAs heterojunction with weakly ordered (In,Ga)P. The interface positions are indicated by vertical dotted lines. Doping levels in the layers are marked by horizontal dashed lines. Multiple recess-etch steps are distinguished by different symbols.[27]

From the carrier distributions in $n-$ and $p-$type (In,Ga)P/GaAs heterostructures, it becomes obvious (see Figs. 1 and 3) that the electron deficit and the hole surplus at the inverted interface are due to the same origin, i.e., negative sheet charges. Such charges raise the conduction and valence band edges and lead to a depletion of electrons and an accumulation of holes. At the normal interface, the electron surplus and the hole deficit are correspondingly caused by positive sheet charges, which lower the band edges. As shown next, the concentration of these sheet charges is related to the degree of (In,Ga)P ordering.

3.   EFFECT OF SPONTANEOUS ORDERING

3.1   GaAs-on-(In,Ga)P Interface

For inverted interfaces of $n-$type heterojunctions with differently ordered (In,Ga)P layers, characteristic depth profiles of the electron density are plotted in Fig. 4. The peak concentration on the GaAs side decreases for interfaces with more strongly ordered (In,Ga)P. The overall electron deficit at this interface is obvious. When the depletion on the (In,Ga)P side becomes dominant at a higher degree of order, the electron accumulation on the GaAs side is missing. For the samples in Fig. 4 with the order parameter η of 0.23, 0.45, and 0.50, the electron deficits are about 3×10^{11}, 9×10^{11}, and 13×10^{11} cm^{-2}, respectively. The density σ_{if} of the related negative sheet charges corresponds apparently to the order parameter η of the (In,Ga)P layer.

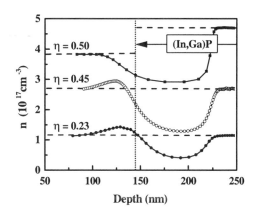

Figure 4. Depth profiles of the electron concentration n for GaAs-on-(In,Ga)P interfaces. Different order parameters η correspond to different MOVPE growth conditions. The depth scales are normalized to an interface position of 145 nm (marked by the vertical dotted line). Horizontal dashed lines denote the doping levels in GaAs and (In,Ga)P.[29]

Due to the strong electron depletion, the GaAs-on-(In,Ga)P interface is often not suitable to determine the conduction band offset ΔE_C.[8,21-23,25,26] In particular, the inverted interface cannot be used to measure ΔE_C in a junction with highly ordered (In,Ga)P. Some of the ΔE_C variations in the literature are likely due to these negative charges, which drastically modify the band structure at the inverted interface. Interestingly, the depletion can be reduced by incorporating between (In,Ga)P and GaAs an interlayer of Ga(As,P), which leads to the formation of (In,Ga)(As,P) and, consequently, less strain at the inverted interface.[26]

3.2   (In,Ga)P-on-GaAs Interface

For (In,Ga)P-on-GaAs interfaces of n–type heterojunctions with differently ordered (In,Ga)P, typical depth profiles of the electron concentration n are depicted in Fig. 5. The common peak/valley structure of isotype heterojunctions is revealed for all studied order parameters. The depletion region on the (In,Ga)P side of the heterointerface remains practically unchanged. The peak carrier concentration on the GaAs side grows with increasing order parameter η.

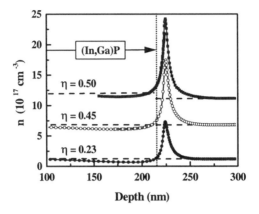

Figure 5. Depth profiles of the electron concentration n for (In,Ga)P-on-GaAs interfaces. Different order parameters η correspond to different MOVPE growth conditions. The depth scales are normalized to an interface position of 215 nm (marked by the vertical dotted line). Horizontal dashed lines denote the doping levels in GaAs and (In,Ga)P.[29]

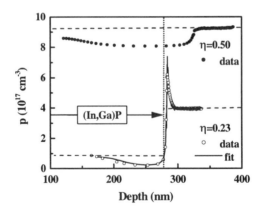

Figure 6. Depth profiles of the hole concentration p for (In,Ga)P-on-GaAs interfaces. The two different order parameters η correspond to different MOVPE growth conditions. The line for the $\eta = 0.23$ sample has been calculated (see, Subsection 4.3). The depth scales are normalized to an interface position of 275 nm (marked by the vertical dotted line). Horizontal dashed lines denote doping levels in GaAs and (In,Ga)P.[29]

When integrating the electron distributions for each curve in Fig. 5, an excess of electrons over the doping level is found, which reaches 1.2×10^{12} cm^{-2} for the interface with highly ordered (In,Ga)P. At the

(In,Ga)P-on-GaAs interface, the concentration of positive sheet charges is apparently related to the order parameter η of the (In,Ga)P layer.

For (In,Ga)P-on-GaAs interfaces of p–type heterojunctions, characteristic depth profiles of the hole density p are plotted in Fig. 6 for two differently ordered (In,Ga)P layers grown by MOVPE on GaAs(001) substrates. Whereas for the weakly ordered sample the peak/valley structure of an isotype heterojuncion is clearly seen, a remarkable hole depletion exists at the normal heterointerface with highly ordered (In,Ga)P. With increasing order parameter η, the hole deficit becomes larger than 1×10^{12} cm^{-2}. Therefore, the band offset ΔE_V for highly ordered (In,Ga)P cannot be determined with p–type (In,Ga)P-on-GaAs interfaces.

It is worth to mention that the sharp peak in Fig. 6 for $\eta = 0.23$ proves the compositional abruptness and the lateral homogeneity of the interface studied. Summarizing the results on the (In,Ga)P-on-GaAs interface, both the hole deficit in p–type and the electron excess in n–type heterojunctions originate from positive interfacial charges.

4.   BAND OFFSETS AND INTERFACIAL CHARGES

4.1   Simulation of Carrier Distributions at Heterointerfaces

If the peak/valley structure of the measured carrier distribution is clearly indicated, information about both the band offset and the interfacial charge can be gained independently.[37,38] The values determined by Kroemer´s method can be used to numerically reconstruct the measured depth profiles of the carrier density. First, the C–V characteristic is calculated by one-dimensional simulations with a Poisson-Schrödinger solver.[45] The calculated depth profile of the carrier density is subsequently obtained from Eqs. (1) and (2).

Calculated carrier distributions are plotted in Fig. 7 for the n–type (In,Ga)P-on-GaAs interface in order to demonstrate the changes, which are induced by separate variations of the sheet charge density σ_{if} and the band offset ΔE_C. For a constant σ_{if} value, the depletion layer on the (In,Ga)P side becomes wider and the peak higher with a larger offset [see Fig. 7(a)]. For a fixed offset, the peak at the interface grows increasing the positive charge density [see Fig. 7(b)], in contrast to the width of the depletion layer on the (In,Ga)P side, which is practically constant.

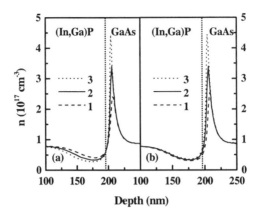

Figure 7. Calculated depth profiles of the electron density n near the (In,Ga)P-on-GaAs interface for (a) $\sigma_{if} = 0$ and ΔE_C values of 0.15 (1), 0.20 (2), and 0.25 eV (3) and (b) $\Delta E_C = 0.20$ eV and interfacial charge densities σ_{if} of -1×10^{11} (1), 0 (2), and $+1\times10^{11}$ cm^{-2} (3). The position of the interface is marked by vertical dotted lines.

4.2   GaAs-on-(In,Ga)P Interface

For the inverted interface of a p–type heterojunction with weakly ordered (In,Ga)P, the depth profile of the hole density is plotted in Fig. 8.

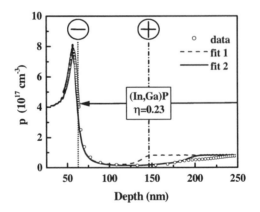

Figure 8. Simulation of the hole distribution at an inverted p–type interface with weakly ordered (In,Ga)P. The experimental p versus depth profile (circles) is reproduced by fits 1 and 2 (see text). The interface position with negative charges and the appearance of a positive sheet charge in (In,Ga)P are marked by vertical lines at 62 and 147 nm, respectively.[29]

The peak/valley structure of the isotype heterojunction is clearly revealed. Fit 1 (dashed line in Fig. 8) is the result of a calculation with a valence band offset ΔE_V of -0.32 eV and an interfacial sheet charge density of

-3×10^{11} cm^{-2}. The simulation is excellent close to the interface. Additional holes are fixed at this interface due to negative charges. It should be noted that practically the same amount of negative charges is found at the inverted interface of n–type heterojunctions with similarly ordered (In,Ga)P (cf. Fig. 1). Both the hole excess in p–type and the electron deficit in n–type heterojunctions originate from the same negative interfacial charges.

Inside the (In,Ga)P layer, there is an additional hole depletion around 150 nm. This decrease of the hole concentration towards the interface does not originate from a spatial variation of the Zn concentration as checked by SIMS measurements.[29] The depletion is most likely linked with a positive sheet charge inside the (In,Ga)P layer. When a density σ_{if} of 3.5×10^{11} cm^{-2} is incorporated at 147 nm, the experimental carrier distribution in Fig. 8 is matched by the calculated depth profile (fit 2). It is therefore evident that the amounts of positive and negative charges inside the (In,Ga)P layer and at the interface, respectively, agree reasonably with each other.

Unfortunately, the electron distribution at the inverted interface of n–type heterojunctions cannot be reconstructed with the same perfection. The reason for this is not clear yet.

4.3   (In,Ga)P-on-GaAs Interface

The simulation for the normal interface of a n–type heterojunction with weakly ordered (In,Ga)P is shown in Fig. 9 (cf. Fig. 1).

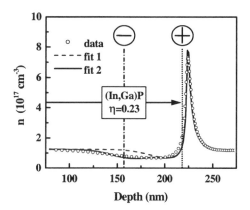

Figure 9. Reconstruction of the electron distribution near a normal n–type interface with weakly ordered (In,Ga)P. The experimental n versus depth profile (circles) is simulated by fits 1 and 2 (see text). The interface position with positive charges and the appearance of a negative sheet charge in (In,Ga)P are marked by vertical lines at 217 and 157 nm, respectively.[29]

Fit 1 (dashed line) is calculated with a conduction band offset ΔE_C of 0.17 eV and a sheet charge density σ_{if} of $+3\times10^{11}$ cm^{-2}. However, the calculation fits the measured electron distribution only on the GaAs side. As for the p–type inverted interface (cf. Fig. 8), carriers are missing inside the (In,Ga)P layer adjacent to the depletion region. The electron deficit around 160 nm does not originate from variations of the Si doping level, but is likely due to an additional negative sheet charge. The depth profile of the electron concentration in Fig. 9 can be perfectly fitted, when a sheet charge density σ_{if} of -2.5×10^{11} cm^{-2} is additionally incorporated at 157 nm (fit 2). This density approaches the concentration of positive sheet charges, which are detected directly at the normal interface.

As an example for the normal interface of a p–type heterostructure, the calculated hole distribution is drawn in Fig. 6 as a line. An excellent fit is obtained with a valence band offset ΔE_V of -0.28 eV and a sheet charge density σ_{if} of $+4.5\times10^{11}$ cm^{-2}.

5.   ORDER-INDUCED POLARIZATION CHARGES

For the investigated MOVPE-grown interfaces, the σ_{if} data are compiled in Fig. 10.

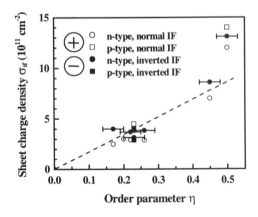

Figure 10. Sheet charge density σ_{if} at normal and inverted (In,Ga)P/GaAs interfaces versus order parameter η. Closed and open symbols denote negative and positive charges, respectively, as observed by the C–V method. For negative sheet charges, error bars are depicted for the order parameter η. The dashed line is a guide to the eye.[29]

The interfacial charge density increases with increasing order parameter η of the (In,Ga)P layer. Positive and negative sheet charges are observed in all heterojunctions at normal and inverted interfaces, respectively. For (In,Ga)P with the same order parameter, the concentration of positive (negative) charges at normal (inverted) interfaces does not depend on the conductivity type of the heterojunctions. For the same sample, the absolute values of the charge densities at the normal and inverted interface are comparable, although the two interfaces are electrically screened.

In summary, the alternating occurrence of positive and negative interfacial charges and the parity of their densities are obvious for the isotype GaAs/(In,Ga)P/GaAs heterojunctions studied. The experimental results can therefore be completely explained by the piezoelectric polarization of ordered (In,Ga)P, which gives rise to the sheet charges at interfaces with GaAs. Ordered (In,Ga)P may exhibit strain-induced piezoelectric polarization.[31,36] The polarization difference between ordered (In,Ga)P and GaAs (no polarization) leads to the opposite sheet charges at normal and inverted (In,Ga)P/GaAs interfaces as observed.

In a simplified model, the (In,Ga)P layers consist of several domains with ordered material in otherwise disordered (In,Ga)P.[46] Additional polarization charges are therefore also expected inside a (In,Ga)P layer, but they are usually screened by free carriers. Adjacent to the (In,Ga)P/GaAs interfaces, however, mirror charges of such domains inside (In,Ga)P can be observed by the $C-V$ method (see Figs. 8 and 9). Since the charge densities inside the (In,Ga)P layer and at both interfaces are practically the same, it is evident that the parameter η of the (In,Ga)P layer does not change during the growth under fixed conditions.

6.   CONCLUSIONS

The order-induced polarization charges at (In,Ga)P/GaAs interfaces have remarkable consequences for the optical and electrical properties of hetero-structures with such interfaces. Because normal and inverted interfaces exhibit opposite charges, the suitable design becomes important for devices. Whereas the interfacial charge density depends on the order parameter η, the charge character (positive or negative) is determined by the growth sequence. Under the MOVPE growth conditions applied, free electron (hole) densities in the 10^{12} cm^{-2} range can be achieved on the GaAs side of the normal (inverted) interfaces with ordered (In,Ga)P without doping (piezoelectric doping). Two-dimensional electron or hole gases and high mobilities are therefore expected in respective field-effect transistors.

Because ordering and related polarization charges have not been considered until now, the variation of the published data on (In,Ga)P/GaAs band offsets is rather large. To determine the band offsets for ordered (In,Ga)P, experimental methods are necessary, which are not affected by the presence of interfacial charges. Using the $C-V$ method on weakly ordered (In,Ga)P ($\eta \approx 0.23$), ΔE_C and ΔE_V are found to be $+(0.20\pm0.03)$ and $-(0.30\pm0.03)$ eV, respectively.[29] The sum of both offsets agrees reasonably well with the measured band gap difference of about 0.50 eV. The interfaces with weakly ordered (In,Ga)P are definitely of type I, in accordance with theoretical predictions.[31] The above values also correspond to $C-V$ results, which have been obtained for interfaces with (In,Ga)P of unknown order.[14,15] The ΔE_V data, in particular, are comparable with calculations for interfaces with disordered (In,Ga)P.[32,34]

The band offsets for interfaces between GaAs and strongly ordered (In,Ga)P are still under discussion. The large carrier deficits at the inverted n–type and normal p–type heterointerfaces inhibit, for example, the simulations of $C-V$ measurements. The other methods discussed in Section 1 are heavily affected by the electric fields, which originate from the polarization charges. Results on inverted p–type interfaces with strongly ordered (In,Ga)P are not available yet. For the normal n–type interface with (In,Ga)P of an order parameter $\eta \approx 0.50$, the electron distribution in Fig. 5 can be fitted with $\Delta E_C = 0.20$ eV. It is therefore suggested that the conduction band offset is not reduced by spontaneous ordering of (In,Ga)P as predicted by theory.[31]

From the experiments described in this chapter, the transition from type I to type II can be excluded for (In,Ga)P/GaAs interfaces, at least for crystallographic (001)-orientation and order parameter $\eta \leq 0.50$. The spatial separation of electron-hole pairs, which has been observed at GaAs-on-(In,Ga)P junctions by photoluminescence experiments, has been attributed to interfaces of type II.[47] However, this effect is probably due to the electric fields generated by the negative sheet charges at the inverted interface. In general, optical investigations on quantum wells in ordered (In,Ga)P should always take into account the electric field induced by opposite polarization charges at the two interfaces.

ACKNOWLEDGMENTS

The contributions of M. Asghar, S. Gramlich, A. Knauer, and H. Kostial are greatly acknowledged. The author would like to thank H. T. Grahn for helpful comments and a careful reading of the manuscript.

REFERENCES

[1] H. Kroemer, J. Vac. Sci. Technol. B1, 126 (1983).
[2] T. Kobayashi, K. Taira, F. Nakamura, and H. Kawai, J. Appl. Phys. 65, 4898 (1989).
[3] M. Razeghi, F. Omnes, M. Defour, P. Maurel, J. Hu, E. Wolk, and D. Pavlidis, Semicond. Sci. Technol. 5, 278 (1990).
[4] T. Oka, K. Ouchi, H. Uchiyama, T. Taniguchi, K. Mochizuki, and T. Nakamura, IEEE Electron. Dev. Lett. 18, 154 (1997).
[5] D. Z. Garbuzov, N. Yu. Antonishkis, A. D. Bondarev, A. B. Gulakov, S. N. Zhigulin, N. I. Katsavets, A. V. Kochergin, and E. V. Rafailov, IEEE J. Quantum Electr. 27, 1531 (1991).
[6] M. Guina, J. Dekker, A. Tukiainen, S. Orsila, M. Saarinen, M. Dumitrescu, P. Sipilä, P. Savolainen, and M. Pessa, J. Appl. Phys. 89, 1151 (2001).
[7] S. D. Gunapala, B. F. Levine, R. A. Logan, T. Tanbun-Ek, and D. A. Humphrey, Appl. Phys. Lett. 57, 1802 (1990).
[8] M. A. Haase, M. J. Hafich, and G. Y. Robinson, Appl. Phys. Lett. 58, 616 (1991).
[9] C. Jelen, S. Slivken, J. Hoff, M. Razeghi, and G. J. Brown, Appl. Phys. Lett. 70, 360 (1997).
[10] J. M. Olson. S. R. Kurtz, A. E. Kibbler, and P. Faine, Appl. Phys. Lett. 56, 623 (1990).
[11] T. Kitatani, Y. Yazawa, S. Watahiki, K. Tamura, J. Minemura, and T. Warabisako, Solar Energy Mat. Solar Cells 50, 221 (1998).
[12] T. Ohori, M. Takechi, M. Suzuki, M. Takikawa, and J. Komeno, J. Crystal Growth 93, 905 (1988).
[13] F. A. J. M. Driessen, G. J. Bauhuis, P. R. Hageman, and L. J. Giling, Appl. Phys. Lett. 65, 714 (1994).
[14] M. A. Rao, E. J. Caine, H. Kroemer, S. I. Long, and D. I. Babic, J. Appl. Phys. 61, 643 (1987).
[15] M. O. Watanabe and Y. Ohba, Appl. Phys. Lett. 50, 906 (1987).
[16] D. Biswas, N. Debbar, P. Bhattacharya, M. Razeghi, M. Defour, and F. Omnes, Appl. Phys. Lett. 56, 833 (1990).
[17] J. B. Lee, S. D. Kwon, I. Kim, Y. H. Cho, and B.-D. Choe, J. Appl. Phys. 71, 5016 (1992).
[18] T. W. Lee, P. A. Houston, R. Kumar, X. F. Yang, G. Hill, M. Hopkinson, and P. A. Claxton, Appl. Phys. Lett. 60, 474 (1992).
[19] S. L. Feng, J. Krynicki, V. Donchev, J. C. Bourgoin, M. DiForte-Poisson, C. Brylinski, S. Delage, H. Blanck, and S. Alaya, Semicond. Sci. Technol. 8, 2092 (1993).
[20] T. H. Lim, T. J. Miller, F. Williamson, and M. I. Nathan, Appl. Phys. Lett. 69, 1599 (1996).
[21] I.-J. Kim, Y.-H. Cho, K.-S. Kim, and B.-D. Choe, Appl. Phys. Lett. 68, 3488 (1996).
[22] J. J. O'Shea, C. M. Reaves, S. P. DenBaars, M. A. Chin, and V. Narayanamurti, Appl. Phys. Lett. 69, 3022 (1996).
[23] T. Kikkawa, K. Imanishi, K. Fukuzawa, T. Nishioka, M. Yokoyama, and H. Tanaka, Inst. Phys. Conf. Ser. 155, 877 (1997).
[24] K. Uchida, T. Arai, and K. Matsumoto, J. Appl. Phys. 81, 771 (1997).
[25] C. Cai, M. I. Nathan, and T. H. Lim, Appl. Phys. Lett. 74, 720 (1999).
[26] Y.-H. Kwon, W. G. Jeong, Y.-H. Cho, and B.-D. Choe, Appl. Phys. Lett. 76, 2379 (2000).
[27] P. Krispin, M. Asghar, A. Knauer, and H. Kostial, "Interface properties of isotype GaAs/(In,Ga)P/GaAs heterojunctions grown by metalorganic-vapour-phase epitaxy on GaAs," J. Crystal Growth 220, 220 (2000), Figs 3-5 used with permission of Elsevier Science.

[28] J. Chen, J. R. Sites, I. L. Spain, M. J. Hafich, and G. Y. Robinson, Appl. Phys. Lett. **58**, 744 (1991).

[29] P. Krispin, A. Knauer, and S. Gramlich, "Electric field-induced redistribution of free carriers at isotype (In,Ga)P/GaAs interfaces," Materials Sci. Engineering B **88**, 129 (2002), Figs 2-7 used with permission of Elsevier Science.

[30] A. Gomyo, T. Suzuki, K. Kobayashi, S. Kawata, I. Hino, and T. Yuasa, Appl. Phys. Lett. **50**, 673 (1987).

[31] S. Froyen, A. Zunger, and A. Mascarenhas, Appl. Phys. Lett. **68**, 2852 (1996).

[32] W. A. Harrison, J. Vac. Sci. Technol. **14**, 1016 (1977).

[33] Y.-C. Ruan and W. Y. Ching, J. Appl. Phys. **62**, 2885 (1987).

[34] C. G. Van de Walle, Phys. Rev. B **39**, 1871 (1989).

[35] M. S. Hybertsen, Materials Sci. Engineering B **14**, 254 (1992).

[36] J.-K. Leong, C. C. Williams, J. M. Olson, and S. Froyen, Appl. Phys. Lett. **69**, 4081 (1996).

[37] H. Kroemer, W.-Y. Chien, J. S. Harris, and D. D. Edwall, Appl. Phys. Lett. **36**, 295 (1980).

[38] H. Kroemer, Appl. Phys. Lett. **46**, 504 (1985).

[39] J. M. Langer and H. Heinrich, Phys. Rev. Lett. **55**, 1414 (1985).

[40] See, for example: P. Blood, J.W. Orton, The Electrical Characterization of Semiconductors: Majority Carriers and Electron States, Academic Press 1992.

[41] P. Ernst, C. Geng, F. Scholz, H. Schweizer, Y. Zhang, and A. Mascarenhas, Appl. Phys. Lett. **67**, 2347 (1995).

[42] P. Ernst, C. Geng, F. Scholz, and H. Schweizer, Phys. Status Solidi (b) **193**, 213 (1996).

[43] See, e.g., T. Hashizume, E. Ikeda, Y. Akatsu, H. Ohno, and H. Hasegawa, Jpn. J. Appl. Phys. Part 2, **23**, L296 (1984).

[44] P. Krispin and A. Knauer, unpublished.

[45] See, e.g., I.-H. Tan, G. L. Snider, L. D. Chang, and E. L. Hu, J. Appl. Phys. **68**, 4071 (1990).

[46] A. Sasaki, K. Tsuchida, Y. Narukawa, Y. Kawakami, Sg. Fujita, Y. Hsu, and G. B. Stringfellow, J. Appl. Phys. **89**, 343 (2001).

[47] Q. Liu, S. Derksen, A. Lindner, F. Scheffer. W. Prost, and F.-J. Tegude, J. Appl. Phys. **77**, 1154 (1995).

Author Index

Subject Index

471